Evolutionary Studies

Series editor
Naruya Saitou, National Institute of Genetics, Mishima, Japan

More information about this series at http://www.springer.com/series/15220

Naruya Saitou
Editor

Evolution of the Human Genome I

The Genome and Genes

Springer

Editor
Naruya Saitou
Division of Population Genetics
National Institute of Genetics
Mishima, Japan

ISSN 2509-484X ISSN 2509-4858 (electronic)
Evolutionary Studies
ISBN 978-4-431-56601-4 ISBN 978-4-431-56603-8 (eBook)
https://doi.org/10.1007/978-4-431-56603-8

Library of Congress Control Number: 2017961845

Printed on acid-free paper

This Springer imprint is published by Springer Nature
The registered company is Springer Japan KK
The registered company address is: Shiroyama Trust Tower, 4-3-1 Toranomon, Minato-ku, Tokyo 105-6005, Japan

Preface

This book is the first volume of the *Evolution of the Human Genome* and is also the first book of the Evolutionary Studies Springer Series. I am the series editor and the editor of this present book. In the fall of 2011, I was asked by Dr. Daisuke Hoshiyama, who was an editor at Springer Japan at that time, to publish a book by Springer written in English. I thus planned one book titled *Evolution of the Human Genome* and asked more than 30 people to contribute to the book. Because of its large size, it was suggested to me that I split the book into two volumes: one dedicated to the diversity of genes in the human genome, and the other dedicated to the diversity of modern humans. After Dr. Hoshiyama left Springer Japan in 2012, Mr. Kaoru Hashimoto handled this book until 2014 and then Dr. Yasutaka Okazaki subsequently took charge of the book. While some authors needed time to finish their chapters, I was asked to start a new Springer series. I proposed Evolutionary Studies and it was accepted by Springer in 2015 (series URL =http://www.springer.com/series/15220). Thus, this book became the first book under the new series. I thank Ms. Aiko Hiraguchi at Springer for overseeing the series and Dr. Okazaki for directly taking this book through the publication process.

The book consists of two parts. Part I is "Overview of the Human Genome", which includes seven chapters. Chapter 1, which I have written, is a brief look at human evolution and the human genome, and is also an introduction to the book. Chapter 2 by Dan Graur is about "rubbish" DNA which consists of the major part of the human genome. Chapter 3 by Satoshi Oota is on GC content heterogeneity of the human genome. Contents of Chaps. 2 and 3 also apply to non-human vertebrate genomes. Protein and RNA coding genes of the human genome are summarized in Chap. 4 by Tadashi Imanishi. Takashi Kitano describes duplicated genes in Chap. 5, and recombinations are discussed by Ludovica Montanucci and Jaume Bertranpetit in Chap. 6, with special reference to linkage disequilibria (LD). Part I ends with Chap. 7, in which copy number variations (CNVs) and microsatellite DNA polymorphism are discussed by Naoko Takezaki.

Part II of this book, "The Human Genome Viewed Through Genes" contains eight chapters with a wide variety of topics covering various gene systems. Yoko

Satta, Yukako Katsura, and Mineyo Iwase discuss genes on X and Y chromosomes in Chap. 8, and Timothy A. Jinam discusses HLA genes from the point of view of human population studies in Chap. 9. Shoji Kawamura and Amanda Melin give an extensive review of opsin genes in primates in general in Chap. 10. Genes involved in morphological phenotypes are discussed in Chap. 11 by Ryosuke Kimura, and transcription factor genes are extensively discussed by Mahoko Takahashi and So Nakagawa in Chap. 12. Ituro Inoue focuses on the genetics of diabetes in Chap. 13, and then he and Hiroto Nakaoka review disease-related genes in Chap. 14. The last chapter of this book, Chap. 15, is not on the genome of humans but of microbes in humans. Genomes of these human superorganisms are amply discussed by Chaochun Wei and Ben Jia.

Although not comprehensive, this book covers a wide variety of genes and genetic systems regarding the human genome. I hope readers will grasp the huge diversity of genes and DNA sequences in the human genome.

Finally, as the editor of the book, I would like to dedicate it to the late Dr. Allan Wilson, a pioneer of human molecular evolutionary studies, who died in 1991 at the age of 56.

Mishima, Japan Naruya Saitou

Contents

Part I
Overview of the Human Genome

Chapter 1
Human Evolution and Human Genome at a Glance

Naruya Saitou

Abstract The long-term human evolution is first explained. Starting from the origin of life, emergence of mammals and evolution of primates and hominoids are outlined, followed by classification of family Hominidae. We then move to the human genome. Its structure in terms of chromosomes and dichotomy of coding and noncoding regions are discussed, followed with general description on various types of mutations such as nucleotide substitutions, insertions and deletions, repeat number changes, and gene duplications.

Keywords Primate evolution · Hominoid classification · Genome sequencing · Structure of human genome · Mutation · Number of protein-coding genes

1.1 Brief Description of Human Evolution

1.1.1 From Origin of Life to Emergence of Primates

The start of this universe is currently estimated to be about 13.7 billion years ago (Hinshaw et al. 2009), and our solar system originated about 4.6 billion years ago (Tilton 1988). The earth was established also at the same time. We do not know the exact age when the primordial life-form emerged on earth, but it happened probably between 4.0 and 3.8 billion years ago (Sleep et al. 1989). The cell is the basic unit of life, and it contains many sorts of molecules, such as proteins, nucleotides, lipids, and carbohydrates. Formation of these molecules is the necessary step for the start of life, called "chemical evolution." Single-cell organisms are majority of the present-day life-forms. There are three major lineages of multicellular eukaryotes: plants, fungi, and animals. Choanoflagellates are phylogenetically the closest single-cell eukaryote species to animals (Ruiz-Trills et al. 2008). Scores of animal phyla were classified into only a few groups based mostly on developmental stages: Porifera (sponges), Radiata,

N. Saitou (✉)
Division of Population Genetics, National Institute of Genetics, Mishima, Japan
e-mail: saitounr@nig.ac.jp

N. Saitou (ed.), *Evolution of the Human Genome I*, Evolutionary Studies,
https://doi.org/10.1007/978-4-431-56603-8_1

Protostomia, and Deuterostomia. One branch of deuterostomes formed notochord at an early stage of their life history, and their descendants are phylum Chordata.

Traditionally, amphioxus was considered to be closer to vertebrates than ascidians. However, genome sequences of ascidian *Ciona intestinalis* (Dehal et al. 2002) and amphioxus (Putnam et al. 2008) now demonstrated that ascidians are closer to vertebrates. Vertebrates emerged about 500 million years ago, and the common ancestor did not have jaw, like present-day jawless vertebrates (Agnatha), lamprey and hagfish. Tetrapods, or land vertebrates, emerged about 400 million years ago. The first tetrapod was amphibian, and then amniotes followed. Reptiles, birds, and mammals are included in amniotes, and dinosaurs flourished during the Mesozoic. The common ancestor of mammals and birds diverged about 300 million years ago, and primordial mammals emerged about 150 million years ago. Monotremes are basal in mammals, and then marsupials and eutherians diverged. Radiation of eutherians started to occur before the Cenozoic started, following the continental drift. Edentates evolved in South America, while elephants, elephant shrew, aardvark, golden mole, hyrax, and tenrec evolved in Africa, and now they are called Afrotheria. It is established that an asteroid hit at the tip of the Yucatan Peninsula 65 million years ago caused mass extinction, including extinction of dinosaurs (Shulte et al. 2010). Disappearance of dinosaurs led creation of open niches, and this prompted the further diversification of mammalian families in each order during early Cenozoic.

1.1.2 Evolution of Primates and Classification of Hominidae Species

Primates probably evolved in tropical forests of Laurasia more than 100 million years ago, and prosimians (Strepsirrhini) and simians (Haplorhini) diverged about 75 million years ago (Steiper and Young 2009). Many present-day prosimians live in Madagascar Island, which separated from the African continent about 50 million years ago. The human lineage belongs to simian and is further classified to Catarrhini (Old World monkeys and hominoids), superfamily Hominoidea (hominoids), family Hominidae, and genus *Homo*. Carl Linnaeus coined Latin binomen *Homo sapiens*, in which the adjective means "clever" and genus name *Homo* means "human."

The phylogenetic position of *Homo sapiens* among the extant hominoids is well established. Hominoids include humans and apes and correspond to superfamily Hominoidea. Chimpanzee (*Pan troglodytes*) and bonobo (*Pan paniscus*) are equally close to humans, followed by gorilla (*Gorilla gorilla*) (e.g., Sibley and Ahlquist 1984, 1987; Saitou 1991; Horai et al. 1995). The close phylogenetic relationship with these African great apes suggests that the human lineage originated somewhere in Africa, as suggested by Darwin (1871). In fact, the oldest fossils considered to be part of the human lineage immediately after the divergence from the chimpanzee/bonobo lineage have so far been found only in Africa.

Chimpanzee, corresponding to genus *Pan*, has traditionally been classified into three subspecies (*P. t. troglodytes*, *P. t. verus*, and *P. t. schweinfurthii*) mainly according to their geographical distribution (Central Africa, West Africa, and East Africa, respectively) (Groves 1997). Mitochondrial DNA sequence comparisons have confirmed the existence of these three lineages (Morin et al. 1992; Gagneux et al. 1999). One group of West African chimpanzees (*P. t. verus*) in Nigeria was shown to have a rather distinct mitochondrial DNA lineage, and a new subspecies status (*P. t. vellerosus*) was proposed (Gonder et al. 1997, 2006). Later, this new subspecies was renamed as *P. t. ellioti* (Oates et al. 2009).

Bonobo, whose Latin binomen is *Pan paniscus*, is distributed to the south of the Congo (Zaire) River and is congeneric with chimpanzee. A hybrid offspring of chimpanzee and bonobo was reported (Vervaecke and van Elsacker 1992). This suggests that the reproductive barrier between these two species is relatively low. It should also be mentioned that Wildman et al. (2003) proposed that chimpanzee and bonobo should be included in the genus *Homo*.

Gorilla, corresponding to genus *Gorilla*, has been traditionally classified into three subspecies, *Gorilla gorilla gorilla*, *G. b. graueri*, and *G. b. beringei*, corresponding to western lowland gorilla, eastern lowland gorilla, and mountain gorilla, respectively (Nowark 1991). However, molecular studies indicated a clear genetic differentiation of the west and east lowland gorilla, while mountain gorilla is genetically close to the eastern lowland gorilla (Garner and Ryder 1996). IUCN Red List of Threatened Species Version 2010.1 (http://www.iucnredlist.org) was cited by Scally et al. (2012) for the current classification of gorilla. According to that, there are two congeneric species; western species (*Gorilla gorilla*) and eastern species (*G. beringei*). Western species comprise two subspecies: western lowland gorillas (*G. g. gorilla*) and Cross River gorillas (*G. g. diehli*). The eastern species is also subclassified into eastern lowland gorillas (*G. b. graueri*) and mountain gorillas (*G. b. beringei*).

Orangutans, corresponding to genus *Pongo*, are much more remotely related to humans than are the African apes. The geographical distribution of the extant orangutan is restricted to the islands of Borneo and Sumatra, and the two populations of which were traditionally classified as two subspecies (*Pongo pygmaeus* and *P. abelii*, respectively), since hybrids are fertile. Because these two subspecies are quite divergent in terms of mitochondrial DNA sequences (Xu and Arnason 1996) and because some chromosomal difference exists, they are now often considered to be congeneric but different species, *Pongo pygmaeus and Pongo abelii*, respectively. The phylogenetic relationship of the gibbons is not yet well known, and even the number of species varies depending on different reports. For example, Gloves (1997) classified gibbons (Hylobatidae) into one genus (*Hylobates*), four subgenera, and 11 species, while Napier and Napier (1985) identified only six species. One example of hybridization was reported between common gibbon and siamang (Myers and Shafer 1979).

1.1.3 Divergence Patterns of Hominoids

Molecular evolutionary studies of primates were started by G. H. F. Nuttall according to Gribbin and Charfas (1982). The phylogenetic relationship of the hominoid species was first shown by Goodman (1962), who indicated a closer relationship of the African apes (chimpanzee and gorilla) with humans than the Asian apes (orangutan and gibbon), using the semiquantitative immunodiffusion method. This discovery was later supported by many molecular data (see Saitou 2005).

In the presence of genetic polymorphisms, a phylogenetic tree reconstructed from a single gene from each species (gene tree) may be different from the true species phylogeny (species tree), even if a large number of nucleotides are used (e.g., see Nei 1987; Saitou 2013). This is especially true when the two speciation events occurred during a short period.

Many studies were conducted with the aim of determining the phylogenetic relationship among humans, chimpanzee, gorilla, and orangutan. Satta et al. (2000), Chen and Li (2001), O'hUigin et al. (2002), and Kitano et al. (2004) compared 34, 53, 51, and 103 genes or DNA segments, respectively, before genome sequences of these organisms were available. In all of these studies, about 40% of the gene trees were different from the species tree in which humans and chimpanzees are clustered. These studies finally established that chimpanzee and bonobo are the closest organisms to us humans.

The human–chimpanzee–gorilla trichotomy problem attracted many researchers for more than 40 years. After the establishment of the branching order among humans, chimpanzee/bonobo, and gorilla, it is now interesting to estimate the speciation times and effective sizes of ancestral species. Hara et al. (2012) analyzed genome data by using an evolutionary model in which mutation rates vary across lineages and chromosomes. The method used was expansion of that proposed by Takahata and Satta (1997). They estimated speciation times of the human lineage from chimpanzee, gorilla, and orangutan to be 5.9–7.6, 7.6–9.7, and 15–19 million years ago, respectively. They also estimated the population size of the common ancestor of humans and chimpanzee, that of human–chimpanzee and gorilla, and that of human–chimpanzee–gorilla and orangutan to be 59,300–75,600, 51,400–66,000, and 159,000–203,000, respectively. These new estimates may be more compatible with those envisaged from paleoanthropological studies.

1.1.4 Sequencing of Hominoid Genomes

Soon after the draft human genome sequencing was finished in 2001, sequencing projects of the chimpanzee genome emerged. One was headed by RIKEN and National Institute of Genetics in Japan, and four other groups from Korea, Taiwan, China, and Germany joined. RIKEN group first sequenced BAC ends of 64,000 chimpanzee BAC clones, and the nucleotide difference between humans and chimpanzee was determined to be 1.23% (Fujiyama et al. 2002). They then used these BES information to collect chimpanzee BAC clones which are orthologous to human chromosome 21 and determined BAC clone sequences to obtain a high-quality 33.3 megabase data for chimpanzee chromosome 22 (International Chimpanzee Chromosome 22 Consortium 2004). However, this was only a slight more than 1% of the chimpanzee genome, and 1 year later, the American group reported draft chimpanzee genome using whole genome shotgun sequencing technique (The Chimpanzee Sequencing and Analysis Consortium 2005).

Because only short nucleotide sequences are initially obtained under shotgun sequencing, chromosome-wide long genomic sequences cannot be obtained. In contrast, BAC clone-based technique can start from minimum BAC clone tiling array, and by sequencing BAC clones independently, we can determine almost full chromosome sequences. Unfortunately, next-generation sequencing methods also produce only short nucleotide sequences, and mammalian genomes which are rich in repeat sequences are still very difficult to produce long and reliable contigs. It should also be noted that chimpanzee chromosome 22 is orthologous to human chromosome 21 (e.g., Yunis and Prakash 1982). This is also true to bonobo, gorilla, and orangutan. However, some human-centric-thinking people decided to rename great ape chromosomes by human orthologous counterpart, and the genome database followed this movement. Therefore, we should be careful for chromosome numbers.

Soon after chimpanzee genome sequencing efforts, the draft genome of rhesus macaque was determined (Rhesus Macaque Genome Sequencing and Analysis Consortium 2006). However, it took several more years to generate genome sequences of orangutan (Locke et al. 2011), gorilla (Scally et al. 2012), bonobo (Prüfer et al. 2012), and gibbon (Carbone et al. 2014). Although these are all draft genomes, we now have genomic sequences of all hominoid species. Genomic diversity of great apes was also studied extensively (Prado-Martinez et al. 2013).

1.2 Structure of the Human Genome

1.2.1 Human Chromosomes and Their Band Structure

There are two sets of genomes in one human cell, and they consist of 22 pairs of autosomal chromosomes and one pair of sex chromosomes (X and Y). Females have two X chromosome and males have one X and one Y chromosomes. In total, one human cell usually has 46 chromosomes ($22 \times 2 + 2$). The total DNA amount of one human genome (haploid) was estimated to be about 3.5×10^{-12} g (Gregory 2005). The molecular weight of one nucleotide pair is 1.08×10^{-21} g, and the total number of base pair of one human genome becomes about 3.28×10^{9}. The X chromosome is much bigger than the Y chromosome; thus one female autosomal cell has a slightly larger amount of DNA than that of male.

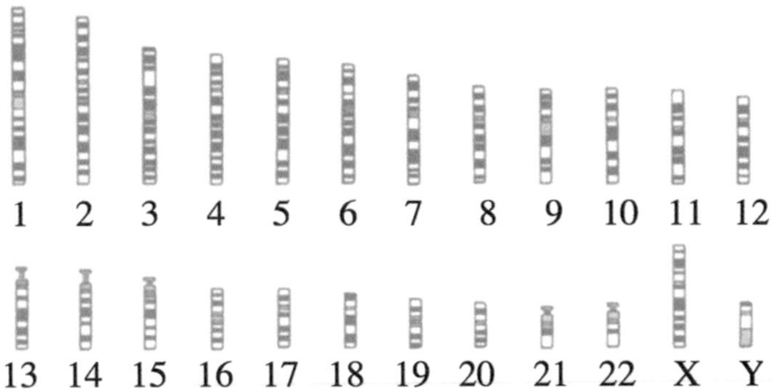

Fig. 1.1 Schematic view of 46 human chromosomes (From Saitou 2013)

Figure 1.1 shows a schematic view of these 46 human chromosomes (from Saitou 2013). These chromosomes are divided into autosomes and sex chromosomes. Autosomes are numbered according to their size, chromosome 1 being the longest. However, the shortest chromosome is not chromosome 22, but chromosome 21, because of a historical reason. There are band patterns in human chromosomes, and these patterns were caused by differential dye concentration depending on the chromosomal locations. The difference is correlated with the GC content of each DNA region, but the molecular mechanism for this is still elusive. Oota (2018; Chap. 3 of this book) discusses the GC content heterogeneity of the human genome.

Determination of nucleotide sequences of the human genome was often called "Human Genome Project" and was a symbol of the genome sequencing efforts of many organisms in the twentieth century. The draft sequences of the human genome were reported by the International Human Genome Sequencing Consortium (2001) and by Celera Genomics (Venter et al. 2001). The completion of the euchromatin region of the human genome was reported 3 years later (International Human Genome Sequencing Consortium 2004). Heterochromatins, which were unable to be sequenced under the Human Genome Project, are rich in repeat sequences. This incomplete situation remains as of 2018, even with the advent of second-generation sequencing technologies.

The majority of the human genome is junk DNA or nonfunctional noncoding sequences, though a significant proportion of junk DNA may be sporadically transcribed (The ENCODE Project Consortium 2012). Only 1.5% or 48 Mb is responsible for coding amino acid sequences. Table 1.1 is the content of the human genome (from Saitou 2013; based on Fig. 7.13 of Brown 2007). If we include introns and pseudogenes, these gene-related regions consist of 38% of the human genome, and the remaining 62% is the intergenic region. The majority of introns and intergenic regions are nonfunctional. The dispersed repeats cover 1400 Mb, or the 44% of the human genome, and they include LINEs and SINEs. Microsatellites or short tandem repeats (STRs) are known to have high mutation rates in terms of repeat number changes, and they consist of 90 Mb. It should be noted that there are

Table 1.1 The content of the human genome

Human genome [3.2 Gb]
Coding regions and related sequences [1.2 Gb]
Exons [48 Mb]
Related sequences (introns, pseudogenes, etc.) [1152 Mb]
Intergenic sequences [2.0 Gb]
Interspersed repeat sequences [1,400 Mb]
LINEs [640 Mb]
SINEs [420 Mb]
LTR elements [250 Mb]
DNA transposons [90 Mb]
Other intergenic sequences [600 Mb]
Microsatellites [90 Mb]
Various sequences [510 Mb]

From Saitou (2013)

also non-repeat, unique sequences in the intergenic regions, and some of them, about ~3% of the human genome, are highly conserved. If we combine these conserved noncoding DNA regions and also conserved protein-coding regions, about 5% of the human genome inherits the important information to shape up humans, and the rest, 95%, are mostly nonfunctional. Graur (2018; Chap. 2 of this book) discusses the functionless rubbish DNA of the human genome.

1.2.2 *Protein Genes of the Human Genome*

The number of protein-coding genes in the human genome was initially estimated to be 30,000–40,000 based on the draft genome data (International Human Genome Sequencing Consortium 2001) but was revised to be 20,000–25,000 based on the finished genome data (International Human Genome Sequencing Consortium 2004). Venter et al. (2001) found a total of 26,383 genes. Imanishi et al. (2004) annotated the 21,037 protein-coding genes based on cDNA sequences, while Clamp et al. (2007) estimated the number of proteincoding genes to be ~20,500. As of June 2016, OMIM (Online Mendelian Inheritance in Man; http://www.ncbi.nlm.nih.gov/omim) contains 23,529 entries including 3425 without molecular basis (only phenotypes such as disease), while PANTHER (Protein Analysis Through Evolutionary Relationships; http://www.pantherdb.org/) and GeneCards (http://www.genecards.org/) contain 20,814 and 21,976 protein-coding genes, respectively. Therefore, the total number of functional protein-coding genes in the human genome may be in the range of 20,000–22,000.

Table 1.2 lists the major protein-coding gene categories in the human genome (from Saitou 2013; based on PANTHER database). The most frequent gene

Table 1.2 Major categories of protein-coding genes in the human genome

Category	No.
Enzyme	6931
Nucleic acid binding	2806
Transcription factor	2179
Receptor	1904
Enzyme modulator	1592
Transporter	1151
Signaling molecule	1260
Cytoskeletal protein	1028
Defense/immunity protein	749
Cell adhesion molecule	715
Extracellular matrix protein	582
Calcium-binding protein	482
Transfer/carrier protein	475
Membrane traffic protein	425
Structural protein	331
Chaperone	224
Cell junction protein	170

From Saitou (2013) with some modification)

category is enzymes; they cover 6931 genes, or 30% of the total human genes. Nucleic acid binding (2806 genes), transcription factors (2179 genes), and receptors (1904 genes) are the next three major categories. Because transcription factors are also DNA binding, ~27% of the human genome is classified as DNA- or RNA-binding proteins. Olfactory receptors (~800) are the most dominant receptors, but more than half are pseudogenes (Nozawa et al. 2007). Imanishi (2018; Chap. 4 of this book) discusses protein-coding genes of the human genome with special reference to alternative splicing.

1.3 RNA-Coding Genes and Gene Expression Control Regions in the Human Genome

When we discuss about genes, they often mean protein-coding genes. However, there is another type of genes which code structural RNA molecules vital for the human cell. Table 1.3 shows the major RNA-coding genes in the human genome (from Saitou 2013). The total number of tRNA genes was estimated to be 625, including 110 predicted pseudogenes (from http://gtrnadb.ucsc.edu), though Ensembl BioMart database (www.ensembl.org/biomart/) gives a slightly different number (128) for tRNA pseudogenes. Known functional tRNA genes in the human

Table 1.3 List of RNA-coding genes in the human genome

Class	No.
tRNA	625
Pseudogene	110
rRNA	535
Pseudogene	179
snRNA	1951
Pseudogene	73
snoRNA	1523
Pseudogene	73
miRNA	1809
Pseudogene	15
Total	6732

From Saitou (2013)

genome are 509 for 20 standard amino acids and selenocysteine. The codon GTT corresponding to Asn has the most frequent 32 tRNA genes followed by GCA for Cys (30 tRNA genes). In contrast, 13 codons do not have their own tRNA genes. There are 535 rRNA genes in the human genome with 179 pseudogenes (www.ensembl.org/biomart/). More abundant RNA genes in the human genome are those for small nuclear RNA (snRNA) and small nucleolar RNA (snoRNA). They are relatively short (~100 bp) sequences, and 1951 snRNA and 1523 snoRNA genes were annotated (www.ensembl.org/biomart/). Genes for microRNA (miRNA) are much shorter (~22 bp) and are also quite abundant (1809 genes) in the human genome (www.ensembl.org/biomart/). There are also genes for small cytoplasmic RNA (scRNA). The total number of these RNA genes is ~7000, about one third of the protein-coding genes. Imanishi (2018; Chap. 4 of this book) discusses RNA-coding genes of the human genome.

There also exist DNA regions that control gene expressions. The ENCODE project published a series of papers on the systematic examination of transcription of the human genome in 2012 (Thurman et al. 2012; Neph et al. 2012; Gerstein et al. 2012; Djebali et al. 2012). They previously conducted a pilot study targeted for the 1% of the human genome (Birney et al. 2007). Their result – a high proportion of the human genome is transcribed – was criticized by van Bakel et al. (2010), who concluded that most "dark matter" transcripts are associated with known genes. Although later ENCODE project experiments used the ChIP-seq technology which is more precise than the ChIP-chip technology, there still seems to exist problems in the interpretation of their data. The ENCODE Project Consortium (2012) assigned biochemical functions for 80% of the human genome. This particular statement implies that these "functional" regions are not junk DNAs. However, the data remain consistent with the view that the nonconserved 80–95% of the human genome is mostly composed of nonfunctional decaying transposons: "junk"

(Eddy 2012; see also Graur et al. 2013). After more than 40 years of classic paper by Ohno (1972), the recognition that our genome has so much junk DNA is essentially correct.

However, there are many noncoding sequences which are evolutionarily highly conserved. They are usually abbreviated as CNSs (conserved noncoding sequences). Bejerano et al. (2004) compared human, mice, and rat genomes and found more than 400 CNSs (their term was "ultraconserved elements") which are at least 200 bp long and identical among three species. Those sequences were often conserved among all vertebrates. Matsunami and Saitou (2013) examined 309 vertebrate paralogous CNSs and found that those may be related to gene expressions in the brain. Takahashi and Saitou (2012) and Babarinde and Saitou (2013) studied lineage-specific CNSs in mammals, and their flanking protein-coding genes were often involved in neuronal system according to GO analysis. Babarinde and Saitou (2016) examined amniote-specific CNSs by comparing the chicken genome and four mammalian species genomes and found that physical distances between CNS and their nearest protein-coding genes were well conserved between human and mouse genomes. Saber et al. (2016) found 1658 Hominidae-specific CNSs by examining genome sequences of humans, chimpanzee, gorilla, and orangutan. Interestingly, in spite of the exact sequence identity among these four species, a high rate of nucleotide substitutions was observed in the common ancestor of Hominidae. This result suggests existence of some kind of positive selection followed by stringent purifying selection for CNSs.

1.4 Various Types of Mutations Which Accumulated During Evolution of the Human Genome

Nucleotide substitutions are mutual interchanges of four kinds of nucleotides or bases. If a substitution is between chemically similar bases, i.e., between purines (adenine and guanine) or pyrimidines (cytosine and thymine), it is called transition. If a substitution is between a purine and a pyrimidine, it is called transversion.

It was predicted that transition should occur in higher frequency than transversion, because transitions have four possible intermediate mispair states, while transversions have only two such states. Absolute rates of mutations for 12 kinds of directions are not easy to estimate, because we need to directly compare parental and offspring genomes, and the rate of fresh or de novo mutations in eukaryotes is usually quite low. Instead, we can compare evolutionarily closely related sequences. Relative mutation rates of six pairs of bases (A<==>G, C<==>T, A<==>T, A<==>C, G<==>T, and G<==>C) can be estimated by comparing many numbers of SNPs (single nucleotide polymorphisms) in one species. However, we need the closely related out-group species when the directionality of mutation comes in. International Chimpanzee Chromosome 22 Consortium (2004) used chimpanzee chromosome 22 sequence as out-group for human chromosome 21 SNP data. Later Oota et al. (2010) expanded this technique to all human chromosomes. (Please also see Oota (2018; Chap. 3 of this book) on the related problem.)

The physical order of nucleotide sequence can be modified through recombination, paralogous gene conversion, or inversion. Chromosomal level changes of DNA sequences are classified into inversion, translocation, and fusion. The largest type of mutation is genome duplication, but the last time this happened in the human lineage was ~500 million years ago.

The length of DNA does not change with nucleotide substitutions, while it is well known that genome sizes vary from organism to organism. It is thus clear that there exist mutations changing length of DNA. They are generically called "insertion" or "deletion" when the DNA length increases or decreases, respectively. When mutational directions are not known, combinations of insertions and deletions may be called gaps or indels. When the gap length is only one, this gap or indel polymorphism may be included as a special case of SNP (single nucleotide polymorphism). In real nucleotide sequence data analysis, insertions and deletions are detected only after multiple alignments of homologous sequences.

A special class of insertions and deletions is repeat number changes. If repeat unit length is very short (less than 10 nucleotides), it is called STRs (short tandem repeats) or microsatellites. In contrast, "minisatellites" or VNTRs (variable number of tandem repeats) have typically repeat unit lengths of 10–100 nucleotides. CNVs and microsatellite DNA polymorphism are discussed by Takezaki (2018; Chap. 7 of this book). Ngai and Saitou (2016) pointed out the effect of perfection status on mutation rates of microsatellites in primates.

Duplication of DNA fragment can happen in any region of chromosomes, but historically duplicated regions containing protein-coding genes were the focus of research on duplication. Therefore, when we mention "gene duplication," nongenic regions may also be included. Under this broad meaning, gene duplication can be classified into four general categories: (1) tandem duplication; (2) RNA-mediated duplication; (3) drift duplication, which was proposed by Ezawa et al. (2011); and (4) genome duplication.

Tandem duplication results in two homologous genes in close proximity with each other in the same chromosome via unequal crossing-over, while RNA-mediated duplication can create duplicate copies, complementary to original RNA molecules, far from the original gene with the help of reverse transcriptase. Kitano (2018; Chap. 5 of this book) discussed duplicated genes.

Recombination is also considered to be a mutation. Montanucci and Bertranpetit (2018; Chap. 6 of this book) discussed recombination in the human genome.

References

Babarinde IA, Saitou N (2013) Heterogeneous tempo and mode of conserved noncoding sequence evolution among four mammalian orders. Genome Biol Evol 5:2330–2343

Babarinde IA, Saitou N (2016) Genomic locations of conserved noncoding sequences and their proximal protein-coding genes in mammalian expression dynamics. Mol Biol Evol 33:1807–1817

Bejerano G et al (2004) Ultraconserved elements in the human genome. Science 304:1321–1325

Birney E et al (2007) Identifi cation and analysis of functional elements in 1% of the human genome by the ENCODE pilot project. Nature 447:799–816

Brown TA (2007) Genomes 3. Garland Science Publishing, New York

Carbone L et al (2014) Gibbon genome and the fast karyotype evolution of small apes. Nature 513:195–201

Chen FC, Li WH (2001) Genomic divergences between humans and other hominoids and the effective population size of the common ancestor of humans and chimpanzees. Am J Hum Genet 68:444–456

Clamp M et al (2007) Distinguishing protein-coding and noncoding genes in the human genome. Proc Natl Acad Sci U S A 104:19428–19433

Darwin C (1871) The descent of man and selection in relation to sex. Appleton, New York

Dehal P et al (2002) The draft genome of *Ciona intestinalis*: insights into chordate and vertebrate origins. Science 298:2157–2167

Djebali S et al (2012) Landscape of transcription in human cells. Nature 489:101–108

Eddy S (2012) The C-value paradox, junk DNA and ENCODE. Curr Biol 22:R898

Ezawa K, Ikeo K, Gojobori T, Saitou N (2011) Evolutionary patterns of recently emerged animal duplogs. Genome Biol Evol 3:1119–1135

Fujiyama A, Watanabe H, Toyoda A, Taylor TD, Itoh T, Tsai SF, Park HS, Yaspo ML, Lehrach H, Chen Z, Fu G, Saitou N, Osoegawa K, de Jong PJ, Suto Y, Hattori M, Sakaki Y (2002) Construction and analysis of a human-chimpanzee comparative clone map. Science 295:131–134

Gagneux P, Wills C, Gerloff U, Tautz D, Morin PA, Boesch C, Fruth B, Hohmann G, Ryder OA, Woodruff DS (1999) Mitochondrial sequences show diverse evolutionary histories of African hominoids. Proc Natl Acad Sci U S A 96:5077–5082

Garner KJ, Ryder O (1996) Mitochondrial DNA diversity in gorillas. Mol Phylogenet Evol 6:39–48

Gerstein MB et al (2012) Architecture of the human regulatory network derived from ENCODE data. Nature 489:91–100

Gonder MK, Oates JF, Disotell TR, Forstner MR (1997) A new west African chimpanzee subspecies? Nature 388:337

Gonder MK, Distill TR, Oates JF (2006) New genetic evidence on the evolution of chimpanzee populations and implications for taxonomy. Int J Primatol 27:1103–1127

Goodman M (1962) Evolution of the immunologic species specificity of human serum proteins. Hum Biol 34:104–150

Graur D (2018) Chapter 2. Rubbish DNA: the functionless fraction of the human genome. In: Saitou N (ed) Evolution of the human genome I. Springer, Tokyo, pp 19–60

Graur D, Zheng Y, Price N, Azevedo RBR, Zufall RA, Elhaik E (2013) On the immortality of television sets: function in the human genome according to the evolution-free gospel of ENCODE. Genome Biol Evol 5:578–590

Gregory TR (ed) (2005) The evolution of the genome. Elsevier Academic, Burlington

Gribbin J, Charfas J (1982) The monkey puzzle. Pantheon Books, New York

Groves CP (1997) Taxonomy and phylogeny of primates. In: Blancher A, Klein J, Socha W (eds) Molecular biology and evolution of blood group and MHC antigens in primates. Springer, Berlin, pp 3–23

Hara Y, Satta Y, Imanishi T (2012) Reconstructing the demographic history of the human lineage using whole-genome sequences from human and three great apes. Genome Biol Evol 4:1133–1145

Hinshaw G et al (2009) Maps, & basic results. Astrophys J Suppl Ser 180:225

Horai S, Hayasaka K, Kondo R, Tsugane K, Takahata N (1995) Recent African origin of modern humans revealed by complete sequences of hominoid mitochondrial DNAs. Proc Natl Acad Sci U S A 92:532–536

Imanishi T (2018) Chapter 4. Protein-coding and non-coding RNA genes. In: Saitou N (ed) Evolution of the human genome I. Springer, Tokyo, pp 93–115

Imanishi T et al (2004) Integrative annotation of 21,037 human genes validated by full-length cDNA clones. PLoS Biol 2:856–875

International Chimpanzee Chromosome 22 Consortium (2004) DNA sequence and comparative analysis of chimpanzee chromosome 22. Nature 429:382–388

International Human Genome Sequencing Consortium (2001) Initial sequencing and analysis of the human genome. Nature 409:860–921

International Human Genome Sequencing Consortium (2004) Finishing the euchromatic sequence of the human genome. Nature 431:931–945

Kitano T (2018) Chapter 5. Duplicated genes. In: Saitou N (ed) Evolution of the human genome I. Springer, Tokyo, pp 117–130

Kitano T, Liu Y-H, Ueda S, Saitou N (2004) Human specific amino acid changes found in 103 protein coding genes. Mol Biol Evol 21:936–944

Locke DP et al (2011) Comparative and demographic analysis of orang-utan genomes. Nature 469:529–533

Matsunami M, Saitou N (2013) Vertebrate paralogous conserved noncoding sequences may be related to gene expressions in brain. Genome Biol Evol 5:140–150

Montanucci L, Bertranpetit J (2018) Chapter 6. Recombination. In: Saitou N (ed) Evolution of the human genome I. Springer, Tokyo, pp 131–142

Morin PA, Moor JJ, Woodruff DS (1992) Identification of chimpanzee subspecies with DNA from hair and allele-specific probes. Proc R Soc Lond B 249:293–297

Myers RH, Shafer DA (1979) Hybrid ape offspring of a mating of gibbon and siamang. Science 205:308–310

Napier JR, Napier PH (1985) The natural history of the primates. MIT Press, Cambridge, MA

Nei M (1987) Molecular evolutionary genetics. Columbia University Press, New York

Neph S et al (2012) An expansive human regulatory lexicon encoded in transcription factor footprints. Nature 489:83–90

Ngai MY, Saitou N (2016) The effect of perfection statuses on mutation rates of microsatellites in primates. Anthropoid Sci. (in press) 124:85–92

Nowark RM (1991) Walker's mammals of the world, vol I, 5th edn. The Johns Hopkins University Press, Baltimore

Nozawa M, Kawahara Y, Nei M (2007) Genomic drift and copy number variation of sensory receptor genes in humans. Proc Natl Acad Sci U S A 104:20421–20426

O'huigin C, Satta Y, Takahata N, Klein J (2002) Contribution of homoplasy and of ancestral polymorphism to the evolution of genes in anthropoid primates. Mol Biol Evol 19:1501–1513

Oates JF, Groves CP, Jenkins PD (2009) The type locality of Pan troglodytes vellerosus (Gray, 1862), and implications for the nomenclature of West African chimpanzees. Primates 50:78–80

Ohno S (1972) So much "junk" DNA in our genome. Brookhaven Symp Biol 23:366–370

Oota S (2018) Chapter 3. GC content heterogeneity. In: Saitou N (ed) Evolution of the human genome I. Springer, Tokyo, pp 61–92

Oota S, Kawamura K, Kawai Y, Saitou N (2010) A new framework for studying the isochore evolution: estimation of the equilibrium GC content based on the temporal mutation rate model

Prado-Martinez et al (2013) Great ape genetic diversity and population history. Nature 499:471–475

Prüfer K et al (2012) The bonobo genome compared with the chimpanzee and human genomes. Nature 486:527–531
Putnam NH et al (2008) The amphioxus genome and the evolution of the chordate karyotype. Nature 453:1064–1071
Rhesus Macaque Genome Sequencing and Analysis Consortium (2006) Evolutionary and biomedical insights from the rhesus macaque genome. Science 316:222–234
Ruiz-Trills I, Roger AI, Burger G, Gray MW, Lang BF (2008) A phylogenomic investigation into the origin of Metazoa. Mol Biol Evol 25:664–672
Saber MM, Babarinde IA, Hettiarachchi N, Saitou N (2016) Emergence and evolution of Hominidae-specific coding and noncoding genomic sequences. Genome Biol Evol. (advance access) 8:2076–2092
Saitou N (1991) Reconstruction of molecular phylogeny of extant hominoids from DNA sequence data. Am J Phys Anthropol 84:75–85
Saitou N (2005) Evolution of hominoids and the search for a genetic basis for creating humanness. Cytogenet Genome Res 108:16–21
Saitou N (2013) Introduction to evolutionary genomics. Springer, Berlin
Satta Y, Klein J, Takahata N (2000) DNA archives and our nearest relative: the trichotomy problem revisited. Mol Phylogenet Evol 14:259–275
Scally A et al (2012) Insights into hominid evolution from the gorilla genome sequence. Nature 483:169–175
Shulte P et al (2010) The Chicxulub asteroid impact and mass extinction at the cretaceous-paleogene boundary. Science 327:1214–1218
Sibley CG, Ahlquist JE (1984) The phylogeny of the hominoid primates, as indicated by DNA-DNA hybridization. J Mol Evol 20:2–15
Sibley CG, Ahlquist JE (1987) DNA hybridization evidence of hominoid phylogeny: results from an expanded data set. J Mol Evol 26:99–121
Sleep NH, Zahnle KJ, Kasting JF, Morowitz HJ (1989) Annihilation of ecosystems by large asteroid impacts on the early earth. Nature 342:139–142
Steiper ME, Young NM (2009) Chapter 74: Primates (primates). In: Hedges SB, Kumar S (eds) The timetree of life. Oxford University Press, Oxford, pp 482–486
Takahashi M, Saitou N (2012) Identification and characterization of lineage-specific highly conserved noncoding sequences in mammalian genomes. Genome Biol Evol 4:641–657
Takahata N, Satta Y (1997) Evolution of the primate lineage leading to modern humans: phylogenetic and demographic inferences from DNA sequences. Proc Natl Acad Sci U S A 94:4811–4815
Takezaki N (2018) Chapter 7. CNVs and microsatellite DNA polymorphism. In: Saitou N (ed) Evolution of the human genome I. Springer, Tokyo, pp 143–155
The Chimpanzee Sequencing and Analysis Consortium (2005) Initial sequence of the chimpanzee genome and comparison with the human genome. Nature 437:69–87
The ENCODE Project Consortium (2012) An integrated encyclopedia of DNA elements in the human genome. Nature 489:57–74
Thurman RE et al (2012) The accessible chromatin landscape of the human genome. Nature 489:75–82
Tilton GG (1988) Age of the solar system. In: Meteorites and the early solar system. University of Arizona Press, Tucson, pp 259–275
van Bakel H, Nislow C, Blencowe BJ, Hughes TR (2010) Most "dark matter" transcripts are associated with known genes. PLoS Biol 8:e1000371
Venter JC et al (2001) The sequence of the human genome. Science 291:1304–1351
Vervaecke H, van Elsacker L (1992) Hybrids between common chimpanzees (*Pan troglodytes*) and pygmy chimpanzees (*Pan paniscus*) in captivity. Mammalia 56:667–669
Wildman DE, Uddin M, Liu G, Li G, Goodman M (2003) Implications of natural selection in shaping 99.4% nonsynonymous DNA identity between humans and chimpanzees: enlarging genus Homo. Proc Natl Acad Sci U S A 100:7181–7188

Xu X, Arnason U (1996) The mitochondrial DNA molecule of Sumatran orangutan and a molecular proposal for two (Bornean and Sumatran) species of orangutan. J Mol Evol 43:431–437

Yunis JJ, Prakash O (1982) The origin of man: a chromosomal pictorial legacy. Science 215:1525–1530

Chapter 2
Rubbish DNA: The Functionless Fraction of the Human Genome

Dan Graur

Abstract Because genomes are products of natural processes rather than "intelligent design," all genomes contain functional and nonfunctional parts. The fraction of the genome that has no biological function is called "rubbish DNA." Rubbish DNA consists of "junk DNA," i.e., the fraction of the genome on which selection does not operate, and "garbage DNA," i.e., sequences that lower the fitness of the organism but exist in the genome because purifying selection is neither omnipotent nor instantaneous. In this chapter, I (1) review the concepts of genomic function and functionlessness from an evolutionary perspective, (2) present a precise nomenclature of genomic function, (3) discuss the evidence for the existence of vast quantities of junk DNA within the human genome, (4) discuss the mutational mechanisms responsible for generating junk DNA, (5) spell out the necessary evolutionary conditions for maintaining junk DNA, (6) outline various methodologies for estimating the functional fraction within the genome, and (7) present a recent estimate for the functional fraction of our genome.

Keywords Human genome · Evolution · Functional DNA · Rubbish DNA · Junk DNA · Garbage DNA · Literal DNA · Indifferent DNA · Pseudogenes · Lazarus DNA · Zombie DNA · Jekyll-to-Hyde DNA

2.1 Introduction

While evolutionary biologists and population geneticists have been comfortable with the concept of genomic functionlessness for more than half a century, classical geneticists and their descendants in the filed of genomics have continued to exist in an imaginary engineered world, in which each and every nucleotide in the genome is assumed to have a function and evolution counts for naught. Under this pre-Darwinian mindset, for instance, Vogel (1964) estimated the human genome

D. Graur (✉)
Department of Biology & Biochemistry, University of Houston, Houston, TX, USA
e-mail: dgraur@gmail.com

N. Saitou (ed.), *Evolution of the Human Genome I*, Evolutionary Studies,
https://doi.org/10.1007/978-4-431-56603-8_2

to contain approximately 6.7 million protein-coding genes. Interestingly, while Vogel (1964) deemed this number to be "disturbingly high," he could not bring himself to admit even a small fraction of "meaningless" DNA. Instead, he postulated one of two possibilities: either the protein-coding genes in humans are 100 times larger than those in bacteria or "systems of higher order which are connected with structural genes in operons and regulate their activity" occupy a much larger part of the genetic material than do protein-coding genes.

Since 1964, the number of protein-coding genes in the human genome has come down considerably, in a process that has been at times extremely humbling. Moreover, scientists discovered that many other species, including plants and unicellular organisms, have more genes than we do (Pertea and Salzberg 2010). Some estimates for the number of protein-coding genes before the Human Genome Project ranged from 100,000 to more than half a million. These estimates were drastically reduced with the publication in 2001 of the draft human genome, in which the number was said to be 26,000–30,000 (Lander et al. 2001). In 2004, with the publication of the "finished" euchromatic sequence of the human genome (International Human Genome Sequencing Consortium 2004), the number was reduced to ~24,500, and in 2007, it further decreased to ~20,500. The lowest ever protein-coding gene count was 18,877 (Pruitt et al. 2009), which agrees quite well with the newest estimate for the number of protein-coding genes in the human nuclear genome (Ezkurdia et al. 2014).

Protein-coding genes turned out to occupy approximately 2% of the human genome. Of course, no one in his right mind thought that all the 98% or so of the human genome that does not encode proteins is functionless. Revisionist claims that equate noncoding DNA with junk DNA (e.g., Hayden 2010) merely reveal that people who are allowed to exhibit their logorrhea in *Nature* and other glam journals are as ignorant as the worst young-earth creationists. The question of genome functionality can be phrased qualitatively (Does the human genome contain functionless parts?) or quantitatively (What proportion of the human genome is functional?). Among people unversed in evolutionary biology (e.g., ENCODE Project Consortium 2012), a misconception exists according to which genomes that are wholly functional can be produced by natural processes. Actually, for the evolutionary process to produce a wholly functional genome, several conditions must be met: (1) the effective population size needs to be enormous—infinite to be precise, (2) the deleterious effects of increasing genome size by even a single nucleotide should be considerable, and (3) the generation time has to be very short. In other words, never! Not even in the commonest of bacterial species population on Earth are these conditions met, let alone in species with small effective population sizes and long generation times such as perennial plants and humans. A genome that is 100% functional is a logical impossibility.

2.2 What Is Function?

Like many words in the English language, "function" has numerous meanings. In biology, there are two main concepts of function: the "selected effect" and "causal role." The selected-effect function, also referred to as the proper biological function, is a historical concept. In other words, it explains the origin, the cause (etiology), and the subsequent evolution of the trait (Millikan 1989; Neander 1991). Accordingly, for a trait, *T*, to have a selected-effect function, *F*, it is necessary and sufficient that the following two conditions hold: (1) *T* originated as a "reproduction" (a copy or a copy of a copy) of some prior trait that performed *F* (or some function similar to *F*, say F') in the past, and (2) *T* exists because of *F* (Millikan 1989). In other words, the selected-effect function of a trait is the effect for which the trait was selected and by which it is maintained. The selected-effect function answers the question: Why does *T* exist?

The causal-role function is ahistorical and nonevolutionary (Cummins 1975; Amundson and Lauder 1994). That is, for a trait, *Q*, to have a causal-role function, *G*, it is necessary and sufficient that *Q* performs *G*. The causal-role function answers the question: What does *Q* do? Most biologists follow Dobzhansky's dictum according to which biological meaning can only be derived from evolutionary context. Hence, with few exceptions, they use the selected-effect concept of function. We note, however, that the causal-role concept may sometimes be useful, for example, as an *ad hoc* device for traits whose evolutionary history and underlying biology are obscure. Furthermore, we note that all selected-effect functions have a causal role, while the vast majority of causal-role functions do not have a selected-effect function. It is, thus, wrong to assume that all causal-role functions are biologically relevant. Doolittle et al. (2014), for instance, prefers to restrict the term "function" to selected-effect function and to refer to causal-role function as "activity."

Using the causal-role concept of function in the biological sciences can lead to bizarre outcomes. For example, while the selected-effect function of the heart can be stated unambiguously to be the pumping of blood, the heart may be assigned many additional causal-role functions, such as adding 300 g to body weight, producing sounds, preventing the pericardium from deflating onto itself, and providing an inspiration for love songs and Hallmark cards (Graur et al. 2013). The thumping noise made by the heart is a favorite of philosophers of science; it is a valuable aid in medical diagnosis, but it is not the evolutionary reason we have a heart. An even greater absurdity of using the causal-role concept of function arises when realizing that every nucleotide in a genome has a causal role—it is replicated! Does that mean that every nucleotide in the genome has evolved for the purpose of being replicated?

Distinguishing between what a genomic element does (its causal-role activity) and why it exists (its selected-effect function) is a very important distinction in biology (Huneman 2013; Brunet and Doolittle 2014). Ignoring this distinction, and assuming that all genomic sites that exhibit a certain biochemical activity are functional, as was done, for instance, by the ENCODE Project Consortium (2012), is equivalent to claiming that following a collision between a car and a pedestrian, a car's hood would be ascribed the "function" of harming the pedestrian, while the pedestrian would have the "function" of denting the car's hood (Hurst 2013).

The main advantage of the selected-effect function definition is that it suggests a clear and conservative method for inferring function in a DNA sequence—only sequences that can be shown to be under selection can be claimed with any degree of confidence to be functional. From an evolutionary viewpoint, a function can be assigned to a DNA sequence if and only if it is possible to destroy it (Graur et al. 2013). All functional entities in the universe can be rendered nonfunctional by the ravages of time, entropy, mutation, and what have you. Unless a genomic functionality is actively protected by selection, it will accumulate deleterious mutations and will cease to be functional. The absurd alternative is to assume that function can be assessed independently of selection, i.e., that no deleterious mutation can ever occur in the region that is deemed to be functional. Such an assumption is akin to claiming that a television set left on and unattended will still be in working condition after a million years because no natural events, such as rust, erosion, static electricity, and the gnawing activity of rodents, can affect it (Graur et al. 2013). A convoluted "rationale" for discarding natural selection as the arbiter of functionality was put forward by Stamatoyannopoulos (2012). This paper should be read as a cautionary tale of how genome biology has been corrupted by medical doctors and other ignoramuses who uncritically use the causal-role concept of function.

Function should always be defined in the present tense. In the absence of prophetic powers, one cannot use the potential for creating a new function as the basis for claiming that a certain genomic element is functional. For example, the fact a handful of transposable elements have been coopted into function cannot be taken as support for the hypothesis that all transposable elements are functional. In this respect, the Aristotelian difference between "potentiality" and "actuality" is crucial.

What is the proper manner in which null hypotheses concerning the functionality or nonfunctionality of a particular genomic element should be phrased? Most science practitioners adhere to Popper's system of demarcation according to which scientific progress is achieved through the falsification of hypotheses that do not withstand logical or empirical tests. Thus, a null hypothesis should be phrased in such a manner as to spell out the conditions for its own refutation. Should one assume lack of functionality as the null hypothesis, or should one assume functionality? Let us consider both cases. A statement to the effect that a genomic element is devoid of a selected-effect function can be easily rejected by showing that the element evolves in a manner that is inconsistent with the expectations of strict neutrality. If, on the other hand, one assumes as the null hypothesis that an element is functional, then failing to find telltale indicators of selection cannot be interpreted as a rejection of the hypothesis, merely as a sign that we have not searched thoroughly enough or that the telltale sign of selection have been erased by subsequent evolutionary events.

There exists a fundamental asymmetry between verifiability and falsifiability in science: scientific hypotheses can never be proven right; they can only be proven wrong. The hypothesis that a certain genomic element is functional can never be rejected and is, hence, unscientific. According to physicist Wolfgang Pauli (quoted in Peierls 1960), a hypothesis that cannot be refuted "is not only not right, it is not even wrong".

2.3 An Evolutionary Classification of Genome Activity and Genome Function

In terms of its biochemical activities, a genome can be divided into three classes: (1) regions that are transcribed and translated, (2) regions that are transcribed but not translated, and (3) regions that are not transcribed (Fig. 2.1). Each of these three classes can be either functional or functionless. Activity has nothing to do with function. A genome segment may be biochemically active yet have no biologically meaningful function. An analogous situation arises when my shoe binds a piece of chewing gum during hot days in Houston. The binding of the chewing gum is an observable activity, but no reasonable person would claim that this is the function of shoes.

Genomic sequences are frequently referred to by their biochemical activity, regardless of whether or not such activity is biologically meaningful. The need for a rigorous evolutionary classification of genomic elements by selected-effect function arises from two erroneous and sometimes deliberately disingenuously equivalencies that are frequently found in the literature. The first equivalency, usually espoused in the medical and popular literature (e.g., Krams and Bromberg 2013; Mehta et al. 2013), misleadingly and inappropriately synonymizes "noncoding DNA"—i.e., all regions in the genome that do not encode proteins—with "junk DNA," i.e., all regions in the genome that are neither functional nor deleterious. The second even more pernicious equivalency transmutes every biochemical

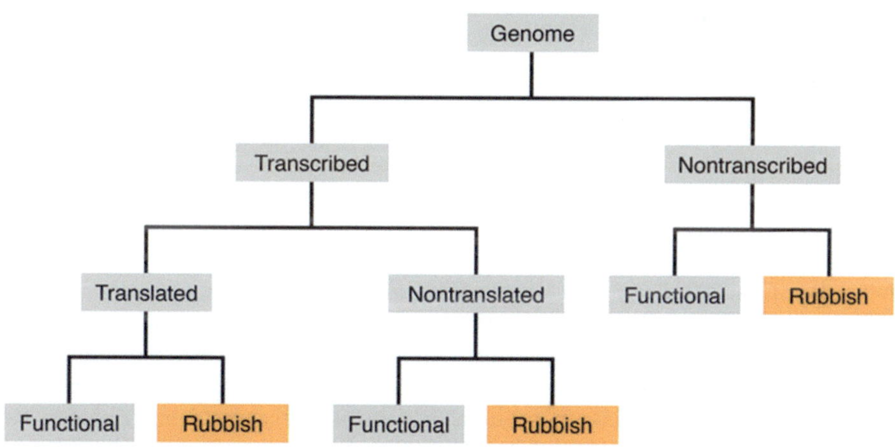

Fig. 2.1 A classification of genomic segments by biochemical activity. Each of the three categories can be functional or functionless (rubbish)

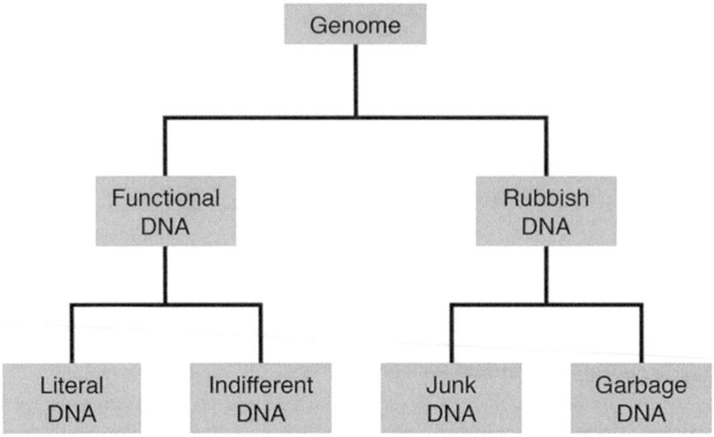

Fig. 2.2 An evolutionary classification of genomic elements according to their selected-effect function (From Graur 2016)

activity into a function (e.g., ENCODE Project Consortium 2012; Sundaram et al. 2014; Kellis et al. 2014).

The classification scheme presented below is based on Graur et al. (2015). The genome is divided into "functional DNA" and "rubbish DNA" (Fig. 2.2). Functional DNA refers to any segment in the genome whose selected-effect function is that for which it was selected and/or by which it is maintained. Most functional sequences in the genome are maintained by purifying selection. Less frequently, functional sequences exhibit telltale signs of either positive or balancing selection. A causal-role activity, such as "low-level noncoding RNA transcription" (e.g., Kellis et al. 2014), is an insufficient attribute of functionality.

Functional DNA is further divided into "literal DNA" and "indifferent DNA." In literal DNA, the order of nucleotides is under selection. Strictly, a DNA element of length l is defined as literal DNA if its function can be performed by a very small subset of the 4^l possible sequences of length l. For example, there are three possible sequences of length 3 that can encode isoleucine according to the standard genetic code, as opposed to the much larger number, 64, of possible three-nucleotide sequences. Functional protein-coding genes, RNA-specifying genes, and untranscribed control elements are included within this category.

Indifferent DNA includes genomic segments that are functional and needed, but the order of nucleotides in their sequences is of little consequence. In other words, indifferent DNA refers to sequences whose main function is being there but whose exact sequence is not important. They serve as spacers, fillers, and protectors against frameshifts and may possess nucleotypic functions, such as determining nucleus size. The third codon position in fourfold degenerate codons may be regarded as a simple example of indifferent DNA; the nucleotide that resides at this position is unimportant, but the position itself needs to be occupied. Thus, indifferent DNA should show no evidence of selection for or against point mutations, but deletions and insertions should be under purifying selection.

Rubbish DNA (Brenner 1998) refers to genomic segments that have no selected-effect function. Rubbish DNA can be further subdivided into "junk DNA" and "garbage DNA." We have written evidence that the term "junk DNA" was already in use in the early 1960s (e.g., Aronson et al. 1960; Ehret and de Haller 1963); however, it was Susumu Ohno (1972, 1973) who formalized its meaning and provided the first evolutionary rationale for its existence. "Junk DNA" refers to a genomic segment on which selection does not operate, and, hence, it evolves neutrally. Of course, some junk DNA may acquire a useful function in the future, although such an even is expected to occur only very rarely. Thus, the "junk" in "junk DNA" is identical in its meaning to the colloquial "junk," such as when a person mentions a "garage full of junk," in which the implications are (1) that the garage fulfills its intended purpose, (2) that the garage contains useless objects, and (3) that in the future some of the useless objects may (or may not) become useful. Of course, as in the case of the garage full of junk, the majority of junk DNA will never acquire a function. Junk DNA and the junk in one's garage are also similar in that "they may be kept for years and years and, then, thrown out a day before becoming useful" (David Wool, personal communication).

The term "junk DNA" has generated a lot of controversy. First, because of linguistic prudery and the fact that "junk" is used euphemistically in off-color contexts, some biologists find the term "junk DNA" "derogatory" and "disrespectful" (e.g., Brosius and Gould 1992). An additional opposition to the term "junk DNA" stems from false teleological reasoning. Many researchers (e.g., Makalowski 2003; Wen et al. 2012) use the term "junk DNA" to denote a piece of DNA that can never, under any evolutionary circumstance, be selected for or against. Since every piece of DNA may become advantageous or deleterious by gain-of-function mutations, this type of reasoning is indefensible. A piece of junk DNA may indeed be coopted into function, but that does not mean that it will be, let alone that it currently has a function. Finally, some opposition to the term is related to the antiscientific practice of assuming functionality as the null hypothesis (Petsko 2003).

Garbage DNA refers to sequences that exist in the genome despite being actively selected against. The reason that detrimental sequences are observable is that selection is neither omnipotent nor rapid. At any slice of evolutionary time, segments of garbage DNA (presumably on their way to becoming extinct) may be found in the genome. The distinction between junk DNA and garbage DNA was suggested by Brenner (1998):

> Some years ago I noticed that there are two kinds of rubbish in the world and that most languages have different words to distinguish them. There is the rubbish we keep, which is junk, and the rubbish we throw away, which is garbage. The excess DNA in our genomes is junk, and it is there because it is harmless, as well as being useless, and because the molecular processes generating extra DNA outpace those getting rid of it. Were the extra DNA to become disadvantageous, it would become subject to selection, just as junk that takes up too much space, or is beginning to smell, is instantly converted to garbage by one's wife, that excellent Darwinian instrument.

Each of the four functional categories described above can be (1) transcribed and translated, (2) transcribed but not translated, or (3) not transcribed. Hence, we may encounter, for instance, junk DNA, literal RNA, and garbage proteins.

2.4 Changes in Functional Affiliation

The affiliation of a DNA segment to a particular functional category may change during evolution. With four functional categories, there are 12 possible such changes. Several such changes are known to occur quite frequently (Fig. 2.3). For example, junk DNA may become garbage DNA if the effective population size increases; the opposite will occur if the effective population size decreases (Ohta 1973). Many of the 12 possible changes have been documented in the literature. Pseudogenes, for instance, represent a change in functional status from literal DNA to junk DNA, while some diseases are caused by either a change from functional DNA to garbage DNA (e.g., Chen et al. 2003) or from junk DNA to garbage DNA (Cho and Brant 2011).

Rubbish DNA mutating to functional DNA may be referred to as "Lazarus DNA," so named after Lazarus of Bethany, the second most famous resurrected corpse in fiction (John 11:38–44; 12:1; 12:9; 12:17). Similarly, functional DNA may mutate to garbage DNA, in which case the term "Jekyll-to-Hyde DNA," based on the fictional transformation of a benevolent entity into a malicious one (Stevenson 1886), was suggested. Garbage DNA may also be derived from junk DNA, for which the term "zombie DNA" seems appropriate (Kolata 2010).

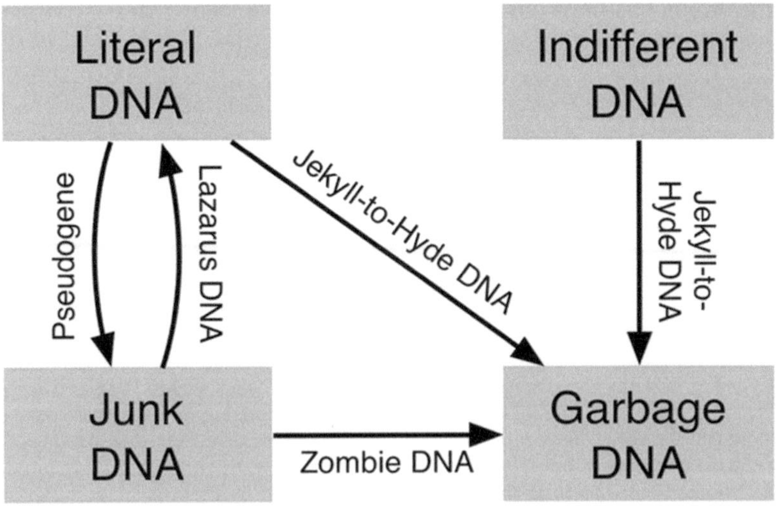

Fig. 2.3 A nomenclature for some possible changes in the functional affiliation of genomic elements (From Graur 2016)

2.5 Genome Size Variation

Because "genome size" and "DNA content" are used inconsistently in the literature (Greilhuber et al. 2005), when referring to the size of the haploid genome, it is advisable to use the unambiguous term "C value" (Swift 1950), where "C" stands for "constant," to denote the fact that the intraspecific variability in haploid genome size is substantially smaller than the interspecific variability.

With very few exceptions, eukaryotes have much larger genomes than prokaryotes. C values in eukaryotes range from less than 4×10^5 bp in the secondary plastid of the chlorarachniophyte *Bigelowiella natans* (Douglas et al. 2001; Gilson et al. 2006) to approximately 1.5×10^{11} bp in the canopy plant *Paris japonica* (Pellicier et al. 2010), close to a 400,000-fold range.

What can explain the huge variation in eukaryotic genome size? Let us first investigate if any genomic compartment correlates with genome size. First, we note that in contradistinction to the situation in prokaryotes, only a miniscule fraction of the eukaryotic genome is occupied by protein-coding sequences. Moreover, the number of protein-coding genes does not correlate with genome size. To illustrate this fact, let us compare the human genome with the genome of a teleost fish called *Takifugu rubripes* (formerly *Fugu rubripes*). At 400 Mb, the *T. rubripes* genome is one of the smallest vertebrate genomes (Aparicio et al. 2002; Noleto et al. 2009). Interestingly, although the length of the *Takifugu* genome is less than about one-eighth that of the human genome, it contains a comparable number of protein-coding genes (Aparicio et al. 2002). It is for this reason that Sydney Brenner regarded the *Takifugu* genome as "the *Reader's Digest* version" of the human genome (quoted in Purves et al. 2004). The small genome size of *Takifugu* can be attributed to a reduction in intron and intergenic lengths, a lack of significant amounts of repetitive sequences, and a very small number of pseudogenes—the genome of *Takifugu rubripes* has no mitochondrial pseudogenes and only 162 nuclear pseudogenes versus at least 15,000 pseudogenes in the human genome (Hazkani-Covo et al. 2010). Many other small-genome organisms have *Takifugu*-like characteristics. For example, the carnivorous bladderwort plant *Utricularia gibba* has a tiny genome, yet it accommodates a typical number of protein-coding genes (~30,000) for a plant (Ibarra-Lachlette et al. 2013).

Can genome size variation be explained by other variables related to protein-coding genes? While there are small differences in the mean mRNA length among different organisms, no correlation exists between this variable and genome size. For instance, mRNAs are slightly longer in multicellular organisms than in protists (1400–2200 bp versus 1200–1500 bp), whereas genome sizes in protists can be much larger than those of multicellular organisms. Similarly, organisms with larger genomes do not always produce larger proteins nor do they have longer introns.

Finally, while there exists a significant positive correlation between the degree of repetition of several RNA-specifying genes and genome size, these genes constitute only a negligible fraction of the eukaryotic genome, such that the variation in the number of RNA-specifying genes cannot explain the variation in genome size.

2.6 The C-Value Paradox as Evidence for the Existence of Junk DNA Within the Human Genome

A genomic paradox refers to the lack of correspondence between a measure of genome size or genome content and the presumed amount of genetic information "needed" by the organism, i.e., its complexity or organismal complexity. Different measures of complexity have been put forward in different fields. However, because researchers ask similar questions about the complexity of their different subjects of research, the answers that they came up with for how to measure complexity in biology, computer science, or finance bear a considerable similarity to one another (Lloyd 2001). Measures of complexity of an entity attempt to answer three questions: (1) How hard is it to describe? (2) How hard is it to create or evolve? (3) What is its degree of organization?

The simplest measures of organismal complexity are the number of cells and the number of cell types (Kaufman 1971). The vast majority of eukaryotic taxa are unicellular, i.e., they have a single cell and a single cell type. In animals, the number of cells varies between 10^3 in some nematodes and 10^{12} in some mammals; the number of cell types varies from 4 and 11 in placozoans and sponges, respectively, to approximately 200 in mammals (Klotzko 2001; Srivastava et al. 2008; Goldberg 2013). The number of cell types in plants is about one order of magnitude smaller than that in mammals (Burgess 1985). Thus, organismal complexity as measured by the number of cell types ranges from 1 to approximately 200, whereas C values range over a 400,000-fold range. Numerous lines of evidence indicate that there is no relationship between genome size and organismal complexity. For example, many unicellular protozoans, which by definition have a single cell and a single cell type, possess much larger genomes than mammals, which possess trillions of cells and 200 or more cell types, and are presumably infinitely more "sophisticated" than protists (Table 2.1). The same goes for the comparison between plants and mammals. Many plants have much larger genomes than mammals while at the same time possessing fewer cell types. Furthermore, organisms that are similar in morphological and anatomical complexity (e.g., flies and locusts or onion and lily) exhibit vastly different C values (Table 2.1). This lack of a positive correlation between organismal complexity and genome size is also evident in comparison with sibling species (i.e., species that are so similar to each other morphologically as to be indistinguishable phenotypically). In insects, protists, bony fishes, amphibians, and flowering plants, many sibling species differ greatly in their C values, even though by definition no difference in organismic complexity exists. The example of sibling species is extremely illuminating since it tells that the same level of complexity can be achieved by vastly different amounts of genomic DNA.

Table 2.1 C values of a few eukaryotic organisms ranked by genome size

Species[a]	C value (Mb)
***Saccharomyces cerevisiae* (baker's yeast)**	**13**
Caenorhabditis elegans (nematode)	78
Ascidia atra (sea squirt)	160
Drosophila melanogaster (fruit fly)	180
***Paramecium aurelia* (ciliate)**	**190**
Oryza sativa (rice)	590
Strongylocentrotus purpuratus (sea urchin)	870
***Gymnosporangium confusum* (rust fungus)**	**893**
Gallus domesticus (chicken)	1200
Lampetra planeri (brook lamprey)	1900
Boa constrictor (snake)	2100
Canis familiaris (dog)	2900
***Homo sapiens* (human)**	**3100**
Nicotiana tabacum (tobacco plant)	3800
Locusta migratoria (migratory locust)	6600
Schistocerca gregaria (desert locust)	9300
Allium cepa (onion)	15,000
***Coscinodiscus asteromphalus* (centric diatom)**	**25,000**
Lilium formosanum (lily)	36,000
Psilotum nudum (skeleton fork fern)	71,000
Amphiuma means (two-toed salamander)	84,000
Pinus resinosa (Canadian red pine)	68,000
Protopterus aethiopicus (marbled lungfish)	130,000
Paris japonica (canopy plant)	150,000

[a]Unicellular eukaryotes and humans are listed in **bold** letters

This lack of correspondence between C values and the presumed complexity of the organisms has become known in the literature as the C-value paradox (Thomas 1971) or the C-value enigma (Gregory 2001). There are two facets to the C-value paradox. The first concerns the undisputable fact that genome size cannot be used as a predictor of organismal complexity. That is, from knowledge of the C value, it is impossible to say whether an organism is unicellular or multicellular, whether the genome contains few or many genes, or whether the organism is made of a few or a couple of hundreds of cell types. The second facet reflects an unmistakably anthropocentric bias. For example, some salamanders have genome sizes that are almost 40 times bigger than that of humans. So, although there is no definition of organismal complexity that shows that salamanders are objectively less complex than humans (Brookfield 2000), it is still very difficult for us humans to accept the fact that our species, from the viewpoint of genome size, does not look at all like the "pinnacle of creation" and the "paragon of animals." Realizing that our genome is so much smaller than those of frogs, sturgeons, shrimp, squids, flatworms, mosses, onions, daffodils, and amoebas can be quite humbling, if not outright insulting.

Moran (2007) referred to the anthropocentric difficulties of some people to accept their reduced genomic status as the "deflated ego problem."

The C-value paradox requires us to explain why some very complex organisms have so much less DNA than less complex ones. Why do humans have less than half the DNA in the genome of a unicellular ciliate? Since it would be illogical to assume that an organism possesses less DNA than the amount required for its vital functions, the logical inference is that many organisms have vast excesses of DNA over their needs. Hence, large genomes must contain unneeded and presumably functionless DNA. This point of view, incidentally, is not very popular with people who regard organisms as epitomes of perfection either because they believe in creationism directed by an omnipotent being or because they lack proper training in evolutionary biology and erroneously believe that natural selection is omnipotent.

As noted by Orgel and Crick (1980), if one assumes that genomes are 100% functional, then one must conclude that the number of genes needed by a salamander is 20 times larger than that in humans. It should be noted that organisms that have much larger genomes than humans are neither rare nor exceptional. For example, more than 200 salamander species have been analyzed thus far, and all their genomes have been found to be 4–35 times larger than the human genome.

These observations pose an insurmountable challenge to any claim that most eukaryotic DNA is functional. The challenge is beautifully illustrated by "the onion test" (Palazzo and Gregory 2014). The domestic onion, *Allium cepa*, is a diploid plant ($2n = 16$) with a haploid genome size of roughly 16 Gb, i.e., approximately five times larger than that of humans. (The example of the onion was chosen arbitrarily, presumably for its shock value—any other species with a large genome could have been chosen for this comparison.) The onion test simply asks: if most DNA is functional, then why does an onion require five times more DNA than a human?

If one assumes that the human genome is entirely functional, then the human genome becomes the Goldilocks of genomes—not too small for the organism who defines itself as the "pinnacle of creation," yet not too big for it to become littered with functionless DNA. The onion genome, on the other hand, would by necessity be assumed to possess junk DNA lest we regard onions as our superiors while the *Drosophila* genome would be interpreted as a sign of diminished organismal complexity. The C-value paradox and the rejection of our singularly exclusive genomic status as Goldilocks inevitably lead us to the conclusion that we possess junk DNA.

2.7 Genetic Mutational Load: Can the Human Genome Be 100% Functional?

Many evolutionary processes can cause a population to have a mean fitness lower than its theoretical maximum. For example, deleterious mutations may occur faster than selection can get rid of them, recombination may break apart favorable combinations of alleles creating less fit combinations, and genetic drift may cause allele frequencies to change in a manner that is antagonistic to the effects of natural selection. Genetic load is defined as the reduction in the mean fitness of a population relative to a population composed entirely of individuals having the maximal fitness. The basic idea of the genetic load was first discussed by Haldane (1937) and later by Muller (1950).

If $w_{\max}$ is the fitness of the fittest possible genotype and $\bar{w}$ is the mean fitness of the actual population, then the genetic load (L) is the proportional reduction in mean fitness relative to highest possible fitness:

$$L = \frac{w_{\max} - \bar{w}}{w_{\max}} \tag{2.1}$$

Here, we shall consider the mutational genetic load, i.e., the reduction in mean population fitness due to deleterious mutations. Haldane (1937) showed that the decrease of fitness in a species as a consequence of recurrent mutation is approximately equal to the total mutation rate multiplied by a factor that ranges from 1 to 2 depending on dominance, inbreeding, and sex linkage. For example, for a recessive mutation, it has been shown that as long as the selective disadvantage of the mutant is larger than the mutation rate and the heterozygote fitness is neither larger nor smaller that the fitness values of the homozygotes, the mutational load is approximately equal to the mutation rate (Kimura 1961; Kimura and Maruyama 1966). Thus

$$L = \mu \tag{2.2}$$

and

$$\bar{w} = (1 - \mu)^n \tag{2.3}$$

where n is the number of loci. In the following we only deal with recessive mutations. We note, however, that under the same conditions, the mutational load for a dominant mutation is approximately twice the mutation rate.

In randomly mating populations at equilibrium, the mutational load does not depend on the strength of the selection against the mutation. This surprising result comes from the fact that alleles under strong selection are relatively rare, but their effects on mean fitness are large, while the alleles under weak purifying selection are common, but their effects on mean fitness are small; the effects of these two types of mutation neatly cancel out. As a result, in order to understand the magnitude of mutation load in randomly mating populations, we conveniently need only know the deleterious mutation rate, not the distribution of fitness effects.

Let us now consider the connection between mutational genetic load and fertility. The mean fertility of a population $(\bar{F})$ is the mean number of offspring born per individual. If the mortality rate before reproduction age is 0 and mean fertility is 1, then the population will remain constant in size from generation to generation. In real populations, however, the mortality rate before reproduction is greater than

0 and, hence, means fertility needs to be larger than 1 to maintain a constant population size. In the general case, for a population to maintain constant size, its mean fertility should be

$$\bar{F} = \frac{1}{w} \tag{2.4}$$

Let us consider the following example from Nei (2013). Assume that there are 10,000 loci in the genome and that the mutation rate is $\mu = 10^{-5}$ per locus per generation. Under the assumption that all mutants are deleterious and recessive, the mutational load is $L = 10,000 \times 10^{-5} = 0.1$, and the mean fitness of the population is $\bar{w} = \left(1 - 10^{-5}\right)^{10,000} \approx 0.90$. Therefore, the average fertility is $\bar{F} = \frac{1}{0.90} = 1.11$. That is, each individual should have on average 1.11 descendants for the population to remain constant in size. For a mammal, like humans, such a load is easily bearable.

Let us now assume that the entire diploid human genome ($2 \times 3 \times 10^9$ bp) is functional and contains functional elements only. If the length of each functional element is the same as that of a bacterial protein-coding gene, i.e., ~1000 nucleotides, then the human genome should consist of approximately three million functional loci. With a mutation rate of 10^5 per locus per generation, the total mutational load would be $L = 30$, and the mean population fitness is 9×10^{-14}. The average individual fertility required to maintain such a population will be $\bar{F} = 1.1 \times 10^{13}$. That is, each individual in the population would have to give birth to 11,000,000,000,000 children, and all but one would die before reproductive age. This number is absurdly high. Muller (1950) suggested that genetic load values cannot exceed $L = 1$. As a matter of fact, he believed that the human genome has no more than 30,000 genes, i.e., a genetic load of $L = 0.3$, an average fitness of $\bar{w} = 0.72$, and an average fertility per individual of $\bar{F} = 1.39$ More recent estimates of mutation rates by Keightley and Eyre-Walker (2000) suggest that humans have a genetic load of between 0.78 and 0.95 depending on whether most of our mutations are recessive or dominant, respectively.

In the above, we assumed that deleterious mutations have an additive effect on fitness. Any factor that increases the number of deleterious mutations removed from the population, such as negative epistasis or inbreeding, will reduce the mutational load. On the other hand, any factor that decreases the efficacy of selection, such as positive epistasis or reduction in effective population size, will increase the mutational genetic load. Be that as it may, let us now consider the implications of the mutational genetic load on the fraction of the genome that is functional.

Studies have shown that the genome of each human newborn carries 56–103 point mutations that are not found in either of the two parental genomes (Xue et al. 2009; Roach et al. 2010; Conrad et al. 2011; Kong et al. 2012). If 80% of the genome is functional, as trumpeted by ENCODE Project Consortium (2012), then 45–82 deleterious mutations arise per generation. For the human population to maintain its current population size under these conditions, each of us should have on average 3×10^{19}–5×10^{35} (30,000,000,000,000,000,000 to 500,000,000,000,000,000,000,000,000,000,000,000) children. This is clearly bonkers. If the human genome consists mostly of junk and indifferent DNA, i.e., if the vast majority of point mutations are neutral, this absurd situation would not arise.

Graur (2017) conducted a study into the mutational load as a function of (1) the number of sites in the genome that are functional, (2) the mutation rate, (3) the fraction of deleterious mutations among all mutations in functional regions, and (4) the maximum tolerable repacement level fertility. These data were used to infer an upper limit on the fraction of the human genome that can be functional. His conclusion was that the functional fraction in the human genome cannot exceed 25%, and is almost certainly much lower.

2.8 Detecting Functionality at the Genome Level

The availability of intraspecific and interspecific genomic sequences has made it possible not only to test whether or not a certain genomic region is subject to selection but also to exhaustively scan the genome for regions likely to have been the target of selection and are, hence, of functional importance. For a given species, such scans allow us to estimate the proportion of the genome that is functional. In the literature, numerous approaches for detecting selection through comparisons of DNA sequences have been proposed (e.g., Nielsen 2005; Andrés et al. 2009; Li 2011; Vitti et al. 2013; Grossman et al. 2013; Lawrie and Petrov 2014). The main difference between tests designed to detect selection at a particular locus and tests involving genome-wide comparisons is that the latter involve multiple tests at multiple loci. Thus, many of the sites that are identified through such methods as having been subjected to selection are expected to be false positives. The statistics must, hence, be adjusted to the number of tests performed, using either standard techniques for multiple comparisons or adjusting significance levels to account for false discovery rates (e.g., Massingham and Goldman 2005).

Some methods for detecting selection at the genomic level require comparisons among species, some rely on intraspecific comparisons, and yet others require both types of data. Some methods are applicable to protein-coding genes only; some are applicable to all sequences. Some are based on comparisons of allele frequencies, some are based on linkage disequilibrium measures, and some rely on population-differentiation measures, such as genetic distances. Some are suitable for detecting purifying selection, and some are suitable for detecting positive or balancing selection. A straightforward method of estimating the functional fraction of a genome is to add up the genomic fractions that are under (1) positive, (2) negative, and (3) balancing selection.

Detecting functional regions subject to purifying selection is relatively straightforward. In interspecific comparisons, homologous genomic regions under purifying selection are expected to be more similar to one another than unselected regions. The rationale for this expectation is that many mutations in functional sequences are deleterious and, hence, weeded out of the population. Thus, purifying selection is primarily observable as highly conserved regions. We note, however, that evolutionary processes other than selection, such as mutational coldspots and gene conversion, can result in sequence conservation (Ahituv et al. 2007).

Balancing selection at the level of the genome is one of the least studied areas in evolutionary genomics. This lack of interest is somewhat unexpected given that a strong association between balancing selection and pathology has been hypothesized for a large number of human diseases, such as sickle-cell anemia, cystic fibrosis, and phenylketonuria. Andrés et al. (2009) devised a method for identifying regions that concomitantly show excessive levels of nucleotide polymorphism (as measured by the number of polymorphic sites in a gene), as well as an excess of alleles at intermediate allele frequencies.

All in all, the main emphasis in the last two decades has been with methods intended to detect positive selection. The principal reason for the emphasis on positive selection at the expense of purifying and balancing selection is that positive Darwinian selection is presupposed to be the primary mechanism of adaptation. Detecting genomic regions that have experienced positive selection requires more nuanced procedures and significantly more data than those required by methods for detecting purifying selection. Moreover, the various methods for detecting positive selection at the level of the genome are known to yield many false positives. Although each method has its own particular strengths and limitations, there are a number of challenges that are shared among all tests. First, deviations from neutrality expectations may be explained by factors other than selection. Demographic events, such as migration, population expansions, and bottlenecks, can often yield signals that mimic selection. This recognition has led some researchers to adopt approaches that explicitly attempt to separate demographic effects from selection effects (Li and Stephan 2006; Excoffier et al. 2009). Second, even when confounding effects are dealt with, the interpretation of selection may not be straightforward. For example, rate-based tests identify as "functional" all regions in which evolutionary rates have been accelerated. Such regions may indeed be subject to positive selection, but the acceleration may also be due to the relaxation of selective constraint due to total or partial nonfunctionalization or to an increase in the rate of mutation. Distinguishing among these possibilities requires a case-by-case analysis—a proposition that is antithetic to the ethos of Big Science genomics and bioinformatics.

Because positive selection leaves a number of footprints on the genome, and each test is designed to pick up on a slightly different signal, researchers sometimes combine multiple metrics into composite tests toward the goal of providing greater power of detection and a finer spatial resolution. Scores of such tests are typically referred to as composite scores. Tests employing composite scores come in two distinct forms. First, some methods calculate a composite score for a contiguous genetic region rather than a single nucleotide site by combining individual scores at all the sites within the region. The motivation for such an approach is that although false positives may occur at any one nucleotide site by chance, a contiguous region of positive markers is unlikely to be spurious and most likely represents a bona fide signature of selection (e.g., Carlson et al. 2005). In another type of method, composite scores are calculated by combining the results of many tests at a single site. The purpose of these methods is to utilize complementary information from different tests in order to provide better spatial resolution and pinpoint selection to the root cause (e.g., Zeng et al. 2006; Grossman et al. 2010).

It is very important to note that regardless of method or combination of methods, there are factors that conspire to underestimate the functional fraction of the genome and factors that conspire to overestimate the functional fraction of the genome. That is, some of the genomic segments identified as functional through telltale signs of positive selections may be false positives, while others may elude detection. For example, functional sequences may be under selection regimes that are difficult to detect, such as positive selection or weak purifying selection. In addition, selection may be difficult to detect as far as recently evolved species-specific elements (e.g., Smith et al. 2004) or very short genetic elements are concerned (e.g., De Gobbi et al. 2006). These factors would cause the fraction of the genome that is under selection and, hence, functional to be underestimated. On the other hand, selective sweeps, background selection, and significant reductions in population size (bottlenecks) would cause an overestimation of the fraction of the genome that is under selection (e.g., Williamson et al. 2007).

2.9 What Proportion of the Human Genome Is Functional?

While it is undisputed that many functional regions within genomes have evolved under complex selective regimes, such as selective sweeps, balancing selection, and recent positive selection, it is widely accepted that purifying selection persisting over long evolutionary times is the most common mode of evolution (Rands et al. 2014). Studies that identify functional sites by using the degree of conservation between sequences from two (or more) species have estimated the proportion of functional nucleotides in the human genome to be 3–15% (Ponting and Hardison 2011; Ward and Kellis 2012). We note, however, that each lineage gains and loses functional elements over time, so the proportion of nucleotides under selection needs to be understood in the context of divergence between species. For example, estimates of constraint between any two species will only include sequences that were present in their common ancestor and that have not been lost, replaced, or nonfunctionalized in the lineages leading up to the genomes of the extant species under study. Functional element turnover is defined as the loss or gain of purifying selection at a particular locus of the genome, when changes in the physical or genetic environment cause a locus to switch from being functional to being nonfunctional or vice versa.

By using genomic data from 12 mammalian species and an estimation model that takes into account functional element turnover, Rands et al. (2014) estimated that 8.2% of the human genome is functional, with a 95% confidence interval of 7.1–9.2%. Because of the difficulties in estimating the functional fraction of a genome, evolutionary biologists treat such numbers as somewhat underestimated. Thus, a claim that 10% or even 15% of the human genome is functional would, thus, be tolerable. On the other hand, a claim that 80% of the human genome is functional (e.g., ENCODE Project Consortium 2012) is misleading in the extreme and logically risible.

Unsurprisingly, in Rands et al.'s (2014) study, constrained coding sequences turned out to be much more evolutionary stable, i.e., experienced least functional element turnover, than constrained noncoding sequences. From among noncoding sequences, the sequences that were most likely to be functional were enhancers and DNase 1 hypersensitivity sites. Transcription factor binding sites, promoters, untranslated regions, and long noncoding RNAs (lncRNAs) contributed little to the functional fraction of the human genome, with lncRNAs exhibiting the most rapid rate of functional element turnover of all the noncoding element types. This finding implies that the vast majority of lncRNAs are devoid of function and represent transcriptional noise.

2.10 How Much Garbage DNA Exists in the Human Genome?

Because humans are diploid organisms and because natural selection is notoriously slow and inefficient in ridding populations of recessive deleterious alleles, the human genome is expected to contain garbage DNA. The amount of garbage DNA, however, should be quite small—many orders of magnitude smaller than the amount of junk DNA.

Deleterious alleles should exhibit a few telltale signs. First, they should be maintained in the population at very low frequencies. The reason for the rarity of deleterious alleles is that at low frequencies, the vast majority of such alleles will be found in heterozygous state, unexposed to purifying selection. Second, deleterious alleles should only very rarely become fixed between populations. Thus, one can identify them by using the ratio of polymorphic alleles to fixed alleles. Third, as shown by Maruyama (1974), deleterious or slightly deleterious allele should, on average, be younger than a neutral allele segregating at the same frequency. The young age of deleterious alleles is due to the fact that although purifying selection is not very efficient, it does eventually eliminate deleterious alleles from the population.

Many studies have shown that human genomes consist of measurable quantities of garbage DNA. Tennessen et al. (2012), for instance, sequenced 15,585 protein-coding genes from 2,440 individuals of European and African ancestry and found that 86% out of the more than 500,000 single-nucleotide variants, had very low allele frequencies of less than 0.5%. Of the 13,595 single-nucleotide variants that each person carries on average, ~43% were missense and nonsense and affected splicing, i.e., affected protein sequence. The rest of the mutations were synonymous. About 47% of all variants (74% of nonsynonymous and 6% of synonymous) were predicted by at least one of several computational methods to be deleterious, and almost all of these deleterious variants (~97%) had very low frequencies. Fu et al. (2013) analyzed 15,336 protein-coding genes from 6,515 individuals and estimated that ~73% of all single-nucleotide variants and ~86% of the variants predicted to be deleterious arose in the past 5,000–10,000 years. Sunyaev et al. (2001) estimated that the average human genotype carries approximately 1,000 deleterious nonsynonymous single-nucleotide variants that together cause a substantial reduction in fitness.

Chun and Fay (2009) approached the problem of distinguishing between deleterious mutations from the massive number of nonfunctional variants that occur within a single genome by using a comparative genomics data set of 32 vertebrate species. They first identified protein-coding sites that were highly conserved. Next they identified amino acid variants in humans at protein sites that are evolutionarily conserved. These amino acids variants are likely to be deleterious. Application of this method to human genomes revealed close to 1,000 deleterious variants per individual, approximately 40% of which were estimated to be at allele frequencies smaller than 5%. Their method also indicated that only a small subset of deleterious mutations can be reliably identified.

So far, we have discussed population genetics and evolutionary methods for predicting the deleteriousness of protein-altering variants based on population properties. There are, however, methods that combine evolutionary and biochemical information to make such inferences (Cooper and Shendure 2011). Nonsense and frameshift mutations are the most obvious candidates, as they are predicted to result in a loss of protein function and are heavily enriched among disease-causal variation. However, this class of variation is not unambiguously deleterious, in some cases allowing functional protein production or resulting in a loss of protein that is apparently not harmful. Considering nonsynonymous variants, the simplest and earliest approaches to estimate deleteriousness were to use discrete biochemical categorizations such as "radical" versus "conservative" amino acid changes. However, there are now numerous more sophisticated approaches to classify nonsynonymous variants on both quantitative and discrete scales. These methods can be divided into "first-principles approaches" and "trained classifiers."

First-principles approaches explicitly define a biological property of deleterious variants and make predictions on the basis of similarity or dissimilarity to that property. For example, first-principles approaches may use the presence of frameshifts in the coding regions as identifiers of deleteriousness (e.g., Sulem et al. 2015). By contrast, trained classifiers generate prediction rules by identifying heuristic combinations of many potentially relevant properties that optimally differentiate a set of true positives from a set of true negatives. First-principles approaches have the advantage of greater interpretability; for example, radical and conservative annotations of amino acid substitutions have a straightforward biochemical interpretation. However, first-principles methods are only as good as the assumptions that they make and do not model all of the relevant factors. Conversely, a trained classifier approach effectively yields a "black box" prediction and will be prone to the biases and errors in the data. However, trained classifiers have the advantage of being specifically tunable to the desired task (such as, predicting disease causality) and are capable of incorporating many sources of information without requiring a detailed understanding of how that information is relevant.

We note that all methods for predicting deleteriousness are prone to estimation error. For example, all methods use multiple sequence alignments and phylogenetic reconstructions. Low qualities of alignment or erroneous phylogenetic tree may result in low-quality inferences. Moreover, the sampling of species is crucial. A sample consisting of sequences that are very similar to one another offers less power of detection, thus increasing the number of false negatives. Conversely, inclusion of distant sequences may increase the number of false positives.

Finally, we note that many methods exploit biochemical data, including amino acid properties (such as charge), sequence information (such as presence of a binding site), and structural information (such as the presence of a β-sheet). The integration of these data with comparative sequence analysis significantly improves predictions of deleteriousness.

2.11 Mutational Origins of Junk DNA

Here, we ask ourselves which among the numerous mutational processes can increase genome size and concomitantly increase the fraction of nonfunctional DNA in the genome. Increases in genome size can be caused by genome duplication, various types of subgenomic duplications, mononucleotide and oligonucleotide insertions, and replicative transposition. Genome size increases can occur either gradually or in a punctuated manner. There are no analogous processes that can cause large and sudden decreases in genome size. Thus, as opposed to genome size increases, which can occur in fits and starts, decreases in genome size can only occur in a gradual manner.

Insertions and subgenomic duplications can increase genome size; however, these processes are expected to increase the size of the genome only very slowly, such that their contribution to genome size is thought to be negligible. Moreover, none of these two processes are expected to alter the fraction of nonfunctional elements in the genome. For example, the probabilities of duplicating a functional sequence or a nonfunctional sequence are proportional to the prevalence of such sequences in the genome, so that following subgenomic duplication, the fraction of junk DNA in the genome neither increases nor decreases.

Genome duplication is the fastest route to genome size increase; in one fell swoop, genome size is doubled. Genome duplication, however, does not increase the fraction of junk DNA in the genome unless the newly created functional redundancy is quickly obliterated by massive gene nonfunctionalization. Of course, in recently formed polyploids, one cannot speak of an increase in the C value, since this value refers to the size of the haploid genome and does not depend on the degree of ploidy. The contribution of gene duplication to the variation in C values only comes into play after the polyploid species has been diploidized and became a cryptopolyploid.

Because of their ability to increase in number, many transposable elements can have profound effects on the size and structure of genomes. Indeed, replicative transposable elements, especially retrotransposons, have the potential to increase their copy number while active, and many are responsible for huge genome size increases during relatively short periods of evolutionary time (e.g., Piegu et al. 2006).

The vast majority of eukaryotic genomes studied to date contain large numbers of transposable elements, and in many species, such as maize (*Zea mays*), the edible frog (*Pelophylax esculentus* formerly known as *Rana esculenta*), and the largest genome sequenced so far, the loblolly pine (*Pinus taeda*), transposable elements constitute the bulk of the genome (e.g., Kovach et al. 2010). Organisms devoid of transposable elements are extremely rare. The only group of organisms that is known to lack transposable elements altogether or to possess only two to three copies of transposable elements are yeast species, such as *Ashbya gossypii*, *Kluyveromyces lactis*, *Zygosaccharomyces rouxii*, and *Schizosaccharomyces octosporus* (Dietrich et al. 2004; Rhind et al. 2011).

Replicative transposition is the only mutational process that can greatly and rapidly increase genome size and at the same time significantly increase the fraction of junk DNA in the genome. The reason for this is that transcription and reverse transcription are very inaccurate methods of copying genetic information, and hence most of the increase in genome size due to transposable elements will consist of nonfunctional elements, i.e., junk DNA.

The claim that most junk DNA is made of transposable elements and their incapacitated descendants yields a quantitative prediction—it is expected that a positive correlation should exist between the genomic fraction inhabited by transposable elements and genome size. We note, however, that estimating the numbers and kinds of transposable elements in a genome is far from trivial or routine. First, different sequenced genomes differ from one another in the quality of the sequences. Because repetitive elements are difficult to sequence and problematic to use in genomic assemblies, the fraction of repetitive DNA is frequently underrepresented in low-quality genome sequences. Second, available algorithms for detecting repeated elements are known to perform with varying degrees of success in different species. The reason is that algorithms use a database of known transposable sequences as their reference, so that the repertoire of transposable elements in one species may be better characterized than that of another species. Third, most transposable elements have neither been coopted into a function by the host nor retained their ability to transpose. Thus, they evolve in an unconstrained fashion, losing their similarity to other members of their transposable element family very rapidly. Algorithms that rely on similarity measures are, hence, unable to identify a significant proportion of dead transposable elements. Finally, most algorithms for detecting repeated transposable elements fail to discover short elements. As a consequence, the fraction of a genome that is taken up by transposable elements is more often than not extremely underestimated. For example, about

50% of the human genome has been identified as derived from transposable elements by using algorithms that rely on a database of consensus element sequences. By using a method that identifies oligonucleotides that are related in sequence space to one another, de Koning et al. (2011) found out that 66–69% of the human genome is derived from repetitive sequences.

Despite the difficulties described above, a positive correlation between total sequence length of transposable elements and genome size is seen in many groups. For example, the Pearson correlation coefficient in a sample of 66 vertebrate genomes was 0.77 (Tang 2015). In other words, ~60% of the variation in genome size could be explained by the variation in total length of transposable elements. In the literature, one can find many reports of positive correlations between the number of transposable element copies per genome and genome size (Kidwell 2002; Biémont and Vieira 2006; Hawkins et al. 2006; Tenaillon et al. 2010; Lynch et al. 2011; Chénais et al. 2012; Chalopin et al. 2015). The rule is simple: large genomes have huge numbers of transposable elements; small genomes have very few (e.g., Roest Crollius et al. 2000; Kovach et al. 2010; Ibarra-Laclette et al. 2013; Kelley et al. 2014).

2.12 Why So Much of the Genome Is Transcribed, or Is It?

The ENCODE Project Consortium (2012) consisted of approximately 500 researchers and cost in excess of 300 million dollars. Its purpose was to identify all functional elements in the human genome. One of its main findings was that 75% of the genome is transcribed—a phenomenon that was originally described 40 years earlier by Comings (1972). Does this observation support the thesis that the human genome is almost entirely functional?

ENCODE systematically catalogued every transcribed piece of DNA as functional. In real life, whether a transcript has a function depends on many additional factors. For example, ENCODE ignored the fact that transcription is fundamentally a stochastic process that is inherently noisy (Raj and van Oudenaarden 2008). Some studies even indicate that 90% of the transcripts generated by RNA polymerase II may represent transcriptional noise (Struhl 2007). In fact, many transcripts generated by transcriptional noise may even associate with ribosomes and be translated (Wilson and Masel 2011). Moreover, ENCODE did not pay any attention to the number of transcripts produced by a DNA sequence. The vast majority of their newly "discovered" polyadenylated and non-polyadenylated RNAs are present at levels below one copy per cell and were found exclusively in the nucleus—never in the cytoplasm (Palazzo and Gregory 2014). According to the methodology of ENCODE, a DNA segment that produces 1,000 transcripts per cell per unit time is counted equivalently to a segment that produces a single RNA transcript in a single cell once in a blue moon.

We note, moreover, that ENCODE used almost exclusively pluripotent stem cells and cancer cells, which are known as transcriptionally permissive environments. In these cells, the components of the RNA polymerase II enzyme complex can increase up to 1000-fold, allowing for high transcription levels from random sequences. In other words, in these cells, transcription of nonfunctional sequences, that is, DNA sequences that lack a *bona fide* promoter, occurs at high rates (Babushok et al. 2007). The use of HeLa cells is particularly a suspect, as these cells have ceased long ago to be representative of human cells. For example, as opposed to humans who have a diploid chromosome number of 46, HeLa cells have a "hypertriploid" chromosome number, i.e., 76–80 regular-size chromosomes, of which 22–25 no longer resemble human chromosomes, as well as a highly variable number of "tiny" chromosomal fragments (Adey et al. 2013). Indeed, HeLa has been recognized as an independent biological species called *Helacyton gartleri* (Van Valen and Maiorana 1991).

The human genome contains many classes of sequences that are known to be abundantly transcribed but are typically devoid of function. Pseudogenes, for instance, have been shown to evolve very rapidly and are mostly subject to no functional constraint. Yet up to one-tenth of all known pseudogenes are transcribed (Pei et al. 2012), and some are even translated, chiefly in tumor cells (Kandouz et al. 2004). Pseudogene transcription is especially prevalent in pluripotent stem, testicular, germline, and cancer cells (Babushok et al. 2007). Unfortunately, because "functional genomics" is a recognized discipline within molecular biology, while "nonfunctional genomics" is only practiced by a handful of "genomic clochards" (Makalowski 2003), pseudogenes have always been looked upon with suspicion and wished away. Gene prediction algorithms, for instance, tend to "resurrect" pseudogenes in silico by annotating many of them as functional genes (Nelson 2004).

Another category of sequences that are devoid of function yet are transcribed is introns. When a human protein-coding gene is transcribed, its primary transcript contains not only functional reading frames but also introns and exonic sequences devoid of reading frames. In fact, only 4% of pre-mRNA sequences is devoted to the coding of proteins; the other 96% is mostly made of noncoding regions. Because introns are transcribed, ENCODE concluded that they are functional. But, are they? Some introns do indeed evolve slightly slower than pseudogenes, although this rate difference can be explained by a minute fraction of intronic sites involved in splicing and other functions. There is a long debate whether or not introns are indispensable components of eukaryotic genome. In one study (Parenteau et al. 2008), 96 introns from 87 yeast genes were removed. Only three of them (3%) seemed to have had a negative effect on growth. Thus, in the majority of cases, introns evolve neutrally, whereas a small fraction of introns are under selective constraint (Ponjavic et al. 2007). Of course, we recognize that some human introns harbor regulatory sequences (Tishkoff et al. 2006), as well as sequences that produce small RNA molecules (Hirose et al. 2003; Zhou et al. 2004). We note, however, that even those few introns under selection are not constrained over their entire length. Hare and Palumbi (2003) compared nine introns from three mammalian species (whale, seal, and human) and found that only about a quarter of their nucleotides exhibit telltale signs of functional constraint. A study of intron 2 of the human *BRCA1* gene revealed that only 300 bp (3% of the length of the intron) is conserved (Wardrop et al. 2005). Thus, the practice of summing up all the lengths of all the introns and adding them to the pile marked "functional" is misleading.

The human genome is also populated by a very large number of transposable elements. Transposable elements, such as LINEs, SINEs, retroviruses, and DNA transposons, may, in fact, account for up to two-thirds of the human genome (de Koning et al. 2011) and for more than 31% of the transcriptome (Faulkner et al. 2009). Both human and mouse had been shown to transcribe *SINE*s (Oler et al. 2012). The phenomenon of *SINE* transcription is particularly evident in carcinoma cell lines, in which multiple copies of *Alu* sequences are detected in the transcriptome (Umylny et al. 2007). Moreover, retrotransposons can initiate transcription on both strands (Denoeud et al. 2007). These transcription initiation sites are subject to almost no evolutionary constraint, casting doubt on their "functionality." Thus, while a handful of transposons have been domesticated into functionality, one cannot assign a "universal utility" for all retrotransposons (Faulkner et al. 2009).

Whether transcribed or not, the majority of transposons in the human genome are merely parasites, parasites of parasites, and dead parasites, whose main "function" appears to be causing frameshifts in reading frames, disabling RNA-specifying sequences, and more often than not littering the genome.

2.13 Hypotheses Concerning the Maintenance of Junk DNA

Three main types of hypotheses have been proposed to explain the C-value paradox. In the first, the selectionist hypothesis, large eukaryotic genomes are assumed to be entirely or almost entirely composed of literal DNA. In the second hypothesis, the nucleotypic hypothesis, eukaryotic genomes are assumed to be almost entirely functional but to contain mostly indifferent DNA and very little literal DNA. In neutralist hypotheses, genomes are assumed to contain large amounts of junk DNA—with bigger genomes containing larger amounts of junk DNA than smaller genomes.

2.14 Selectionist Hypotheses

Selectionist claims to the effect that genomes are entirely or almost entirely functional are usually made in the context of the human genome. For wholly unscientific reasons, humans are frequently assumed to occupy a privileged position, against which all other creatures are measured. In the literature, it frequently seems that as far as humans are concerned, the equations of population genetics are suspended, and evolution abides by a different set of rules. Thus, while no one will ever insist that ferns, salamanders, and lungfish, which have vastly larger genomes than humans, are devoid of junk DNA, one can frequently encounter National Institute of Health bureaucrats denying that human junk DNA exists (Zimmer 2015). It is in the human context, therefore, that we shall examine whether or not selectionist claims hold water.

The first selectionist hypothesis at the level of the genome was put forward by Zuckerkandl (1976), who asserted that there is very little nonfunctional DNA in the genome; the vast majority of the genome performs essential functions, such as gene regulation, protection against mutations, maintenance of chromosome structure, and the binding of proteins. Consequently, the excess DNA in large genomes is only apparent. In the many years since the publication of this article, this theory was rejected multiple times, and the paper was forgotten. The ENCODE Project Consortium (2012) resurrected the selectionist hypothesis (without acknowledging its originator), but their conclusion that the human genome is almost 100% functional was reached through various nefarious means, such as by employing the "causal role" definition of biological function and then applying it inconsistently to different biochemical properties, by committing a logical fallacy known as "affirming the consequent," by failing to appreciate the crucial difference between "junk DNA" and "garbage DNA," by using analytical methods that yield false positives and inflate estimates of functionality, by favoring statistical sensitivity over specificity, and by emphasizing statistical significance over the magnitude of the effect (Eddy 2012; Doolittle 2013; Graur et al. 2013; Niu and Jiang 2013; Brunet and Doolittle 2014; Doolittle et al. 2014; Palazzo and Gregory 2014).

Given that the original selectionist hypothesis has been appropriately relegated to the dustbin of history, it is difficult to rationalize the contemporary resurrection of such theories. Explaining the motives behind the pronouncements of ENCODE Project Consortium (2012) would require the combined skills of specialists in the pathologies of ignorance, pseudoscience, and self-aggrandizement.

2.15 Nucleotypic and Nucleoskeletal Hypotheses

A variety of cellular, organismal, and ecological parameters have been reported to correlate with C values. Unfortunately, these relationships are never universal; they are apparent only in limited taxonomic contexts. For example, although genome size is inversely correlated with metabolic rate in both mammals and birds, no such relationship is found in amphibians. Many correlates of genome size were described in plants (Greilhuber and Leitch 2013) but have no applicability outside this kingdom. For example, species with large genomes flower early in the spring, while later-flowering species have progressively smaller genomes. Additional examples concern weediness (the ability to invade arable lands) and invasiveness (the ability to colonize new environments), which were both found to be negatively correlated with genome size. An intriguing hypothesis for explaining genome size in plants casts phosphorus as a key player. Phosphorus is an important ingredient in the synthesis of nucleic acids (DNA and RNA). Yet despite its being the twelfth most abundant element in the environment, it is not readily accessible to plants and is often present in such low amounts that it may be considered a limiting nutrient for DNA synthesis. Because large genomes require increased supplies of phosphorus, it has been hypothesized that polyploids and plants with large genomes are at a selective disadvantage in phosphorus-limited environments (Hanson et al. 2003). Phosphorus-depleted soils should, accordingly, be populated by species with small genome sizes. Indeed, plants that live in mineral-poor environments seem to have particularly small genomes. Moreover, the smallest plant genome reported so far was found in a family of carnivorous plants that grow in nutrient-poor environments (Greilhuber et al. 2006). Some experimental support for the phosphorus hypothesis has been obtained by Šmarda et al. (2013) in their long-term fertilization experiment with 74 vascular plant species.

So far, the most universal correlate of genome size is cell size. For over a hundred years, cytologists have been aware of a positive correlation between the volume of the nucleus and the volume of the cytoplasm. These observations led to the concept of the nucleocytoplasmic ratio, according to which the ratio of the nuclear volume to that of the cytoplasm is constant, reflecting the need to balance nuclear and cytoplasmic processes. Indeed, neither nuclear-DNA content nor varied growth conditions, nor drug treatments, could alter the nucleocytoplasmic ratio, and deviations from it are associated with disease (Jorgensen et al. 2007; Neumann and Nurse 2007).

Given that a positive correlation exists between genome size and nuclear volume and that cytoplasm volume and cell volume are similarly correlated, the nucleocytoplasmic ratio is frequently presented as a positive correlation between genome size and cell size. A rough correlation between C value and cell size has been noted in some of the earliest studies of genome size evolution (Mirsky and Ris 1951) and has since been confirmed in many groups of animals, plants, and protists. Indeed, a positive correlation with cell size is a most general feature of genome size, and the relationships between C value and cell size have been claimed to rank among the most fundamental rules of eukaryote cell biology (Cavalier-Smith 2005). We note, however, that the best correlations are found among closely related taxa. For example, while a significant correlation between pollen size and genome size was found in a sample of 16 wind-pollinated grass species, a large-scale analysis of 464 angiosperms failed to confirm the correlation (Greilhuber and Leitch 2013). These findings make it clear that genome size evolution cannot be understood without reference to the particular biology of the organisms under study. Finally, we note that in many taxa, cell size and genome size do not seem to be correlated (e.g., Starostova et al. 2008, 2009).

Two basic explanations for the correlation between genome size and cell size have been put forward in the literature: the coincidence and the nucleotypic hypothesis. Under the coincidence hypothesis, most DNA is assumed to be junk, and genome size is assumed to increase through mutation pressure. The increase in the amount of DNA in the genome is not, however, a boundless process. At a certain point, the genome becomes too large and too costly to maintain, and any further increases in genome size will be deleterious and selected against. In the coincidence hypothesis, it is assumed that bigger cells can tolerate the accumulation of more DNA (Pagel and Johnstone 1992). In other words, the correlation between genome size and cell size is purely coincidental. The cell and the nucleus are envisioned as finite containers to be filled with DNA.

Under the nucleotypic hypothesis, the genome is assumed to have a nucleotypic function, i.e., to affect the phenotype in a manner that is dependent on its length but independent of its sequence. As a consequence, genome size may be under secondary selection owing to its nucleotypic effects. Let us assume, for instance, that genome size affects flowering time. Then, the selection for earlier or later flowering times will result in an indirect selection on genome size.

Cavalier-Smith (1978, 1982, 1985, 2005) envisioned the DNA as a "nucleoskeleton" around which the nucleus is assembled, so that the total amount of DNA in a cell as well as its degree of packaging exerts a strong effect on nucleus size and subsequently on cell size. According to this nucleoskeletal hypothesis, the DNA is not only a carrier of genetic information but also a structural material element—a nucleoskeleton that maintains the volume of the nucleus at a size proportional to the volume of the cytoplasm. Since larger cells require larger nuclei, selection for a particular cell volume will secondarily result in selection for a particular genome size. Thus, the correlation between genome size and cell size arises through a process of coevolution in which nuclear size is adjusted to match alterations in cell size.

In nucleotypic hypotheses, the entire genome is assumed to be functional, although only a small portion of it is assumed to be literal DNA. Thus, according to this hypothesis, most DNA is indifferent DNA, whose length is maintained by selection, while its nucleotide composition changes at random. With this hypothesis, the driving force, according to the nucleotypic hypothesis, is selection for an optimal cell size. For example, if a large cell size becomes adaptively favorable due to changes in the environment, then there would also be positive selection for a corresponding increase in nuclear volume, which in turn will be achieved primarily through either increases in the amount of indifferent DNA or modifications in the degree of DNA condensation.

At this point, we need to raise two questions: (1) does cell size matter? (2) does genome size affect cell size? The answer to the first question is that cells do have an optimal size, i.e., they need to be not too big and not too small. Based on studies in *C. elegans* and other systems, it has been argued that cell size is limited by the physical properties of its components (Marshall et al. 2012). For example, in order to proliferate, a cell has to divide, and for faithful cell division, the molecular machinery, such as the centrosome (the organelle that serves as the main microtubule organizing center) and the mitotic spindle, must be constructed at the right position and with the correct size. This may not be accomplished in extremely large or extremely small cells due to the physical properties of microtubules and chromosomes. If a cell exceeds the upper size limit, its centrosome and mitotic spindle may not be able to position themselves at the center of the cell, leading to nonsymmetrical cell division. Moreover, in such a case, microtubules may not reach the cell cortex, potentially leading to insufficient spindle elongation. If a cell falls below a lower size limit, its centrosome may not be able to position itself in a stable manner at the center of the cell due to the excess elastic forces of the microtubules. In addition, there may not be sufficient space for accurate chromosome segregation.

We do not know at present what the upper or lower limits of cell size are. We do, however, know that some cells are so extremely large that they most certainly exceed whatever the theoretical upper limit for cell size may be. Three such examples are the shelled amoeba, *Gromia sphaerica*, which can reach a diameter of 3 cm; ostrich eggs, which can reach 15 cm in diameter and can weigh more than 1 kg; and the record holders, unicellular seaweeds belonging to genus *Caulerpa*, whose tubular stolon may extend to a length of 3 or more meters. These enormous examples cannot, however, be taken as evidence that an upper limit for cell size does not exist. What these examples show is that there exist molecular and cellular devices for escaping the consequences of large cell size during cell division. None of these enormous cells undergo regular binary fission; they either divide by cleaving only a small portion of their mass (e.g., bird eggs) or by becoming multinucleate for at least part of their life cycle and producing large numbers of diminutive progeny or gametes (e.g., *Gromia* and *Caulerpa*).

As to the second question—does genome size affect cell size?—the evidence is quite thin. First, the correlation between cell size and genome size is imperfect at best. Second, there is evidence for contributors to cell size other than DNA. Levy and Heald (2010) studied the regulation of nuclear size in two related frog species: *Xenopus laevis* and *X. tropicalis*. *X. laevis* is a larger animal than *X. tropicalis*, and its cells are tetraploid. *X. tropicalis* is smaller and its cells are diploid. The two species also differ in another aspect: the cells and nuclei of *X. laevis* are larger. Because *Xenopus* nuclei can be assembled in a test tube using chromatin (DNA–protein complexes) and extracts of egg cytoplasm, one can test whether or not DNA has a role in determining cell size. Levy and Heald (2010) added sperm chromatin from either *X. laevis* or *X. tropicalis* to egg extracts from either *X. laevis* or *X. tropicalis*. They found that although both extracts can trigger assembly of the nuclear envelope around the chromatin, the cytoplasmic extract from *X. laevis* forms larger nuclei than the *X. tropicalis* extract, regardless of the DNA used. This indicates that one or more cytoplasmic factors determine nuclear size, while DNA content seems to have no discernable effect. Can we, therefore, state that the nucleoskeletal hypothesis has been invalidated? My answer is that it may be too early to discard the nucleoskeletal hypothesis, although one may certainly call into question its universality.

2.16 The Neutralist Hypothesis

As explained previously, a scientific hypothesis should spell out the conditions for its refutation. As far as the C-value paradox is concerned, the simplest scientific hypothesis is that the fraction of DNA that looks superfluous is indeed superfluous. The assumption that a vast fraction of the genome evolves in a neutral fashion means that this DNA does not tax the metabolic system of eukaryotes to any great extent and that the cost (e.g., in energy, time, and nutrients) of maintaining and replicating large amounts of nonfunctional DNA is negligible.

The first researcher to suggest that part of the genome may lack function was Darlington (1937), who recognized the difficulties in getting rid of redundant DNA because of its linkage to functional genes: "It must be recognized that the shedding of redundant DNA within a chromosome is under one particularly severe restriction, a restriction imposed by its contiguity, its linkage with DNA whose information is anything but dispensable." The theme of redundancy was later adopted by Rees and Jones (1972) and by Ohno (1972), the latter being credited with popularizing the notion that much of the genome in eukaryotes consists of junk DNA. In particular, Ohno (1972) emphasized the interconnected themes of gene duplication and trial and error in genome evolution. Gene duplication can alleviate the constraints imposed by natural selection by allowing one copy to maintain its original function while the other accumulates mutations. Only very rarely will these mutations turn out to be beneficial. Most of the time, however, one copy will be degraded into a pseudogene: "The creation of every new gene must have been accompanied by many other redundant copies joining the ranks of silent DNA base sequences." "Triumphs as well as failures of nature's past experiments appear to be contained in our genome." The discovery of transposable elements, and more importantly the observation that vast majority of transposable elements are nonfunctional and highly degraded by mutation, added support for the junk DNA hypothesis, as did the discovery of other nonfunctional genomic elements, such as pseudogenes, introns, and highly repetitive DNA.

Under the neutralist hypothesis, we need to address three issues. First, we need to ask what mutational processes can create junk DNA, i.e., what does junk DNA consist of? Second, we need to identify the evolutionary driving forces that can maintain junk DNA. Third, we need to explain why different organisms possess vastly different quantities of junk DNA. The first question is relatively easy to answer. Notwithstanding the fact that gene and genome duplications can create redundancies that may ultimately result in junk DNA and that satellite DNA can too contribute to the nonfunctional DNA fraction, there is currently no doubt whatsoever that the bulk of junk DNA is derived from transposable elements. By estimating the relative contribution of the major types of transposable elements, genomes can be classified into four main categories: (1) genomes in which DNA transposons predominate (e.g., *Amphioxus*, *Ciona*, most teleost fish, *Xenopus*), (2) genomes in which *LINE*s and *SINE*s predominate (e.g., lamprey, elephant shark, *Takifugu*, coelacanth, chicken, mammals), (3) genomes with a predominance of LTR retrotransposons (e.g., the tunicate *Oikopleura*), and (4) genomes in which no particular transposable element type predominates (e.g., *Tetraodon*, stickleback, reptiles, zebra finch). Some genomes are particularly poor in DNA transposable elements and contain almost exclusively retroelements (e.g., elephant shark, coelacanth, birds, mammals); however, there are no genomes that contain almost exclusively DNA transposable elements (Chalopin et al. 2015).

2.17 Selfish DNA

The "selfish DNA" hypothesis attempts to explain how superfluous and even deleterious elements can multiply within genomes and spread within populations. Selfish DNA (Doolittle and Sapienza 1980; Orgel and Crick 1980) is a term that applies to DNA sequences that have two distinct properties: (1) the DNA sequences can form additional copies of themselves within the genome, and (2) the DNA sequences either do not make any specific contribution to the fitness of the host organism or are actually detrimental. Some selfish DNA also engage in transmission ratio distortion and horizontal gene transfer—two processes through which they can further increase their frequency in the population.

The vast majority of selfish DNA consists of active class I and class II transposable elements that have not been domesticated, i.e., not coopted into function. A minor fraction of selfish DNA consists of promiscuous DNA as well as tandemly repeated sequences. As an approximation, in the following, we shall use the terms "selfish DNA," "selfish DNA elements," and "transposable elements" interchangeably.

Because of their ability to increase in number, selfish DNA elements can have profound effects on the size and structure of genomes. Two main classes of hypotheses have been put forward to explain the long-term persistence of selfish DNA in the genome. One class of hypotheses proposes that the process reflects two independent equilibria. At the genome level, a balance is achieved between transposition or retroposition, on the one hand, and the mechanisms utilized by the host to restrain transposable element activity, on the other. At the population level, a balance is achieved between the rate with which new copies of transposable elements are created and the efficiency with which negative selection gets rid of genotypes carrying transposable elements. The efficiency of selection, in turn, depends on population genetic parameters, such as effective population size (Brookfield and Badge 1997; Le Rouzic et al. 2007; Levin and Moran 2011). Under this class of hypotheses, a genome is assumed to be a closed system, in which the activity of transposable elements is counteracted by such intrinsic entities as small interfering RNAs, PIWI-interacting RNAs (piRNAs), DNA methylation, and histone modifications. In this closed system, transposable elements employ various evasive strategies, such as preferential insertion into regions transcribed by RNA polymerase II. Sooner or later, the system comprised of the host genome, and the transposable elements reach a stable equilibrium, after which nothing much happens until internal or external perturbations, such as mutations or environmental stress, disrupt the equilibrium, at which point either a burst of transposition is unleashed or the transposable elements become forever incapacitated.

The discovery that some transposable elements, such as the *P* element of *Drosophila*, are able to colonize new genomes by means of horizontal transfer (Daniels et al. 1990) unveiled an additional way for transposable elements to persist over evolutionary time. Horizontal escape of an active transposable element into a new genome would allow the element to evade a seemingly inevitable extinction in its original host lineage resulting from host elimination or inactivation due to mutational decay (Hartl et al. 1997).

Although the inherent ability of transposable elements to integrate into the genome suggested a proclivity for horizontal transfer (Kidwell 1992), the extent to which such processes affected a broad range of transposable elements and their hosts was not clear. Schaack et al. (2010) revealed more than 200 cases of horizontal transfer involving all known types of transposable elements, which may mean that virtually all transposable elements can horizontally transfer.

From a genome-wide study across *Drosophila* species, it was estimated that approximately one transfer event per transposable element family occurs every 20 million years (Bartolome et al. 2009). In several instances, we have evidence for horizontal transfer of transposable elements among extremely distant taxa, including at least 12 movements across phyla. So far, all these "long jumps" were found to involve DNA transposons, suggesting one of two possibilities: either DNA elements are better adapted to invade genomes than RNA elements or the preponderance of DNA elements represents a case of ascertainment bias due to the fact that DNA elements are studied more intensively in an evolutionary context, while the research on RNA elements is almost exclusively focused on a narrow taxonomic range of so-called model organisms.

It is becoming increasingly clear that the life cycle of a transposable element family is akin to a birth-and-death process in that it starts when an active copy colonizes a novel host genome and it ends when all copies of the transposable element family are lost or inactivated by chance through the accumulation of disabling mutations or by negative selection in a process which may be driven by host-defense mechanisms or by the fact that each transposable element contributes negatively to the fitness of the organism.

There are two major ways for transposable elements to escape extinction: the first is to horizontally transfer to a new host genome prior to extinction; the second is to inflict minimal fitness harm. Like other parasites, it is possible that transposable elements will make use of different strategies at different times. Each strategy has a phylogenetic signature. In cases in which horizontal transfer is frequent, there should be dramatic incongruence between the phylogeny of the transposable element family and that of their various host species. In these cases, horizontal transfer might allow the transposable element to colonize a new genome in which host suppression mechanisms are inefficient.

In cases where a transposable element family has persisted for long periods in a host lineage, the reduced frequency of horizontal transfer can be inferred from the similarity between the phylogenies of the transposable element and the hosts. Persistence could be achieved, for instance, through self-regulatory mechanisms that limit copy number or by evolving targeting preference for insertion into "safe havens" in the genome, such as, for instance, through preferential transposition into high copy number genes or heterochromatin. The *LINE-1* element of mammals provides an exceptional example of endurance, having persisted and diversified over the past 100 million years with virtually no evidence of horizontal transfer.

2.18 The Mutational Hazard Hypothesis: A Nearly Neutralist Hypothesis

So far, we have provided plausible explanations for the persistence of junk DNA within genomes. We did not, however, address the question of genome size disparity: Why do certain eukaryotes possess minute quantities of junk DNA, whereas the genome of many eukaryotes is made almost entirely of nonfunctional DNA. A possible explanation may be that some genomes have not been invaded by transposable elements, while others have been invaded many times. Another explanation may be that some genomes have been invaded by very inefficient transposable elements while others by very prolific ones. A third explanation may be that some genomes are extremely efficient at warding off selfish DNA, while others are more permissive. To the best of my knowledge, none of these three hypotheses has been tested. In the following, we present a hypothesis that uses differences in selection intensity and selection efficacy to explain the observed difference between organisms with high junk DNA content and those with low junk DNA content.

The mutational-hazard theory (Lynch and Conery 2003; Lynch 2006) postulates that virtually all increases in genome size in eukaryotes impose a slight fitness reduction. Thus, eukaryotic genomes are assumed to contain varying amounts of excess DNA that behaves like junk DNA in species with large effective population sizes and like garbage DNA in species with small effective population sizes.

The fate of a slightly deleterious allele is determined by the interplay between purifying selection and random genetic drift. Purifying selection acts by decreasing the frequency of the slightly deleterious allele at a rate that depends upon its selective disadvantage, s, where $s < 0$. Random genetic drift changes the allele frequencies in a nondirectional manner at a mean rate that is proportional to $1/N_e$, where N_e is the effective population size. Thus, whether a mutation that increases genome size is selected against or has the same probability of becoming fixed in the population as a neutral mutation is determined by its selective disadvantage relative to the effective size of the population into which it is introduced. Lynch and Conery (2003) argued that the ineffectiveness of selection in species with low N_e is the key to understanding genome size evolution, and the main prediction of the mutational-hazard theory is that large genomes will be found in species with small effective population sizes. How can we test this hypothesis?

First, we must ascertain that two mutational requirements are met. The first requirement is that mutations resulting in genome increases will outnumber mutations resulting in genome diminution. In many prokaryotes, for instance, this condition is not met. As a consequence, random genetic drift causes the genomes of many prokaryotes with small effective population sizes to get smaller rather than larger. In most eukaryotes, the proliferation of transposable elements overwhelms all other mutations, and, hence, at the mutation level, the condition that genome increases outnumber genome decreases is met.

The second mutational requirement is that the addition of each transposable element to the genome should on average decrease the fitness of its carrier. Two types of empirical data support this assumption. First, it has been shown that in *Drosophila melanogaster*, each transposable element insertion decreases the fitness of an individual by 0.4% (Pasyukova et al. 2004). Second, it has been shown that bottlenecks affect the diversity and level of activity of transposable elements, with populations that had experienced population size reductions having an increased level of both transposable element activity and transposable element diversity (Lockton et al. 2008; Picot et al. 2008).

After determining that the mutational requirements of the mutational-hazard hypothesis are met, we need to ascertain that the postulated relationship between genome size and effective population size (N_e) is supported by empirical data. In principle, N_e can be estimated directly by monitoring the variance of allele frequency changes across generations, as this has an expected value $\frac{p(1-p)}{2N_e}$, where p is the initial allele frequency. In practice, however, as the expected changes in allele frequency from generation to generation are extremely small unless N_e is tiny, this approach is difficult to put into practice because errors in estimating p will overwhelm the true change in p unless the sample size is enormous. As a consequence, most attempts to estimate N_e have taken a circuitous route, the most popular being to indirectly infer effective population size from the levels of within-population variation at nucleotide sites assumed to evolve in a neutral fashion. The logic underlying this approach is that if μ is the rate of neutral point mutations per generation per site and if N_e is roughly constant, then an equilibrium level of variation will be reached in which the mutational input to variation, 2μ, is matched by the mutational loss via genetic drift, $1/(2N_e)$. At equilibrium, the nucleotide site diversity is expected to be approximately equal to the ratio of these two forces, i.e., $4N_e\mu$ for a diploid population and $2N_e\mu$ for haploids. Thus, to estimate effective population size, one needs to empirically determine the degree of nucleotide site diversity, which is a relatively straightforward process, and the rate of mutation, μ, which is a slightly more difficult thing to do.

By measuring nucleotide site variation at synonymous sites and taking into account the contribution from mutation, average N_e estimates turned out to be ~10^5 for vertebrates, ~10^6 for invertebrates and land plants, ~10^7 for unicellular eukaryotes, and more than 10^8 for free living prokaryotes (Lynch et al. 2011). Although crude, these estimates imply that the power of random genetic drift is substantially elevated in eukaryotes—e.g., at least three orders of magnitude in large multicellular species relative to prokaryotes. It is also clear that the genetic effective sizes of populations are generally far below actual population sizes.

At present there is insufficient data on effective population sizes from a sufficiently diverse sample of taxa to test the mutational-hazard hypothesis directly. There is, however, indirect evidence supporting it. The mutation rate per generation is expected to be higher in species with low effective population sizes than in species with large effective population sizes. The mutation-hazard hypothesis asserts that organisms with low effective population sizes should have larger genomes than organisms with large effective population sizes. Thus, support for the mutational-hazard hypothesis can be obtained indirectly by showing that a positive correlation exists between mutation rate and genome size. Such a correlation has indeed been reported by Lynch (2010).

2.19 Genome Size and Bottlenecks: The Simultaneous Accumulation of *Alus*, Pseudogenes, and *Numts* Within Primate Genomes

As far as slightly deleterious mutations are concerned, the importance of random genetic drift relative to selection becomes especially pronounced during profound reductions of population size. If genome size can be shown to have increased concomitantly with reductions in population size, then the mutational-hazard theory would gain evidential support.

Gherman et al. (2007) compared the evolution of *numt* insertions in primate genomes and compared it with the insertions of two other classes of nonfunctional elements, *Alu*s and pseudogenes (Britten 1994; Bailey et al. 2003; Ohshima et al. 2003). These elements are unlikely to be functional, since their rate of evolution indicates a complete lack of selective constraint and they possess no positional, transcriptional, or translational features that might indicate a beneficial function subsequent to their integration into the nuclear genome. Using sequence analysis and fossil dating, Gherman et al. (2007) showed that a probable burst of integration of *Alu*s, pseudogenes, and *numt*s in the primate lineage occurred close to the prosimian–anthropoid split, which coincided with a major climatic event called the Paleocene–Eocene Thermal Maximum (~56 million years ago). During this event, which lasted for about 70,000 years, average global temperatures increased by approximately 6 °C, massive amounts of carbon were added on the atmosphere, the climate became much wetter, sea levels rose, and many taxa went extinct or suffered decreases in population size. Thus, the increase in primate genomes can be largely accounted for by a population bottleneck and the subsequent neutral fixation of slightly deleterious nonfunctional insertions. The fact that three classes of nonfunctional elements that use vastly different mechanisms of multiplying in the genome increased their numbers simultaneously effectively rules out selectionist explanations. These findings suggest that human and primate genomic architecture, with its abundance of repetitive elements, arose primarily by evolutionary happenstance.

2.20 Is It Junk DNA or Is It Indifferent DNA?

Distinguishing between neutralist and nucleoskeletal explanations has been quite difficult. Pagel and Johnstone (1992) proposed two expectations derived from each of the two theories. According to these authors, a major cost of junk DNA is the time required to replicate it. Organisms that develop at a slower pace may therefore be able to "tolerate" greater amounts of junk DNA, and thus a negative correlation across species between genome size and developmental rate is predicted. In contrast, the prediction of the nucleoskeletal hypothesis is for a positive correlation between genome size and cell size. Unfortunately, organisms with large cells also tend to develop slowly, whereas faster-growing organisms typically have smaller cells. Thus, according to the skeletal DNA hypothesis, a negative correlation between developmental rate and the C value is also expected. However, according to the nucleotypic hypothesis, the relation between developmental rate and genome size occurs only secondarily, as a result of the relationship between developmental rate and cell size.

Pagel and Johnstone (1992) studied 24 salamander species. The size of the nuclear genome was found to be negatively correlated with developmental rate, even after the effects of nuclear and cytoplasmic volume have been removed. However, the correlations between genome size, on the one hand, and nuclear and cytoplasmic volumes, on the other, become statistically insignificant once the effects of developmental rates have been removed. These results support the hypothesis that most of the DNA in salamanders is junk rather than indifferent DNA. Whether Pagel and Johnstone's results represent a true phenomenon or one restricted to *Salamandra* is still a controversial subject (Gregory 2003), especially since in eukaryotes, the "cost" of replicating DNA may not be correlated with genome size.

References

Adey A et al (2013) The haplotype-resolved genome and epigenome of the aneuploid HeLa cancer cell line. Nature 500:207–211

Ahituv N, Zhu Y, Visel A, Holt A, Afzal V, Pennacchio LA, Rubin EM (2007) Deletion of ultraconserved elements yields viable mice. PLoS Biol 5:e234

Amundson R, Lauder GV (1994) Function without purpose. Biol Philos 9:443–469

Andrés AM et al (2009) Targets of balancing selection in the human genome. Mol Biol Evol 26:2755–2764

Aparicio S et al (2002) Whole-genome shotgun assembly and analysis of the genome of *Fugu rubripes*. Science 297:1301–1310

Aronson AI et al (1960) Biophysics. In: Year book 59: Carnegie Institution of Washington. Lord Baltimore Press, Baltimore, pp 229–289
Avarello R, Pedicini A, Caiulo A, Zuffardi O, Fraccaro M (1992) Evidence for an ancestral alphoid domain on the long arm of human chromosome 2. Hum Genet 89:247–249
Babushok DV et al (2007) A novel testis ubiquitin-binding protein gene arose by exon shuffling in hominoids. Genome Res 17:1129–1138
Bailey JA, Liu G, Eichler EE (2003) An *Alu* transposition model for the origin and expansion of human segmental duplications. Am J Hum Genet 73:823–834
Bartolomé C, Bello X, Maside X (2009) Widespread evidence for horizontal transfer of transposable elements across *Drosophila* genomes. Genome Biol 10:R22
Biémont C, Vieira C (2006) Junk DNA as an evolutionary force. Nature 443:521–524
Brenner S (1998) Refuge of spandrels. Curr Biol 8:R669
Britten RJ (1994) Evidence that most *Alu* sequences were inserted in a process that ceased about 30 million years ago. Proc Natl Acad Sci U S A 91:6148–6150
Brookfield JF (2000) Genomic sequencing: the complexity conundrum. Curr Biol 10:R514–R515
Brookfield JFY, Badge RM (1997) Population genetics models of transposable elements. Genetica 100:281–294
Brosius J, Gould SJ (1992) On "genomenclature": a comprehensive (and respectful) taxonomy for pseudogenes and other "junk DNA". Proc Natl Acad Sci U S A 89:10706–10710
Brunet TDP, Doolittle WF (2014) Getting "function" right. Proc Natl Acad Sci U S A 111:E3365
Burgess J (1985) An introduction to plant cell development. Cambridge University Press, Cambridge
Carlson CS, Thomas DJ, Eberle MA, Swanson JE, Livingston RJ, Rieder MJ, Nickerson DA (2005) Genomic regions exhibiting positive selection identified from dense genotype data. Genome Res 15:1553–1565
Cavalier-Smith T (1978) Nuclear volume control by nucleoskeletal DNA, selection for cell volume and cell growth rate, and the solution of the DNA C-value paradox. J Cell Sci 34:247–278
Cavalier-Smith T (1982) Skeletal DNA and the evolution of genome size. Annu Rev Biophys Bioeng 11:273–302
Cavalier-Smith T (ed) (1985) The evolution of genome size. Wiley, New York
Cavalier-Smith T (2005) Economy, speed and size matter: evolutionary forces driving nuclear genome miniaturization and expansion. Ann Bot 95:147–175
Chalopin D, Naville M, Plard F, Galiana D, Volff JN (2015) Comparative analysis of transposable elements highlights mobilome diversity and evolution in vertebrates. Genome Biol Evol 7:567–580
Chen YH et al (2003) KCNQ1 gain-of-function mutation in familial atrial fibrillation. Science 299:251–254
Chénais B, Caruso A, Hiard S, Casse N (2012) The impact of transposable elements on eukaryotic genomes: from genome size increase to genetic adaptation to stressful environments. Gene 509:7–15
Cho JH, Brant SR (2011) Recent insights into the genetics of inflammatory bowel disease. Gastroenterology 140:1704–1712
Chun S, Fay JC (2009) Identification of deleterious mutations within three human genomes. Genome Res 19:1553–1561
Comings DE (1972) The structure and function of chromatin. Adv Hum Genet 3:237–431
Conrad DF et al (2011) Variation in genome-wide mutation rates within and between human families. Nat Genet 43:712–714
Cooper GM, Shendure J (2011) Needles in stacks of needles: finding disease-causal. Nat Rev Genet 12:628–640
Cummins R (1975) Functional analysis. J Philos 72:741–765

Daniels SB, Peterson KR, Strausbaugh LD, Kidwell MG, Chovnick A (1990) Evidence for horizontal transmission of the *P* transposable element between *Drosophila* species. Genetics 124:339–355

Darlington CD (1937) Recent advances in cytology, 2nd edn. Blakiston, Philadelphia

De Gobbi M et al (2006) A regulatory SNP causes a human genetic disease by creating a new transcriptional promoter. Science 312:1215–1217

Denoeud F et al (2007) Prominent use of distal 5′ transcription start sites and discovery of a large number of additional exons in ENCODE regions. Genome Res 17:746–759

Dietrich FS et al (2004) The *Ashbya gossypii* genome as a tool for mapping the ancient *Saccharomyces cerevisiae* genome. Science 304:304–307

Doolittle WF (2013) Is junk DNA bunk? A critique of ENCODE. Proc Natl Acad Sci U S A 110:5294–5300

Doolittle WF, Sapienza C (1980) Selfish genes, the phenotype paradigm and genome evolution. Nature 284:601–603

Doolittle WF, Brunet TDP, Linquist S, Gregory TR (2014) Distinguishing between "function" and "effect" in genome biology. Genome Biol Evol 6:1234–1237

Douglas S et al (2001) The highly reduced genome of an enslaved algal nucleus. Nature 410:1091–1096

Eddy SR (2012) The C-value paradox, junk DNA and ENCODE. Curr Biol 22:R898–R899

Ehret CF, de Haller G (1963) Origin, development, and maturation of organelles and organelle systems of the cell surface in *Paramecium.* J Ultrastruct Res 23(Supplement 6):1–42

ENCODE Project Consortium (2012) An integrated encyclopedia of DNA elements in the human genome. Nature 489:57–74

Excoffier L, Hofer T, Foll M (2009) Detecting loci under selection in a hierarchically structured population. Heredity 103:285–298

Ezkurdia I et al (2014) Multiple evidence strands suggest that there may be as few as 19,000 human protein-coding genes. Hum Mol Genet 23:5866–5878

Faulkner GJ et al (2009) The regulated retrotransposon transcriptome of mammalian cells. Nat Genet 41:563–571

Fu W et al (2013) Analysis of 6,515 exomes reveals the recent origin of most human protein-coding variants. Nature 493:216–220

Gherman A et al (2007) Population bottlenecks as a potential major shaping force of human genome architecture. PLoS Genet 3:e119

Gilson PR, Su V, Slamovits CH, Reith ME, Keeling PJ, McFadden GI (2006) Complete nucleotide sequence of the chlorarachniophyte nucleomorph: nature's smallest nucleus. Proc Natl Acad Sci U S A 103:9566–9571

Goldberg WM (2013) The biology of reefs and reef organisms. University of Chicago Press, Chicago

Graur D (2016) Molecular and genome evolution. Sinauer Associates, Sunderland

Graur D (2017) An upper limit on the functional fraction of the human genome. Genome Biol Evol 9:1880–1885

Graur D, Zheng Y, Price N, Azevedo RB, Zufall RA, Elhaik E (2013) On the immortality of television sets: "function" in the human genome according to the evolution-free gospel of ENCODE. Genome Biol Evol 5:578–590

Graur D, Zheng Y, Azevedo RB (2015) An evolutionary classification of genomic function. Genome Biol Evol 7:642–645

Gregory TR (2001) Coincidence, coevolution, or causation? DNA content, cell size, and the C-value enigma. Biol Rev 76:65–101

Gregory TR (2003) Variation across amphibian species in the size of the nuclear genome supports a pluralistic, hierarchical approach to the C-value enigma. Biol J Linn Soc 79:329–339

Greilhuber J, Leitch IJ (2013) Genome size and the phenotype. In: Leitch IJ, Greilhuber J, Dolezel J, Wendel J (eds) Plant genome diversity, vol 2. Springer, Wien, pp 323–344

Greilhuber J, Dolezel J, Lysák MA, Bennett MD (2005) The origin, evolution and proposed stabilization of the terms 'genome size' and 'C-value' to describe nuclear DNA contents. Ann Bot 95:255–260

Greilhuber J, Borsch T, Müller K, Worberg A, Porembski S, Barthlott W (2006) Smallest angiosperm genomes found in Lentibulariaceae, with chromosomes of bacterial size. Plant Biol 8:770–777

Grossman SR et al (2010) A composite of multiple signals distinguishes causal variants in regions of positive selection. Science 327:883–886

Grossman SR et al (2013) Identifying recent adaptations in large-scale genomic data. Cell 152:703–713

Haldane JBS (1937) The effect of variation on fitness. Am Nat 71:337–349

Hanson L, Brown RL, Boyd A, Johnson MA, Bennett MD (2003) First nuclear DNA C-values for 28 angiosperm genera. Ann Bot 91:31–38

Hare MP, Palumbi SR (2003) High intron sequence conservation across three mammalian orders suggests functional constraints. Mol Biol Evol 20:969–978

Hartl DL, Lohe AR, Lozovskaya ER (1997) Modern thoughts on an ancyent marinere: function, evolution, regulation. Annu Rev Genet 31:337–358

Hawkins JS, Kim H, Nason JD, Wing RA, Wendel JF (2006) Differential lineage-specific amplification of transposable elements is responsible for genome size variation in *Gossypium*. Genome Res 16:1252–1261

Hayden EC (2010) Life is complicated. Nature 464:664–667

Hazkani-Covo E, Zeller RM, Martin W (2010) Molecular poltergeists: mitochondrial DNA copies (*numts*) in sequenced nuclear genomes. PLoS Genet 6:e1000834

Hirose T, Shu MD, Steitz JA (2003) Splicing-dependent and –independent modes of assembly for intron-encoded box C/D snoRNPs in mammalian cells. Mol Cell 12:113–123

Huneman P (ed) (2013) Functions: selection and mechanisms. Springer, Dordrecht

Hurst LD (2013) Open questions: a logic (or lack thereof) of genome organization. BMC Biol 11:58

Ibarra-Laclette E et al (2013) Architecture and evolution of a minute plant genome. Nature 498:94–98

International Human Genome Sequencing Consortium (2004) Finishing the euchromatic sequence of the human genome. Nature 431:931–945

Jorgensen P, Edgington NP, Schneider BL, Rupes I, Tyers M, Futcher B (2007) The size of the nucleus increases as yeast cells grow. Mol Biol Cell 18:3523–3532

Kandouz M, Bier A, Carystinos GD, Alaoui-Jamali MA, Batist G (2004) *Connexin43* pseudogene is expressed in tumor cells and inhibits growth. Oncogene 23:4763–4770

Kaufman SA (1971) Gene regulation networks: a theory for their global structures and behaviours. In: Moscona AA, Monroy A (eds) Current topics in developmental biology. Academic, New York, pp 145–182

Kauffman SA (1993) The origins of order: self-organization and selection in evolution. Oxford University Press, Oxford

Keightley PD, Eyre-Walker A (2000) Deleterious mutations and the evolution of sex. Science 209:331–333

Kelley JL et al (2014) Compact genome of the Antarctic midge is likely an adaptation to an extreme environment. Nat Commun 5:4611

Kellis M et al (2014) Defining functional DNA elements in the human genome. Proc Natl Acad Sci U S A 111:6131–6138

Kidwell MG (1992) Horizontal transfer of *P*-elements and other short inverted repeat transposons. Genetica 86:275–286

Kidwell MG (2002) Transposable elements and the evolution of genome size in eukaryotes. Genetica 115:49–63

Kimura M (1961) Some calculations on the mutational load. Jap J Genet 36:S179–S190

Kimura M, Maruyama T (1966) The mutational load with interactions in fitness. Genetics 54:1337–1351

Klotzko AJ (ed) (2001) The cloning sourcebook. Oxford University Press, Oxford

Kolata G (2010) Reanimated 'junk' DNA is found to cause disease. New York Times http://www.nytimes.com/2010/08/20/science/20gene.html?_r=0

Kong A et al (2012) Rate of *de novo* mutations and the importance of father's age to disease risk. Nature 488:471–475

de Koning AP, Gu W, Castoe TA, Batzer MA, Pollock DD (2011) Repetitive elements may comprise over two-thirds of the human genome. PLoS Genet 7:e1002384

Kovach A et al (2010) The *Pinus taeda* genome is characterized by diverse and highly diverged repetitive sequences. BMC Genomics 11:420

Krams SM, Bromberg JS (2013) ENCODE: life, the universe and everything. Am J Transplant 13:245

Lander ES et al (2001) Initial sequencing and analysis of the human genome. Nature 409:860–921

Lawrie DS, Petrov DA (2014) Comparative population genomics: power and principles for the inference of functionality. Trends Genet 30:133–139

Le Rouzic A, Boutin TS, Capy P (2007) Long-term evolution of transposable elements. Proc Nat Acad Sci 104:19375–19380

Levin HL, Moran JV (2011) Dynamic interactions between transposable elements and their hosts. Nature Rev Genet 12:615–627

Levy DL, Heald R (2010) Nuclear size is regulated by importin α and Ntf2 in *Xenopus*. Cell 143:288–298

Li H, Stephan W (2006) Inferring the demographic history and rate of adaptive substitution in *Drosophila*. PLoS Genet 2:e166

Li X et al (2011) Chromosome size in diploid eukaryotic species centers on the average length with a conserved boundary. Mol Biol Evol 28:1901–1911

Lloyd S (2001) Measures of complexity: a nonexhaustive list. IEEE Control Syst Mag 21(4):7–8

Lockton S, Ross-Ibarra J, Gaut BS (2008) Demography and weak selection drive patterns of transposable element diversity in natural populations of *Arabidopsis lyrata*. Proc Natl Acad Sci U S A 105:13965–13970

Lynch M (2006) The origins of eukaryotic gene structure. Mol Biol Evol 23:450–468

Lynch M (2010) Evolution of the mutation rate. Trends Genet 26:345–352

Lynch M, Conery JS (2003) The origins of genome complexity. Science 302:1401–1404

Lynch M, Bobay L-M, Catania F, Gout J-F, Rho M (2011) The repatterning of eukaryotic genomes by random genetic drift. Annu Rev Genomics Hum Genet 12:347–366

Makalowski W (2003) Not junk after all. Science 300:1246–1247

Marshall WF et al (2012) What determines cell size? BMC Biol 10:101

Maruyama T (1974) The age of a rare mutant gene in a large population. Am J Hum Genet 26:669–673

Massingham T, Goldman N (2005) Detecting amino acid sites under positive selection and purifying selection. Genetics 169:1753–1762

Mehta G, Jalan R, Mookerjee RP (2013) Cracking the ENCODE: from transcription to therapeutics. Hepatology 57:2532–2535

Millikan RG (1989) In defense of proper functions. Philos Sci 56:288–302

Mirsky AE, Ris H (1951) The desoxyribonucleic acid content of animal cells and its evolutionary significance. J Gen Physiol 34:451–462

Moran LA (2007) The deflated ego problem. http://sandwalk.blogspot.com/2007/05/deflated-ego-problem.html

Muller HJ (1950) Our load of mutations. Am J Hum Genet 2:111–176

Neander K (1991) Functions as selected effects: the conceptual analyst's defense. Philos Sci 58:168–184

Nei M (2013) Mutation driven evolution. Oxford University Press, Oxford

Nelson DR (2004) "Frankenstein genes," or the *Mad Magazine* version of the human pseudogenome. Hum Genomics 1:310–316

Neumann FR, Nurse P (2007) Nuclear size control in fission yeast. J Cell Biol 179:593–600

Nielsen R (2005) Molecular signatures of natural selection. Annu Rev Genet 39:197–218

Niu DK, Jiang L (2013) Can ENCODE tell us how much junk DNA we carry in our genome? Biochem Biophys Res Commun 430:1340–1343

Noleto RB, de Souza Fonseca Guimarães F, Paludo KS, Vicari MR, Artoni RF, Cestari MM (2009) Genome size evaluation in Tetraodontiform fishes from the Neotropical region. Mar Biotechnol 11:680–685

Ohno S (1972) So much "junk" DNA in our genome. Brookhaven Symp Biol 23:366–370

Ohno S (1973) Evolutional reason for having so much junk DNA. In: Pfeiffer RA (ed) Modern aspects of cytogenetics: constitutive heterochromatin in man. F. K. Schattauer Verlag, Stuttgart, pp 169–180

Ohshima K, Hattori M, Yada T, Gojobori T, Sakaki Y, Okada N (2003) Whole-genome screening indicates a possible burst of formation of processed pseudogenes and *Alu* repeats by particular *L1* subfamilies in ancestral primates. Genome Biol 4:R74

Ohta T (1973) Slightly deleterious mutant substitutions in evolution. Nature 246:96–98

Oler AJ, Traina-Dorge S, Derbes RS, Canella D, Cairns BR, Roy-Engel AM (2012) *Alu* expression in human cell lines and their retrotranspositional potential. Mob DNA 3:11

Orgel LE, Crick FHC (1980) Selfish DNA: the ultimate parasite. Nature 284:604–607

Pagel M, Johnstone RA (1992) Variation across species in the size of the nuclear genome supports the junk-DNA explanation for the C-value paradox. Proc Roy Soc 249B:119–124

Palazzo AF, Gregory TR (2014) The case for junk DNA. PLoS Genet 10:e1004351

Parenteau J et al (2008) Deletion of many yeast introns reveals a minority of genes that require splicing for function. Mol Biol Cell 19:1932–1941

Pasyukova EG, Nuzhdin SV, Morozova TV, Mackay TF (2004) Accumulation of transposable elements in the genome of *Drosophila melanogaster* is associated with a decrease in fitness. J Hered 95:284–290

Pei B et al (2012) The GENCODE pseudogene resource. Genome Biol 13:R51

Peierls R (1960) Wolfgang Ernst Pauli, 1900-1958. Biogr Mem Fellows R Soc 5:174–192

Pellicier J, Fay MF, Leitch IJ (2010) The largest eukaryotic genome of them all? Bot J Linn Soc 164:10–15

Pertea M, Salzberg SL (2010) Between a chicken and a grape: estimating the number of human genes. Genome Biol 11:206

Petsko GA (2003) Funky, not junky. Genome Biol 4:104

Picot S, Wallau GL, Loreto EL, Heredia FO, Hua-Van A, Capy P (2008) The *mariner* transposable element in natural populations of *Drosophila simulans*. Heredity 101:53–59

Piegu B et al (2006) Doubling genome size without polyploidization: dynamics of retrotransposition-driven genomic expansions in *Oryza australiensis*, a wild relative of rice. Genome Res 16:1262–1269

Ponjavic J, Ponting CP, Lunter G (2007) Functionality or transcriptional noise? Evidence for selection within long noncoding RNAs. Genome Res 17:556–565

Ponting CP, Hardison RC (2011) What fraction of the human genome is functional? Genome Res 21:1769–1776

Pruitt KD, Tatusova T, Klimke W, Maglott DR (2009) NCBI reference sequences: current status, policy and new initiatives. Nucleic Acids Res 37:D32–D36

Purves WK, Sadava D, Orians GH, Heller HC (2004) Life: the science of biology, 7th edn. Sinauer, Sunderland

Raj A, van Oudenaarden A (2008) Stochastic gene expression and its consequences. Cell 135:216–226

Rands CM, Meader S, Ponting CP, Lunter G (2014) PLoS Genet 10:e1004525

Rees H, Jones RN (1972) The origin of the wide species variation in nuclear DNA content. Int Rev Cytol 32:53–92

Rhind N et al (2011) Comparative functional genomics of the fission yeasts. Science 332:930–936
Roach JC et al (2010) Analysis of genetic inheritance in a family quartet by whole-genome sequencing. Science 328:636–639
Roest Crollius H et al (2000) Characterization and repeat analysis of the compact genome of the freshwater pufferfish *Tetraodon nigroviridis*. Genome Res 10:939–949
Schaack S, Gilbert C, Feschotte C (2010) Promiscuous DNA: horizontal transfer of transposable elements and why it matters for eukaryotic evolution. Trends Ecol Evol 25:537–546
Šmarda P, Hejcman M, Březinová A, Horová L, Steigerová H, Zedek F, Bureš P, Hejcmanová P, Schellberg J (2013) Effect of phosphorus availability on the selection of species with different ploidy levels and genome sizes in a long-term grassland fertilization experiment. New Phytol 200:911–921
Smith NG, Brandström M, Ellegren H (2004) Evidence for turnover of functional noncoding DNA in mammalian genome evolution. Genomics 84:806–813
Srivastava M et al (2008) The *Trichoplax* genome and the nature of placozoans. Nature 454:955–960
Stamatoyannopoulos JA (2012) What does our genome encode? Genome Res 22:1602–1611
Starostova Z, Kratochvil L, Flajshans M (2008) Cell size does not always correspond to genome size: phylogenetic analysis in geckos questions optimal DNA theories of genome size evolution. Zoology 111:377–384
Starostova Z, Kubicka L, Konarzewski M, Kozlowski J, Kratochvil L (2009) Cell size but not genome size affects scaling of metabolic rate in eyelid geckos. Am Nat 174:E100–E105
Stevenson RL (1886) Strange case of Dr. Jekyll and Mr. Hyde. Charles Scribner's Sons, New York. https://books.google.com/books?id=xZHv4CWgWaEC&printsec=frontcover&dq=jekyll+hyde&hl=en&sa=X&ei=uW6LVL2KKsGpgwSM4oHgCQ&ved=0CB4Q6wEwAA#v=onepage&q=jekyll%20hyde&f=false
Struhl K (2007) Transcriptional noise and the fidelity of initiation by RNA polymerase II. Nat Struct Mol Biol 14:103–105
Sulem P et al (2015) Identification of a large set of rare complete human knockouts. Nat Genet. https://doi.org/10.1038/ng.3243
Sundaram V et al (2014) Widespread contribution of transposable elements to the innovation of gene regulatory networks. Genome Res 24:1963–1976
Sunyaev S, Ramensky V, Koch I, Lathe W, Kondrashov AS, Bork P (2001) Prediction of deleterious human alleles. Hum Mol Genet 10:591–597
Swift H (1950) The constancy of deoxyribose nucleic acid in plant nuclei. Proc Natl Acad Sci U S A 36:643–654
Tang D (2015) Repetitive elements in vertebrate genomes. http://davetang.org/muse/2014/05/22/repetitive-elements-in-vertebrate-genomes/
Tenaillon MI, Hollister JD, Gaut BS (2010) A triptych of the evolution of plant transposable elements. Trends Plant Sci 15:471–478
Tennessen JA et al (2012) Evolution and functional impact of rare coding variation from deep sequencing of human exomes. Science 337:64–69
Thomas CA (1971) The genetic organization of chromosomes. Annu Rev Genet 5:237–256
Tishkoff SA et al (2006) Convergent adaptation of human lactase persistence in Africa and Europe. Nat Genet 39:31–40
Umylny B, Presting G, Ward WS (2007) Evidence of *Alu* and *B1* expression in dbEST. Syst Biol Reprod Med 53:207–218
Van Valen LM, Maiorana VC (1991) Hela, a new microbial species. Evol Theor 10:71–74
Vitti JJ, Grossman SR, Sabeti PC (2013) Detecting natural selection in genomic data. Annu Rev Genet 47:97–120
Vogel F (1964) A preliminary estimate of the number of human genes. Nature 201:847
Ward LD, Kellis M (2012) Evidence of abundant purifying selection in humans for recently acquired regulatory functions. Science 337:1675–1678

Wardrop SL, kConFab Investigators, Brown MA (2005) Identification of two evolutionarily conserved and functional regulatory elements in intron 2 of the human BRCA1 gene. Genomics 86:316–328

Wen Y-Z, Zheng L-L, Qu L-H, Ayala FJ, Lun Z-R (2012) Pseudogenes are not pseudo any more. RNA Biol 9:27–32

Williamson SH, Hubisz MJ, Clark AG, Payseur BA, Bustamante CD, Nielsen R (2007) Localizing recent adaptive evolution in the human genome. PLoS Genet 3:e90

Wilson BA, Masel J (2011) Putatively noncoding transcripts show extensive association with ribosomes. Genome Biol Evol 3:1245–1252

Xue Y et al (2009) Human Y chromosome base-substitution mutation rate measured by direct sequencing in a deep-rooting pedigree. Curr Biol 19:1453–1457

Zeng K, Fu Y-X, Shi S, Wu C-I (2006) Statistical tests for detecting positive selection by utilizing high-frequency variants. Genetics 174:1431–1439

Zhou H, Zhao J, Yu CH, Luo QJ, Chen YQ, Xiao Y, Qu LH (2004) Identification of a novel box C/D snoRNA from mouse nucleolar cDNA library. Gene 327:99–105

Zimmer C (2015) Is most of our DNA garbage? New York Times http://www.nytimes.com/2015/03/08/magazine/is-most-of-our-dna-garbage.html?_r=0

Zuckerkandl E (1976) Gene control in eukaryotes and the C-value paradox: "excess" DNA as an impediment to transcription of coding sequences. J Mol Evol 9:73–104

Chapter 3
GC Content Heterogeneity

Satoshi Oota

Abstract Conventional evolutionary theories have been mainly constructed based on coding and its surrounding regions. The genome evolution with its genomic context is scarcely covered by sophisticated evolutionary models. GC content studies have been elaborated long before the "genome sequencing era," mainly with bacterial genomes. After the intrachromosomal guanine+cytosine (GC) content heterogeneity (isochore) was found as a universal genomic context, various models were proposed to explain the complex and enigmatic isochore evolution. So far, however, no single model can explain the existing data without flaws. I introduced a totally new framework to elucidate evolution of the genomic context: the nonlinear dynamic model or $f(x)$ framework, which potentially expands power of existing evolutionary models.

Keywords Isochore · Direction of mutations · GC-biased gene conversion

3.1 Brief History of the GC Content Studies

In late 1970s, Frederick Sanger (Sanger et al. 1973, 1977; Sanger and Coulson 1975) and Walter Gilbert (Gilbert and Maxam 1973; Maxam and Gilbert 1977) made a breakthrough in efficient DNA sequencing, which has potential to decode the whole-genome sequence of mammals. Following their achievements, many researchers plunged into the frenzy "genome sequencing era" in the end of the twentieth century (Karolchik et al. 2003; Scientists 2009).

Meanwhile, there had been another stream of the genome study (Belozersky and Spirin 1958; Sueoka 1961; Clay et al. 2003). Before the whole-genome sequence era, researchers have already studied genomes with rather gross approaches (Sueoka et al. 1959; Davis 1998). Since genome is a set of gigantic molecules carrying genetic information, it has measurable physical properties (in terms of

S. Oota (✉)
RIKEN Bioresource Center, Tsukuba, Japan
e-mail: oota@riken.jp

N. Saitou (ed.), *Evolution of the Human Genome I*, Evolutionary Studies,
https://doi.org/10.1007/978-4-431-56603-8_3

mechanics and thermodynamics) reflecting sequence-level genomic information. By fragmenting the genome into short segments with deoxyribonuclease (Bernardi et al. 1973), we can measure their mass distribution or a kind of "mass spectrum" by using a density gradient centrifugation (Cortadas et al. 1977; Macaya et al. 1978). From the obtained mass spectra and thermal melting points, it is possible to estimate composition of the fragmented genome associated with (rough) positional information (Tikchonenko et al. 1981). Since the segmented genome sequences are double stranded with complementary pairs, we can directly measure genomic composition as equal amounts of the complimentary nucleotide pairs: adenine (A) for thymine (T) and guanine (G) for cytosine (C) and verse versa.

Many researchers had already tried this approach from the end of 1950s. Sueoka was one of the pioneers who noticed evolutionary significance of the genome composition (Sueoka 1961). By comparing between various bacteria, protozoa, and algae, he concluded that there are two kinds of directed mutation pressures in the genome. Meanwhile, Bernardi and his colleagues carried out extensive comparative analysis of mammalian genome compositions (Bernardi et al. 1985). Uniqueness of their insight was that they tagged the genomic DNA compositions with approximate locations across different mammalian species, i.e., they performed the comparative genome analysis without massive sequence data. They studied the genomic DNA composition with a temporal (comparative analysis with various organisms) and spatial (genomic locations) manner. And they made an outstanding discovery: mammals have spatial heterogeneity of the GC content at the genomic level (Bernardi and Bernardi 1986; Bernardi et al. 1988).

As their predecessors have reported (Belozersky and Spirin 1958; Sueoka 1961, 1962), it has already been known that compositional (GC content) variation between/within species exists (Nekrutenko and Li 2000; Wu et al. 2012). For example, certain bacterial genomes have high GC content, and genic regions have a tendency to be high GC (Galtier et al. 2001). But Bernerdi's point of view was somewhat different from theirs. Bernardi mainly focused on compositional difference within the genome. In other words, he boldly claimed that "nonfunctional" intergenic regions also have GC content heterogeneity, which had been supposed to be "junk" DNA (Ohno 1972; Ohno and Yomo 1991).

Bernardi designated this GC content heterogeneity as isopycnic isochores or "isochore" (Bernardi et al. 1985).

3.2 Three Hypotheses on the Isochore Evolution

Right after Bernardi's discovery, "the whole-genome sequencing era" arrived (Almasy 2012). This ignited elaborate studies on the isochore evolution: with genome sequence data, we can use exact locations of genomic regions associated with GC content (Liberles 2001).

One of the first striking discoveries of the GC content heterogeneity was clear and drastic structure of human major histocompatibility (MHC) regions (Stephens

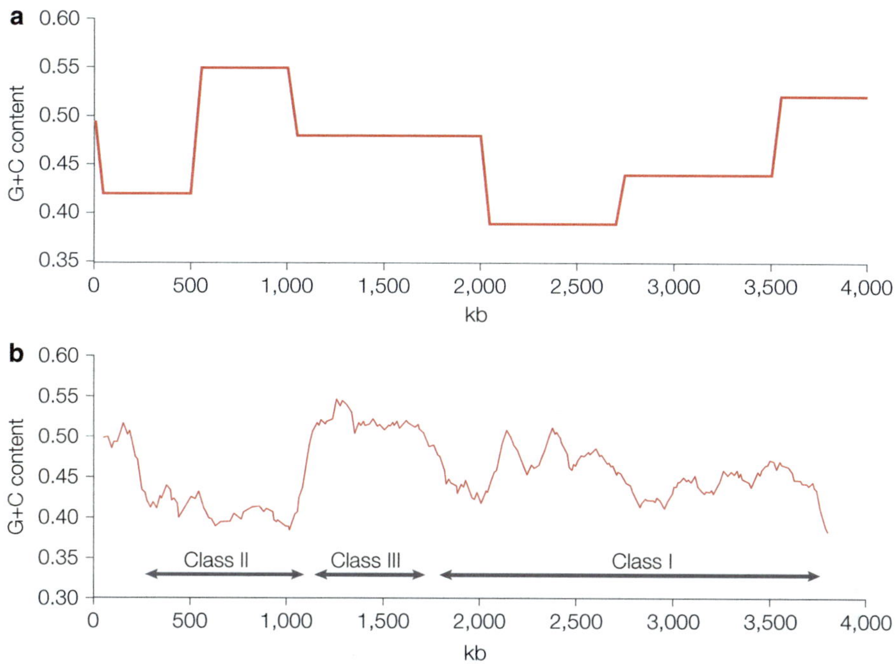

Fig. 3.1 (**a**) A typical isochore structure corresponding to actual GC content heterogeneity in the human major histocompatibility (MHC) regions. (**b**) Here three isochore classes are revealed corresponding to the MHC classes. However, this kind of visually clear isochore is relatively rare in mammalian genome (Eyre-Walker and Hurst 2001)

et al. 1999), which exactly followed what they expected as isochore (Eyre-Walker and Hurst 2001) (Fig. 3.1). However, before long, a lot of discoveries were made with rich genome sequence data, which greatly bewildered isochore researchers.

Considering that intergenic regions are supposed to be "junk (Ohno 1972)," mutations are randomly and uniformly accumulated, and no structure cannot be generated at the whole-genome level. But such very structure has been discovered as isochore.

Surprisingly, the GC content heterogeneity (when exists) spans across intergenic and genic regions (Eyre-Walker and Hurst 2001; Vinogradov 2003a, b). Even in coding sequences, "remnant" of isochore can be observed as strong correlation of GC content values between coding sequences and their flanking genomic regions, as well as untranslated regions (UTR) and introns. In other words, it looks as if ubiquitous isochore structure exists as a kind of "background" of genomic structure and was "destroyed" due to strong selection of coding sequences (CDS).

How did such GC content heterogeneity emerge and why is it maintained to date? There are mainly three hypotheses on the isochore evolution.

3.2.1 The Thermodynamic Stability Hypothesis (The Selection Model)

Bernardi claimed that isochore is a result of strong selection due to high body temperature that can disrupt fragile DNA's hydrogen bonds (Bernardi et al. 1985): since GC base pairs with three hydrogen bonds are stronger than AT base pairs with two hydrogen bonds owing to the stacking interactions (Note that it is actually not due to the triple hydrogen bonds of the GC base pair despite of general understanding; see Yakovchuk et al. 2006). Since this thermodynamics theory is intuitively acceptable, Bernardi's theory won a leading position in the isochore evolution. But before long the floods of genome data generated by the emerging genome sequencing technology threatened Bernard's triumph (Galtier et al. 2001). The "isochore" structure was also discovered in part of reptiles (Hamada et al. 2003) and teleost fishes (Costantini et al. 2007; Melodelima and Gautier 2008), which are definitely cold-blooded animals. Meanwhile, some warm-blooded animals appeared to lack GC-rich or GC-poor isochore: e.g., the opossum (*Monodelphis domestica*) (Mikkelsen et al. 2007) and the platypus (*Ornithorhynchus anatinus*) (Warren et al. 2008; Costantini et al. 2009), respectively. After all, some amniotes have isochore and some do not (see Fig. 3.2). Today, the thermodynamic stability hypothesis is considered to be revised to fit the new observations.

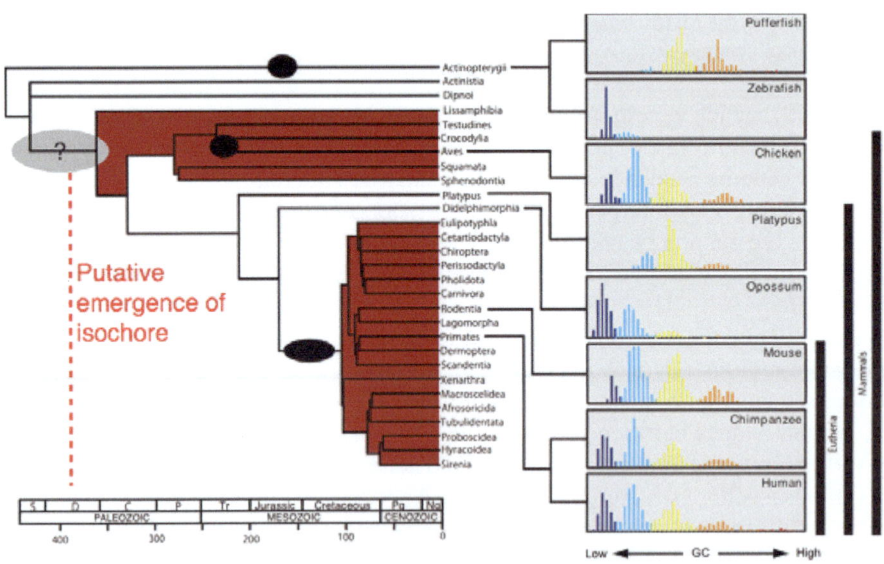

Fig. 3.2 A phylogenetic tree of amniotes. The tree was generated by integrating several "linearized-time" trees (Hedges et al. 2006). It is obvious that species that have GC-rich isochore are polyphyletic. This means that we have to assume very complicated isochore evolution: e.g., multiple emergence and/or lineage-specific disappearances of GC-rich isochore (Oota et al. 2010)

3.2.2 *The GC-Biased Gene Conversion Hypothesis*

Biased gene conversion (BGC) is a recombination-associated segregation distortion or, in our context, BGC favoring G and C over A and T (GC-biased gene conversion, gBGC) (Holmquist 1989; Eyre-Walker 1993). The gBGC is a powerful and really tempting concept: many phenomena associated with isochore evolution can be explained by this model, without introducing complicated selection (Duret and Arndt 2008).

Allelic gene conversion occurs via the heteroduplex structure of DNA strands during homologous recombination (Chen et al. 2007). When DNA molecules are damaged, the mismatches are immediately repaired using information of an intact strand as a template. But in the heteroduplex structure, we have no "parent" or template between the both strands. The mismatches have to be repaired without the genuine strand information. Usually, the repair is carried out blindly and randomly. But for some reason, this fixation has a tendency toward G or C, which is interpreted as a nature of DNA repair machinery (Brown and Jiricny 1988; Lesecque et al. 2013). Therefore, when heteroduplex-associated homologous recombination sufficiently occurs, the genomic regions are prone to GC rich across the extended segments.

This gBGC model was thought to be a versatile theory to explain the isochore evolution (Montoya-Burgos et al. 2003). Ironically, this versatility began to cripple its reputation: if the gBGC model is genuine, isochore should be somehow associated with local recombination rates. More specifically saying, if there is recombination, there should be GC-rich isochore. As mentioned above, with today's rich genome sequence data, some amniotes appeared to have no isochore (Fujita et al. 2011), while gBGC is quite universal across various species (Pessia et al. 2012). This is a strong counterexample against the gBGC hypothesis.

There is another perspective to criticize the gBGC hypothesis. Gene conversion is literally a "convenient" concept for evolution. We can virtually explain any kind of evolutionary process by using gene conversion as an "excuse." However, even the gene conversion could not fully explain the isochore evolution. We should seriously take into account this issue.

3.2.3 *The Mutation Bias Hypothesis*

The mutation bias hypothesis actually has the twofold structure, which makes this model somehow confusable but still more plausible over the others.

It is well known that mutation rate is heterogeneous across the genome (Nachman and Crowell 2000; Kvikstad and Duret 2014). Note that "mutation" in this context is not "observed" diversity, but (estimated) fresh mutations in the germ line (Nachman and Crowell 2000). Due to the another nature of repair machinery (different from one mentioned in the gBGC section), the mis-corporation can be prone to G or C, resulting in GC-rich regions with higher mutation rate. Regarding the DNA repair machinery, Wolfe et al. (1989) claimed as follows: in a somatic cell, there is a free nucleotide pool, whose composition is associated with replication timing (Phear and Meuth 1989; Nachman and Crowell 2000; Eyre-Walker and Hurst 2001). As a result, G and C nucleotides are preferentially mis-incorporated into DNA when the nucleotide pool is GC rich. For example, if the nucleotide pool is AT rich in the early replication, while the nucleotide pool is GC rich in the late replication, the late replication regions will be prone to GC rich.

Meanwhile, it is also reported that the replication timing is associated with mutation patterns (see Fig. 3.3) (Holmquist 1989; Stamatoyannopoulos et al. 2009): i.e., the early replication regions during S phase are more secure against mutations, while late replication regions are more mutation prone. It is obvious that a later replication region has disadvantage for the mutagen attacks, resulting in GC rich in the above context.

We should note that, however, some of the data that support the mutation bias hypothesis are based on somatic mutations (Wolfe et al. 1989) because it is difficult to observe the fresh mutations in the germ line. Furthermore, more importantly, the

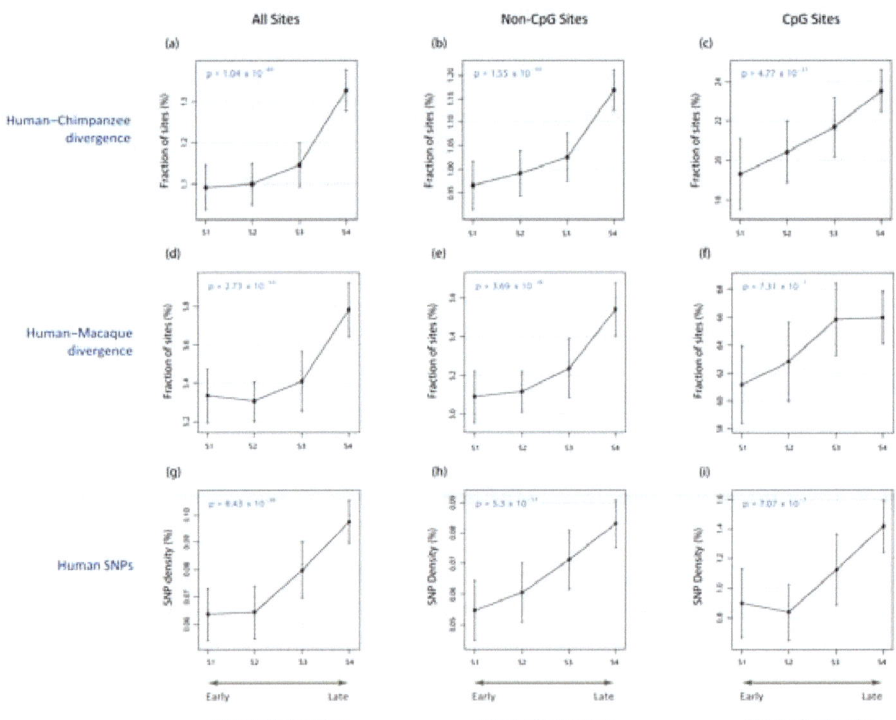

Fig. 3.3 Replication time-dependence of evolutionary divergence and human SNP density (Stamatoyannopoulos et al. 2009)

mutation bias is a quite universal phenomenon, and again, this model also cannot explain why some amniotes have no GC-rich isochore.

3.3 Definition of Isochore: As a Framework for the Genomic Evolutions

When we consider GC content heterogeneity, "isochore" is evidently an important concept. While some researches have already noticed the interspecific heterogeneity of GC content (Belozersky and Spirin 1958; Sueoka 1961, 1962; Lio and Goldman 1998; Clay et al. 2003) before the discovery of isochore in the mammalian genomes, its intrachromosomal characteristics had long been overlooked.

As I mentioned above, the term "isochore" in this context was coined by Bernardi et al. in 1981 with a strong evolutionary paradigm (Cuny et al. 1981; Bernardi et al. 1985). They believed that this genomic structure was formed by thermodynamic stability during evolution. Their hypothesis is called "the selection model," but more precisely speaking, what they essentially claimed is that isochore was formed by selection in a positive feedback manner (Fryxell and Zuckerkandl 2000). When amniotes turned warm-blooded, their genome got persistently exposed to high-temperature environment, which is an adversity for fragile DNA molecules. Assuming that the genome, a blueprint of an organism, should be intact as much as possible for its fitness and any kind of mutations should be harmful, the persistent high body temperature is nothing but malignant mutagen (Donatsch et al. 1982; King and Wild 1983). While the G-C pair is bound by three hydrogen bonds, the A-T pair is bound by only two hydrogen bonds, leading to an intuitive speculation that the G-C pair is more stable than the A-T pair. As a result, the G-C pair is supposed to be robust against undesirable mutations. However, we should note that since the hydrogen bond is not so stable, this naïve idea is not accurate to describe actual thermodynamic stability (Yakovchuk et al. 2006).

Still it is true that the G-C pair is thermodynamically resistant. For example, paired DNA molecules are relatively stable at room temperature. When the temperature reaches a melting point, the strand pair will separate. The melting point is determined by several factors: the DNA sequence length, the extent of mismatch in the pair, and the GC content. Higher GC content causes higher resistance for the environmental temperature: i.e., resulting in a higher melting point (Mandel et al. 1968; Vinogradov 2003a). Organisms living in extreme conditions (extremophile), like *Thermus thermophilus*, have typically GC-rich genomes (Guagliardi et al. 1997; McDonald 2001). However, today we know some cold-blooded organisms that harbor isochore in their genomes (Hamada et al. 2003). So at least some revision is required in the selection model to adapt the hypothesis to the observation.

Furthermore, it is no longer acceptable today that "isochore" is literally mosaic structure of GC content bounded by "GC content cliffs." In the genome era, owing to the advanced sequencing technology, we can easily "compute" GC content associated with precise genomic positions at nucleotide level. Some genomic regions, like MHC regions, beautifully show the mosaic structure of GC content (Pavlicek et al. 2002). Meanwhile, such "axiomatic" observation is rather exceptional, and, in most genomic regions, typical GC content heterogeneity is vaguer or sometimes dimmer than expected before.

We should note that this does not mean the death of isochore. The null hypothesis of randomness in GC content heterogeneity can be easily tested (Haiminen and Mannila 2007). It is reasonable to conclude that the GC content heterogeneity surely exists in certain amniote lineages while it does not exist in the other lineages. Meantime the "isochore" is no longer the isochore in the original sense. Today isochore means statistically significant GC content heterogeneity across the genome, but not a clear mosaic structure bounded by GC content cliffs at visual perception level (Li 2002).

3.4 Controversy Over the Isochore Evolution

So far no evidence was found to exclusively support one of the above three hypotheses, suggesting that the isochore evolution harbors a complicated evolutionary mechanism (Oota et al. 2010).

Furthermore, there are several controversies over the isochore evolution. As I mentioned before, GC-rich isochore was observed only in certain lineages, mammals, birds, and part of reptiles. Currently, the minimum consensus on the isochore evolution is that GC-rich isochore emerged in an ancestor of amniotes. But even this basal consensus is threatened by new discoveries: when we map species that harbor GC-rich isochore to a phylogeny of amniotes, the enigma is inevitable (Fig. 3.2).

This situation is very hard to interpret in terms of the conventional molecular evolution framework: implicitly or explicitly, we typically use an idea of "Occam's razor" to assess the evolution (Smith 1980). For example, if we consider two characters derived from a common ancestor, we assume a minimum evolution (Rzhetsky and Nei 1993) between them. In other words, we usually choose to deny any detours in an evolutionary pathway. Or we assume that an organism evolves along a nearly shortest path.

This assumption is fairly reasonable. If an evolutionary pathway extremely winds in environmental changes, the organism will need to pay extra cost comparing with its competitors that chooses a shorter path (Kopp and Matuszewski 2014). Besides that, the "Occam's razor" is widely accepted in natural science, and we usually had no reason to refute this assumption.

In case of isochore, however, it is difficult to model its evolution with a simple process: i.e., we need to assume multiple emergences or independent disappearances of isochore (Fig. 3.2) at least. This kind of "nonlinear" transition cannot be treated with typical linear dynamics or a stationary Markov process, which is a convenient toolkit for the molecular evolution.

To make matters worse, this is not all for the enigmatic isochore evolution. Mutation patterns of GC content show a stunningly rapid convergence of GC content values: i.e., many researchers reported that GC content of the mammalian genome is not yet at the equilibrium and still in a converging state (Webster et al. 2003). This estimation is supported by both mutation and substitution patterns, and their estimated equilibria are fairly close to each other (The vanishing isochore theory, Duret et al. 2002). But nobody had mentioned how rapid the convergence is. Oota et al. roughly estimated the convergence rate by using human SNP data comparing with its closest relative, chimpanzee, and found that it is far more rapid than we expected (Fig. 3.4) (Oota et al. 2010). If we assume the neutrality of the intergenic regions (which is supposed to be a fair assumption), the convergence of GC content is too rapid to observe the extant isochore.

Of course, the average coalescence time of randomly selected pairs of SNP in the population has huge standard deviation (around $2N_e$) because the probability of coalescence can be approximated by the standard exponential distribution (see Fig. 3.5). But still there is potential incongruence between the estimation and the observation. How can we reconcile them?

Associated with this problem, we also have another issue: the vanishing isochore theory (Duret et al. 2002). As described above, it is almost certain that GC content of the mammalian genome is converging toward an equilibrium according to several reports (Duret et al. 2002; Smith et al. 2002; Webster et al. 2003). So isochore seems to be vanishing. But we should note that this inference is based on insights on relatively short-term evolution. The precedent studies mostly used the

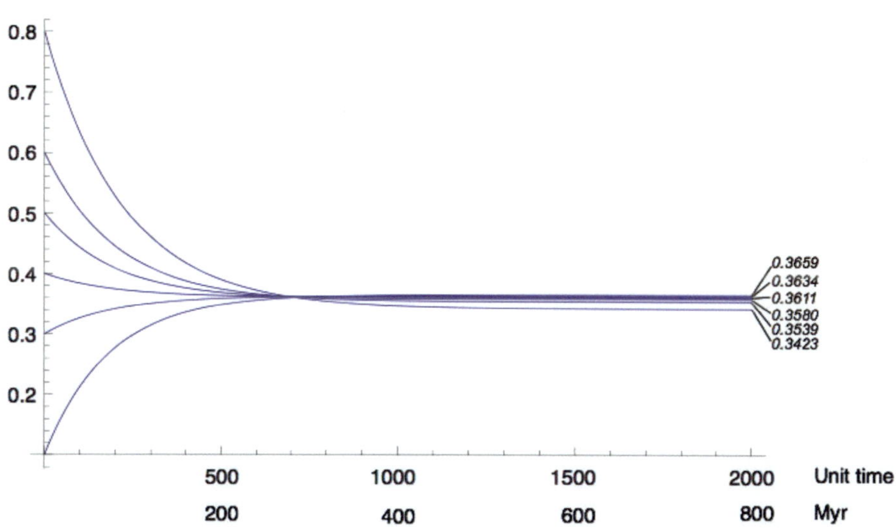

Fig. 3.4 Convergence toward the equilibrium of GC content based on the conventional framework (stationary Markov model or "constant model" Oota et al. 2010)

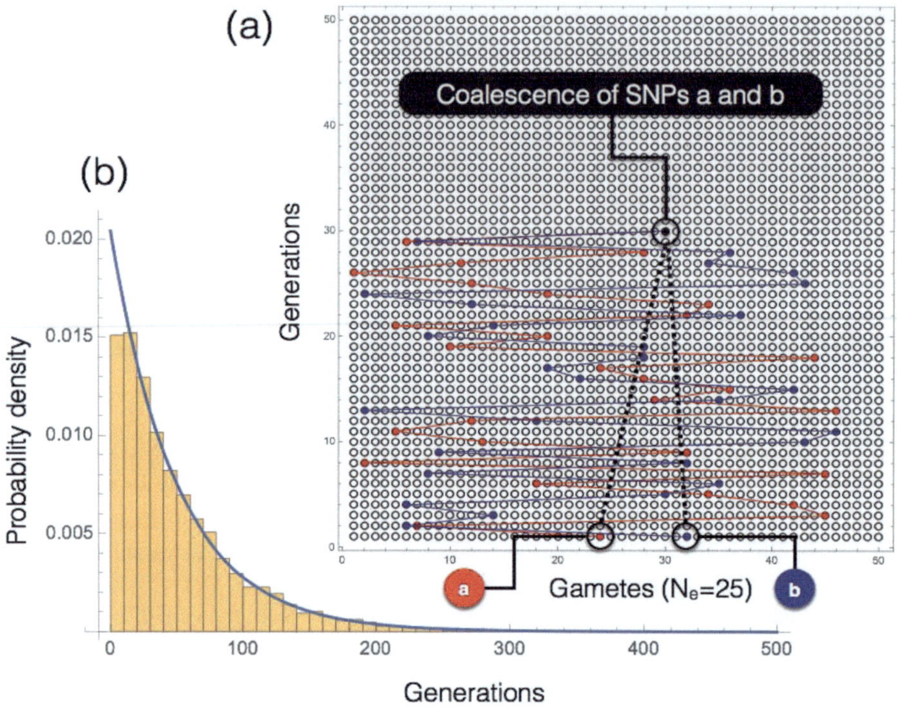

Fig. 3.5 A brief explanation average coalescence time of randomly sampled two SNPs. (**a**) A snapshot of simulation with effective population size (N_e) = 25; no selection is considered. (**b**) A distribution of the simulated coalescence time (histogram) and the theoretical distribution (solid line)

parsimonious method to estimate ancestral states, which are essential factors to elucidate the isochore evolution. Gu and Li (2006) published a paper to review the vanishing isochore theory: they reviewed the parsimonious approaches, which may be too naive to treat the long-term evolution. They used the maximum likelihood method instead and derived surprising results: in some mammalian linages (e.g., rabbit), isochore is rather emerging.

There is an issue to be noted on their approach. While the parsimonious method uses the minimum evolution principle (Gascuel et al. 2001), which is a kind of implicit evolutionary model, the maximum likelihood requires explicit evolutionary models: transition probability matrices (at implementation level, the model is expressed by transition rate matrices). And the matrices are derived from known closely related extant data. So after all we use the "parsimonious" approach to make a seed of the maximum likelihood inference. The maximum likelihood method can compensate a shortcoming of the maximum parsimony method, which is prone to omit part of evolutionary processes in the long-term evolution. Since the maximum likelihood method stochastically takes into account all possible ancestral states, its estimation is quite robust (Felsenstein 1981). Therefore, as long as given models are correct (or reasonable at least), we can rely on the estimates.

However exactly here we have a rather fundamental issue to use the maximum likelihood method for the isochore evolution, especially to evaluate the vanishing isochore theory. A stationary Markov chain can have its own equilibrium distribution. Therefore, as long as we use the typical Markov framework, we always have an equilibrium of GC content, which superficially supports the vanishing isochore theory.

In the approach of Gu and Li (2006), they use plural evolutionary models attributed to different lineages (or branches) of a phylogenetic tree (Gu and Li 2006). This is quite a clever approach to dig out hidden characteristics embedded in the long-term evolution. According to their sound (at least as sound as possible at present) approach, emergence of GC-rich isochore is strongly supported in some lineages. We should note that they do not totally deny the vanishing isochore theory. They suggested that the isochore evolution may be more complex than we thought. In fact, this complexity is fairly consistent with the polyphyletic characteristics of isochore (Oota et al. 2010).

Considering those issues, it is quite hard to explain the isochore evolution with a conventional model alone. There should be a mechanism that generates the GC content heterogeneity across the genome, but we may still lack a sufficient penetration depth with the traditional framework.

3.5 A New Framework with Nonlinear Dynamics

We introduce a typical framework to analyze nonlinear dynamics. Instead of directly constructing a rate matrix by using closely related extant genomic sequences, we focused on relationships between temporally neighboring two states: $f(x)$ and $f(x+1)$. Since we are interested in GC content values, we treat a state as GC content at step x: $f(x)=\mathrm{GC}(x)$.

GC content is a positional concept by nature, and it is trivial to treat GC content as a function of positional information like GC(x), where x is a positional information in the genome (note that x is not a physical position itself, but an abstract value associated with genomic coordinates). Meanwhile GC content itself is determined by a local mutation rate. This sounds a bit queer, but is explained as follows: GC content always has its own bin size due to definition. With alignable (comparable) sequences in the bin, we can compute mutation rates from changes of GC to AT (GC⇒AT) and from AT to GC (AT⇒GC). Note that those mutation rates in this context are standardized by the number of GC and AT sites, respectively. Since the comparable sequences have their common ancestor, we can estimate their divergence time or coalescence time. This means that we can estimate standardized GC⇒AT and AT⇒GC mutation rates separately. This class of mutation rate is often called **the per base pair rates of GC⇒AT (*u*) and AT⇒GC (*v*) mutations** (Webster et al. 2003).

With the per base pair rates of u and v mutations, the number of G+C sites at time $t + 1$ is determined in a recursive way as follows (Oota et al. 2010):

$$\bar{N}_{\mathrm{GC}}(t) = N_{\mathrm{GC}}(0) + \bar{N}_{\mathrm{AT}\to\mathrm{GC}}(t) - \bar{N}_{\mathrm{GC}\to\mathrm{AT}}(t) \tag{3.1}$$

where $\bar{N}_{\mathrm{AT}\to\mathrm{GC}}(t)$ and $\bar{N}_{\mathrm{GC}\to\mathrm{AT}}(t)$ are the expected accumulated numbers of AT⇒GC and GC⇒AT changes at time t, respectively. Note that unit time corresponds to $2N_e \times 20$ years if the SNP data are subject to the neutral evolution, where N_e is an expected effective population size of the human lineage (see Fig. 3.5). The expected total numbers of AT⇒GC and GC⇒AT changes at time t are described as follows:

$$\bar{N}_{\mathrm{GC}\to\mathrm{AT}}(t) = u\sum_{x=0}^{t-1}\bar{N}_{\mathrm{GC}}(x)$$

and

$$\bar{N}_{\mathrm{AT}\to\mathrm{GC}}(t) = v\sum_{x=0}^{t-1}\bar{N}_{\mathrm{AT}}(x)$$

where $N_{\mathrm{AT}}(t)$ and $N_{\mathrm{GC}}(t)$ are expected numbers of AT and GC sites at time t, respectively. Equation 3.1 can be transformed to a recursive expression:

$$\bar{N}_{\mathrm{GC}}(t) = \bar{N}_{\mathrm{GC}}(0) - (u+v)\sum_{x=0}^{t-1}\bar{N}_{\mathrm{GC}}(x) + vLt \tag{3.2}$$

where L is the total number of sites. Therefore, an expecting GC content at time t is

$$\hat{G}C(t) = \hat{G}C(0) - (u+v)\sum_{x=0}^{t-1}\hat{G}C(x) + vt \tag{3.3}$$

With the per base pair rates of u and v mutations, the number of G+C sites at time $t + 1$ is determined in a recursive way as follows:

$$\begin{aligned} N_{\mathrm{GC}}(t+1) &= vN_{\mathrm{AT}}(t) - uN_{\mathrm{GC}}(t) + N_{\mathrm{GC}}(t) \\ &= v(L - N_{\mathrm{GC}}(t)) - uN_{\mathrm{GC}}(t) + N_{\mathrm{GC}}(t) \\ &= (-u - v + 1)N_{\mathrm{GC}}(t) + Lv \end{aligned} \tag{3.4}$$

where L is the length of the sequence. Therefore, GC content in the sequence at time $t + 1$ is

$$\mathrm{GC}(t+1) = \mathrm{GC}(t)(-u - v + 1) + v \tag{3.5}$$

We call this model "the constant model" because rates u and v are given as constant. With GC($t+1$) = GC(t), we can immediately obtain the equilibrium of GC content f^* as follows:

$$f^* = \frac{u}{u+v} \tag{3.6}$$

which is identical to the formula that Sueoka showed in a different framework (Sueoka 1988).

In the conventional Markov framework, the per base pair rates of u and v mutations are constant: i.e., a rate matrix is time independent. However, this expression is somewhat confusable. More precisely speaking, they can be temporally constant, but spatially variable (Oota et al. 2010). In fact, it is easy to show the per base pair rates u and v are spatially heterogeneous across the genome. So like GC content, the per base pair rates u and v can be expressed as functions of position in the genome, like $u(x)$ and $v(x)$, where x is the positional information described above.

Now we can associate the per base pair rates of u and v mutations with GC content by using the positional information x. Therefore, we can express the per base pair rates of u and v mutations as a functions of GC content, which is associated with the positional information x.

From the relationships between the per base pair rates of u and v mutations and GC content (Fig. 3.6), they are spatially variable as suggested by the previous works (Nachman and Crowell 2000; Smith et al. 2002; Berlin et al. 2006; Kvikstad and Duret 2014). The important point here is that GC content can be mapped to corresponding per base pair rate of u and v of the mutations: i.e., u and v at time t can be expressed as a function of local GC content at time t.

As described in Eq. 3.5, the per base pair rates of u and v mutations is determinants of temporal GC content of time t. Meanwhile, by using human SNP data and the chimpanzee genome as a reference, it is easily derived that the u and v spatially vary depending on the local GC content (Fig. 3.7). This observation raises a complex situation: i.e., we need to handle a variable that depends on another variable in a recursive manner.

The significance of the rate heterogeneity is very intriguing. For now, it is difficult to describe a biological reason that such mutation rate heterogeneity exists in the genome. In this chapter, I do not detail the mechanism harbored this phenomenon. Instead, I focus on a feasible model that explains observed SNP data to acquire the rate heterogeneity.

Since the SNP are actually alleles, its significance is associated with the sample size. If we can obtain denser SNP data, we can acquire more informative mutation patterns. Regarding relationships to GC content, there were few SNP data that fall into extremely low and high GC content classes. When we could not obtain dense SNP data, this kind of analysis was virtually impossible. Today, owing to the rich SNP data, the rate heterogeneity (U-shaped curve in Fig. 3.7) turned to be substantial according to the confidence interval based on the bootstrap resampling.

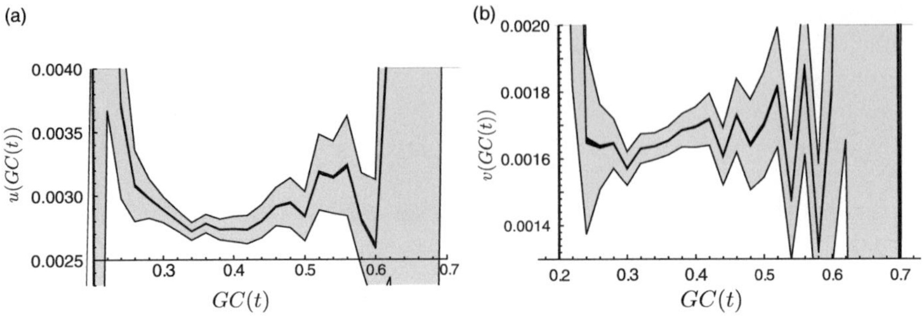

Fig. 3.6 The U-shaped relationships between GC content and the per base pair rates of GC⇒AT (u) and AT⇒GC (v) mutations in human chromosome 3 (HSA3). The relationships were obtained mutation patterns based on observed SNP data and comparison with the chimpanzee genome (Fig. 3.7). Confidence interval (CI) was calculated with bootstrap sampled data from HSA3 (1000 replicates) (Oota et al. 2010)

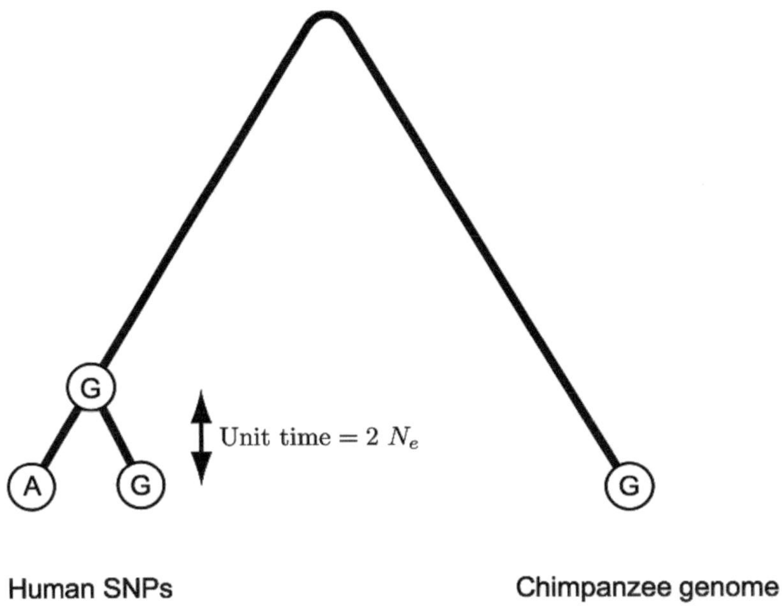

Fig. 3.7 Inference of ancestral states at coalescence of Human SNPs with the chimpanzee genome as a reference (Oota et al. 2010)

This spatial rate heterogeneity triggers a new question: is the assumption that the per base pair rates of u and v mutations temporally constant is acceptable? The genome is a mosaic of segmented sequences that experienced various histories. This local GC content-dependent bias actually drives the convergence of GC content toward the equilibrium (Karro et al. 2008): i.e., genomic regions that have higher and lower GC content have higher mutation rates. Meanwhile, the U-shaped relationships are observed in all the chromosomes with relatively small variance as shown in Fig. 3.6. Therefore, we can regard the mutation rate-GC content correspondences as a universal characteristic of the genome.

With currently available data, it is virtually impossible to directly model the per base pair rates of u and v mutations as a function of time t, even if the u and v vary depending on time. However, assuming that the above U-shaped relationships between GC content and the per base pair rates of mutations are universal across the genome, the per base pair rates of u and v mutations can be presented as a function of temporal GC content GC(t): i.e.,

$$u = u(\mathrm{GC}(t))$$

and

$$v = v(\mathrm{GC}(t)).$$

Since the u and v are determinants of temporal GC content at the next state (i.e., time $t + 1$), we can formulate this relationship as a recursive manner as follows:

$$\mathrm{GC}(t+1) = \mathrm{GC}(t)(-(u(\mathrm{GC}(t)) + v(\mathrm{GC}(t)))) + v(\mathrm{GC}(t)) + \mathrm{GC}(t). \quad (3.7)$$

GC content at arbitrary time t can be obtained if we know GC content at time 0: i.e., GC(0). Note that GC(0) is the present local GC content. We call this model "the variable model" because u and v are variable as functions of GC content at time t: GC(t).

We also call this new approach "the $f(x)$ framework," which is a typical way to analyze nonlinear dynamics (Rabinovich and Abarbanel 1998; Dokoumetzidis et al. 2001).

If we consider long-term evolution, we need to extrapolate this framework from diversity level to divergence level. It looks crude, but this framework depends on only two kinds of parameters: $u = u(\mathrm{GC}(t))$ and $v = v(\mathrm{GC}(t))$. So only assumption that we need here is universal characteristics of the u-shaped relationships (as shown in Fig. 3.7) over time. This assumption implies that possible mutation patterns are already included in the genome: or more precisely speaking, the temporal mutation pattern space is included in the spatial genome space. This "*time-space conversion*" is the key concept of the variable model.

Interestingly, the $f(x)$ framework provides results consistent to the conventional framework under certain conditions (Fig. 3.8). The conventional framework (actually the stationary Markov framework) and the $f(x)$ framework give almost the same equilibrium GC content (Oota et al. 2010), supporting the vanishing isochore theory (Duret et al. 2002). Note that their rates of convergence are also similar to each other.

In this regard, the variable model is a natural extension of the constant model. Or the constant model is a special case of the variable model, in which $u(\mathrm{GC}(t))$ and v

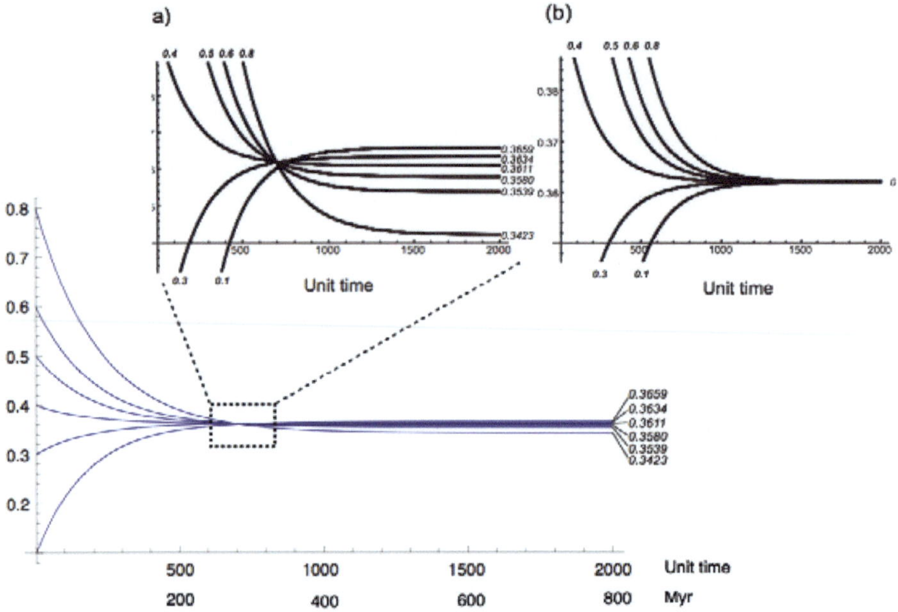

Fig. 3.8 Convergence of GC content computed by (**a**) the conventional framework (the constant model) and (**b**) the $f(x)$ framework (the variable model). Their results are fairly consistent except for subtle difference: in the variable model, the exact convergence is guaranteed in Class I and Class II (Oota et al. 2010)

(GC(t)) are uniform functions. In this case, the above U-shaped relationships are to be flat.

Figure 3.9 shows a typical asymptotic convergence in the $f(x)$ framework. However, this behavior needs a certain condition. Depending on relationships between $f(x)$ and $f(x + 1)$ or $f(\mathrm{GC}(t))$ and $f(\mathrm{GC}(t + 1))$ in our case, the GC content transition can drastically change, which is a characteristic of a nonlinear dynamics (Bishop 2012). The asymptotic convergence is guaranteed only if

$$\left| \frac{df\left(\widehat{\mathrm{GC}(t_e)}\right)}{d\widehat{\mathrm{GC}(t_e)}} \right| < 1 \tag{3.8}$$

holds, where t_e is the minimum time to reach $\mathrm{GC}(t_e+1) = \mathrm{GC}(t_e)$: i.e., an equilibrium. We call this situation, which is consistent with the conventional framework, Class I (Fig. 3.9). Note that this condition is required for existence of a fixed point, which actually does not immediately mean "equilibrium" in the context of evolution. Therefore, the inequality (Eq. 3.8) is a *requisite condition*. However, there are little reasons to distinguish between a fixed point and the equilibrium in this class.

Meanwhile, in the $f(x)$ framework, we can theoretically meet "unprecedented" behaviors of GC content transition. A subtle change of relationships between GC(t) and GC(t + 1) leads to typical nonlinear behaviors.

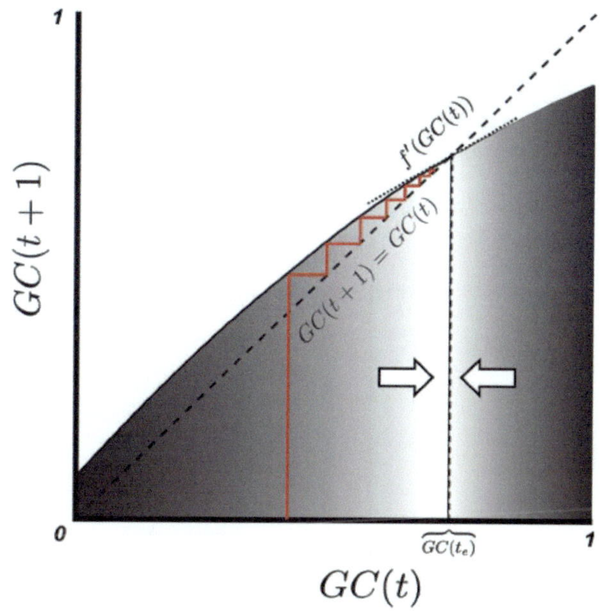

Fig. 3.9 "Asymptotic" convergence of GC content at Class I. In the $f(x)$ framework, this class gives results consistent to the conventional stationary Markov framework (Oota et al. 2010)

Figure 3.10 shows an oscillatory convergence of GC content. This occurs only if

$$-1 < \frac{df\left(\widehat{\mathrm{GC}(t_e)}\right)}{d\widehat{\mathrm{GC}(t_e)}} < 0 \tag{3.9}$$

holds. Note that the curvature of Fig. 3.10 is exaggerated for convenience. The "oscillatory" means asymptotic convergence in reciprocal directions: i.e., GC content gradually approaches an equilibrium point from both sides. This means that this convergence itself can potentially generate local GC content heterogeneity before reaching an equilibrium state. We call this transition pattern Class II.

Figure 3.11 shows multiple convergence of GC content. This occurs only if

$$1 < \frac{df\left(\widehat{\mathrm{GC}(t_e)}\right)}{d\widehat{\mathrm{GC}(t_e)}} \tag{3.10}$$

holds. This situation is very intriguing. In Fig. 3.11, only three fixed points (potential equilibria) are shown for convenience, but the number of fixed points depends on a shape of $\mathrm{GC}(t+1)=f(\mathrm{GC}(t))$. As shown in below, results of our analysis in the $f(x)$ framework are fairly consistent with the currently observed isochore classes (Duret et al. 1995). We call this situation Class III (Oota et al. 2010).

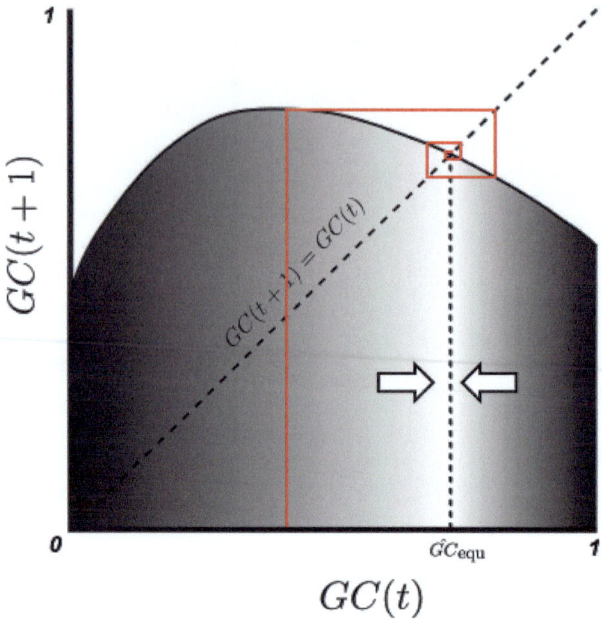

Fig. 3.10 An oscillatory convergence of GC content (Class II) (Oota et al. 2010)

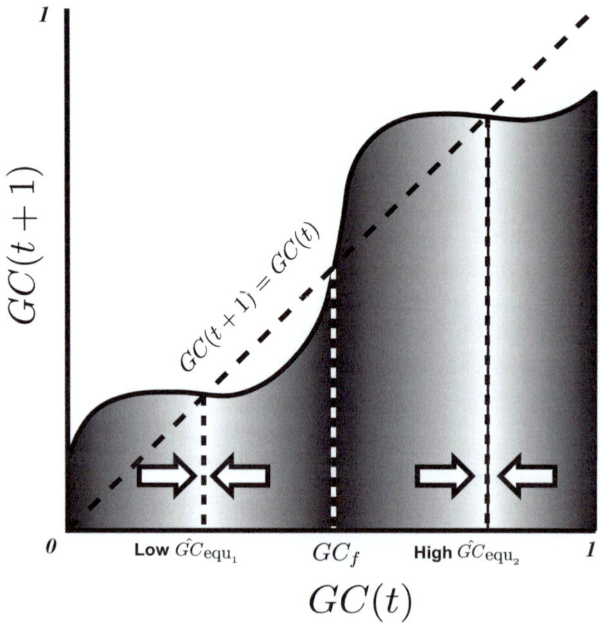

Fig. 3.11 Attraction and repulsion of GC content transition around plural fixed points. Three fixed points are presented here (Class III) (Oota et al. 2010)

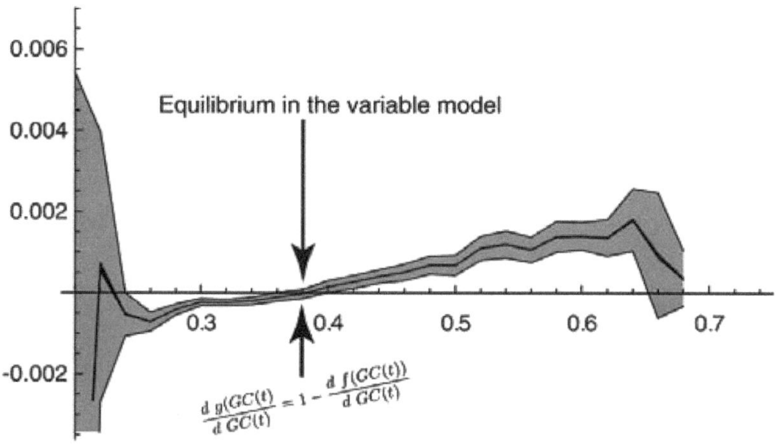

Fig. 3.12 An example of relationships between GC(t) and g(GC(t)) = GC(t)–f(GC(t)) = GC(t)–GC(t + 1) in the variable model by using a data set generated with human chromosome 21 (HSA21) and the chimpanzee genome. The upper and lower arrows indicate an equilibrium GC value and the slope of g(GC(t)) at the fixed point, respectively. Shaded regions indicate 95% confidence interval based on 1000 bootstrap resampling from the HSA21 data (Oota et al. 2010)

It is almost straightforward to apply the variable model to actual data with above conditions. Figure 3.12 shows an intersection between GC($t+1$) = = GC(t) and g (GC(t)) = GC(t) − f(GC(t)) = GC(t) − GC($t+1$). Here we use $g(x)$ instead of $f(x)$ to visualize the intersection: since difference between GC(t) and GC(t + 1) is very subtle in the actual data, it was difficult to perceptually demonstrate the relationships without this conversion. Note that the confidence interval (CI) is very small around the intersection: i.e., the potential equilibrium is stable across the chromosome.

In this "$g(x)$" framework, a derivative of g(GC(t_e)) with respect to GC(t_e) is the determinant of the GC content transitional behaviors. In human chromosome 21 (HSA21), an estimated GC content equilibrium is approximately 0.4 (Fig. 3.12), which is consistent with one by the conventional framework. The majority of the data of HSA21 belong to Class I (Oota et al. 2010).

Meanwhile, when we review genome-wide data, it is relatively easy to find the other classes.

Figure 3.13 shows relationships of fixed points (potential equilibria), derivatives at the fixed points, and corresponding GC content values in human chromosome 1 (HSA1). While most fixed points belong to Classes I and II, an considerable number of fixed points belong to Class III. We can see three clusters of fixed points (a, b, and c), suggesting multiple equilibria in HSA1. Meanwhile, there is a high frequency of fixed points in extremely low and high GC content classes (d). For now, however, we cannot exclude a possibility that they are just artifacts due to the small sample size. We excluded such suspicious data from the current analysis. In the future, rich SNP data may overcome this problem.

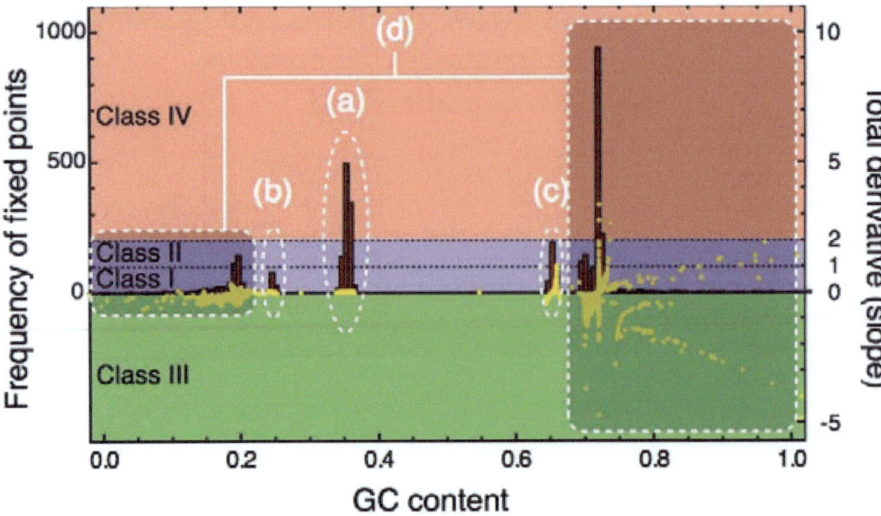

Fig. 3.13 Distribution of fixed points of $f(\mathrm{GC}(t))$ in chromosome 1 (HSA1). Left axis, frequencies of fixed points of $f(\mathrm{GC}(t))$ in HSA1 (histogram); right axis, the derivative of $g(\mathrm{GC}(t))$ with respect to $\mathrm{GC}(t)$ at t_e (fixed points of $g(\mathrm{GC}(t))$) (yellow dots). Classes I and II, "attractive" fixed points or potential equilibria; Class III, "divergent" fixed points (green). Note that, in actual computation, the fixed points were given by roots of $g(\mathrm{GC}(t)) = 0$. Therefore, the three classes in terms of behaviors of GC content transition were determined as follows: *(1)* $0 \leq g'(\mathrm{GC}(t_e)) < 2$: convergence, suggesting an equilibrium (*purple*); *(2)* $g'(\mathrm{GC}(t_e)) > 2$ or $g'(\mathrm{GC}(t_e)) < 0$: divergence (green). The constant model shows equilibria around GC content ~0.4, which are consistent with those of the constant model with substitution patterns of GC_3 (Webster et al. 2003; Oota et al. 2010). Note that CpG contexts are included in those data

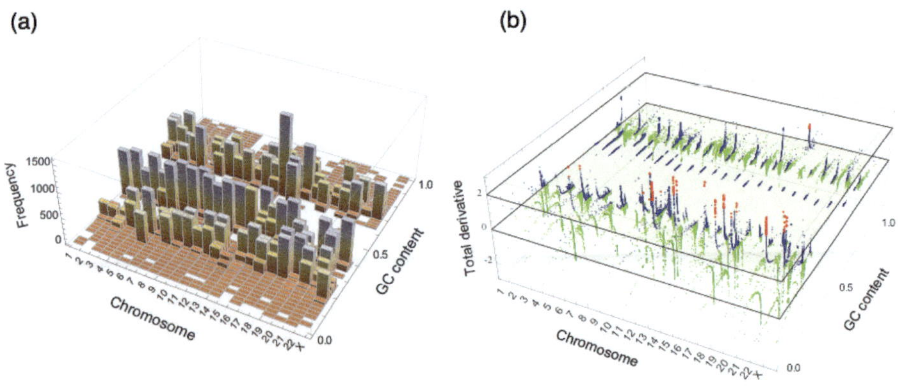

Fig. 3.14 Distribution of GC content (**a**) and fixed points of $\mathrm{g}(\mathrm{GC}(t))$ (**b**) in human chromosomes 1, 2, 3, 4, 5, 6, 7, 8, 9, 10, 11, 12, 13, 14, 15, 16, 17, 18, 19, 20, 21, 22, and X. Frequency, the number of segments associated with GC content values; total derivative, the derivative at a fixed point of $g(\mathrm{GC}(t))$ with respect to $\mathrm{GC}(t)$. Blue, Classes I and II; green, Class III; red, Class IV

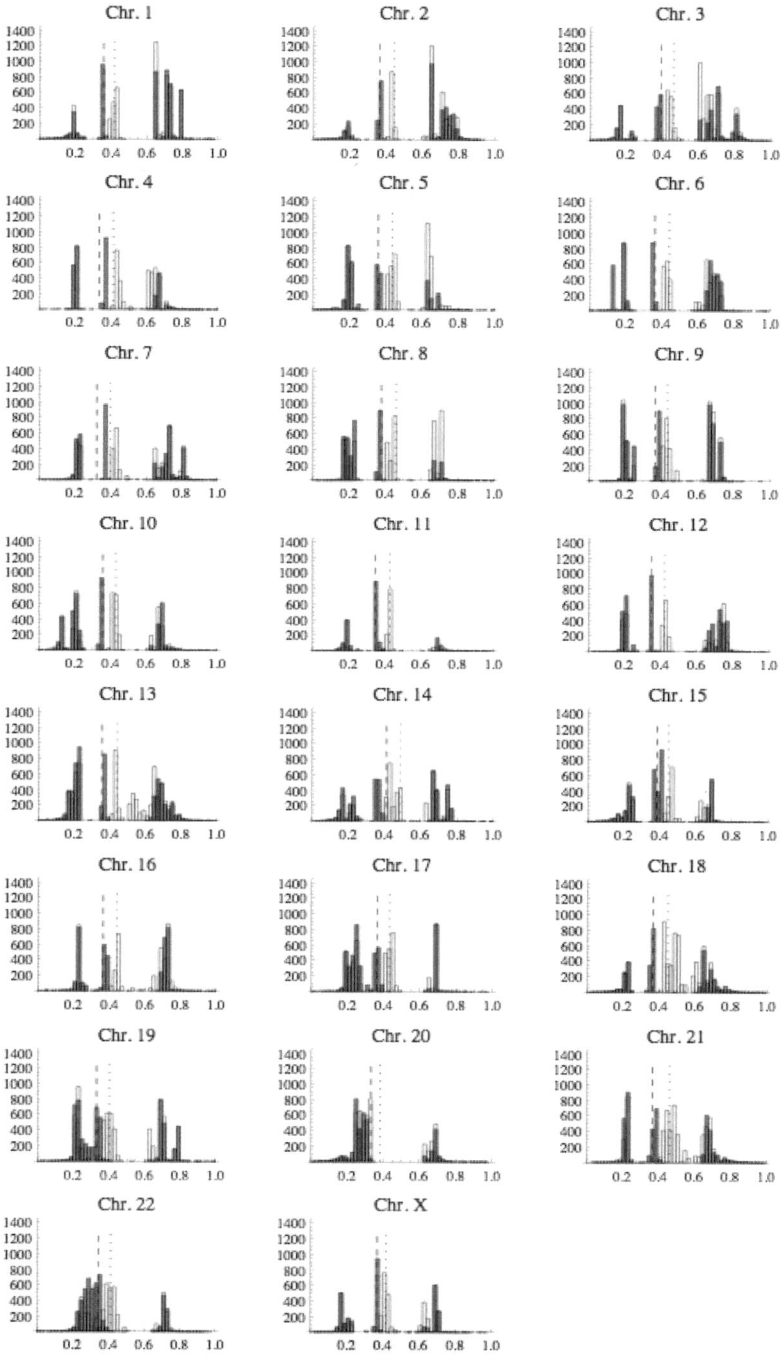

Fig. 3.15 Distribution of fixed points of $g(\mathrm{GC}(t))$ in human chromosomes (open bars, all SNP data; shaded bars, non-CpG SNP data). Subsets of the fixed points are potential equilibria. Abscissa and ordinate are GC content and frequency of fixed points, respectively. Dashed and dotted lines indicate estimated GC content equilibria in each chromosome under the constant model by using all SNP data and non-CpG SNP data, respectively (Oota et al. 2010)

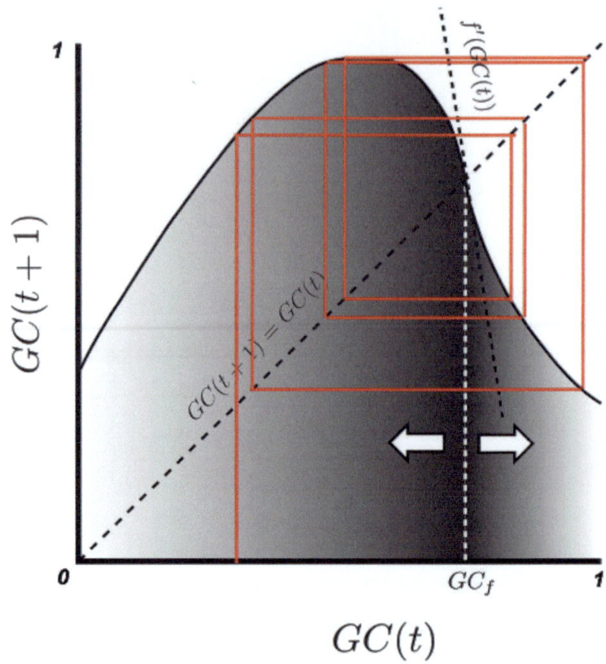

Fig. 3.16 A schematic representation of a chaotic behavior of GC content transitions. Red lines represent the transitions of GC content. Diagonal broken lines represent GC(t+1) = GC(t), which is a requisite condition in the $f(x)$ framework for a fixed point (potential equilibrium) (Oota et al. 2010)

Figure 3.14 shows an overall landscape of fixed points across the human genome. All the distributions share a common characteristic: three-four clusters of fixed points are observed, suggesting multiple equilibria. Since CpG dinucleotide context is subject to higher mutation rates (Fryxell and Moon 2005; Walser et al. 2008), we also generated two kinds of data sets: i.e., with and without the CpG dinucleotide context (shaded and open bars, respectively in Fig. 3.15). Without the CpG dinucleotide context, estimated equilibria shifted toward high GC values, which fairly matches results of prior researches (Smith et al. 2002; Webster et al. 2003).

In the $f(x)$ framework, there exists the fourth class (Fig. 3.16) only if

$$2 \le \frac{dg\left(\mathrm{GC}(t)\right)}{d\mathrm{GC}(t)} \text{ or } \frac{dg\left(\mathrm{GC}(t)\right)}{d\mathrm{GC}(t)} \le 0 \tag{3.11}$$

holds. This Class IV is totally different from the other classes. First of all, Class IV has no equilibria. In this class, GC content can drastically change between neighboring steps, revealing almost random-like behaviors. Note that this is actually not "random" in terms of mathematics, but definitely deterministic. In other words, a simple rule expressed by the Eq. 3.5 generates this complicated transitions under the above condition (Eq. 3.11). This kind of dynamics is called "chaos," if its period has virtually infinite length (Strizhak and Pojman 1996).

This nonlinear dynamics implicates a lot of things (Smyrlis and Papageorgiou 1991). Here I would like to point out just one of them: even complicated phenomena may harbor a fundamental principle: i.e., they may just follow a simple rule

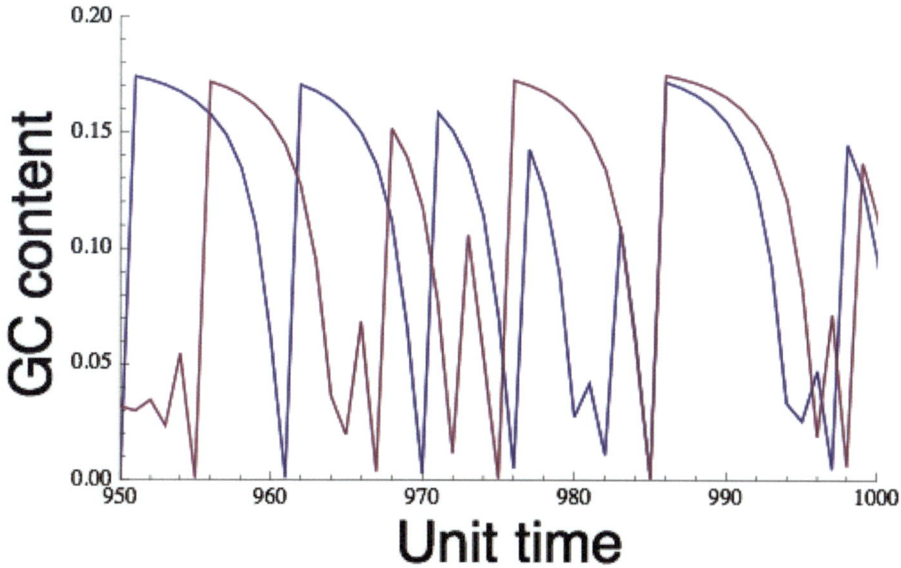

Fig. 3.17 Results of simulation with Class IV data that belong to clusters d in Fig. 3.13. Subtle difference of initial values led to drastically diverged results, suggesting that the Class IV data may "shuffle" or "rest" converging GC content toward multiple equilibria

(Oestreicher 2007). In fact, this kind of "chaos" was observed in various fields (Mazzocchi 2008).

In our current data set, we could identify no Class IV cases except for several fixed points that belong to clusters d in Fig. 3.13, which we ignored. Figure 3.17 reveals results of simulation by using the Class IV data.

Considering that many data belong to Class III (multiple convergence toward fixed points or potential equilibria), existence of the Class IV data seems to justify that the *f*(*x*) framework explains the enigmatic isochore evolution. Namely, conjunction of two kinds of dynamics may exist: convergence toward multiple equilibria and occasional shuffles and/or "resets" of the asymptotic transition course by the chaotic behaviors.

This prediction will be validated by high-density SNP data available in future.

3.6 General Discussion on an Evolutionary Study Based on Nonlinear Dynamics

The concept of molecular clock is often mentioned as an example of the triumph of molecular evolution over traditional morphology-based evolution studies (Ayala 1999; Bromham and Penny 2003; Hipsley and Muller 2014). While the molecular clock is a definitely powerful indicator to associate an ancestral state with a physical time, we need a strong assumption: persistent constancy of the evolutionary rate (Kimura 1983; Takahata 2007). This is a fundamental issue in the molecular evolution because we have to implicitly assume that the molecules have been subject to nonfunctional changes during the evolution despite of the observed evolutionary changes (otherwise the molecular clock doesn't "tick") (Nei 2005).

In the very same context (Gingerich 1986), the molecular evolution itself is based on an implicit but distinctive assumption: basically, evolution at molecular level is treated as a stationary Markov process (Siegel 1976; Kaehler et al. 2014). The linearity in this context is that a small change results in a small change and a large change results in a large change. Therefore, we can infer and/or predict an evolutionary process with an intuitive manner, at least. For example, the molecular clock is a very understandable concept as an analogy of radioactive decay of isotopes (Ayala 1999), which is a reliable temporal measure. But unlike the behavior of isotopes, the determinants of the molecular clock are far unclear. In other words, the molecular clock itself is far more indirect for measuring time. We have to consider complicated mechanism of the biological system (or phenotypes) associated with the evolutionary changes. When some representative molecules are almost constantly changing (Zuckerkandl 1987, 2012), we usually use the neutral evolution principle (Kimura 1968) to explain the constancy. This seems a sound assumption, but may not be always true (Charlesworth and Eyre-Walker 2007).

An important issue here is that there may be some potential flaws in the current framework of evolutionary studies (Rodriguez-Trelles et al. 2002; Koonin 2007; Dean 2010). The most prominent one is that the current framework is virtually the "open-loop" framework (Fudenberg and Levine 1988) mainly due to a lack of experimental evidences (Orzack 2012). But this is one of the natures of evolutionary studies. The second one is insufficient soundness in inference of the ancestral states. For this inference, usually we construct an evolutionary model by using extant closely related species data (Kosiol and Goldman 2011; Verbyla et al. 2013; Mir and Schober 2014). This is a typical example of an extrapolative way that we often apply to the molecular evolution. We sample data from reachable evidences (including ancient DNA sequences such as Green et al. 2006) and construct a transition probability (or rate) matrix (Gu and Li 1998; Baier et al. 1999). We should note that we do not use any explicit hypotheses or "models" to generate the transition rate matrix to be extrapolated, except for the parsimonious inference (Zhaxybayeva et al. 2007).

To apply the extrapolation, we again need to assume that uniform properties of long-term evolutionary behaviors: i.e., a *stationary model* (Gu and Li 1998; Kosiol and Goldman 2011). That is, something occurred yesterday is expected to be occurring for some million years.

This issue can be expanded to a more extreme case: polymorphism is obviously "seeds" of the long-term evolution (Sawyer and Hartl 1992; Barton 1995). But they are different kinds of phenomena. More specifically describing, in terms of the isochore evolution, can we directly associate substitution rate with mutation rate? This question has been already answered by Kimura (1983). Actually they are identical if the exact neutrality holds.

But what if the neutrality does not hold? What if intergenic regions of the genome are subject to evolutionary dynamics essentially different from genic (or coding) regions? Is it possible to perform certain "calibration" to make the extrapolation precise (Subramanian and Lambert 2011)?

At present, it is hard to reveal evidences to elucidate these problems except for several cases (Andolfatto 2005). It is not so surprising that intergenic regions have their own evolutionary traits (Komiya et al. 1995). We should note that the molecular evolution has been constructed mainly by using data sampled from coding regions and their surrounding genomic regions, which may have rather exceptionally behaved in the evolutionary pathways.

3.7 An Impact of Big Data and Extensive Simulation

In the molecular evolution, researchers typically rely on molecular data sampled from extant (living) organisms. This claim sounds trivial. Traditional (non-molecular) evolutionary studies, however, do not always follow this paradigm: e.g., archeologists mainly use excavated fossil records to study evolutionary changes of morphological traits (Norell and Novacek 1992). In this case, we rely on "in situ" data with geological evidence, which can tag the data with physical time (Kirchner and Weil 2000). Although it is considerably difficult to acquire intact and/or informative fossil records, it is certain that this traditional method is the most direct way to elucidate morphological evolution (O'Higgins 2000; Giribet 2003): we can use radiocarbon dating (Wood et al. 2013) as I mentioned above. By contrast, the molecular evolution in fact heavily relies on the comparative analysis of extant data, which may lead to erroneous inference due to the indirect estimation. After all, what we are doing with the "modern" molecular-based methodology is nothing but extrapolation from extant data to extinct events (Dietrich 1994; Koonin 2009).

It is obvious that the most of advantage of the molecular evolutionary studies resides in its objectivity and quantitativity based on highly precise (and *discrete*, i.e. nonambiguous) molecular data (Kell and Lurie-Luke 2015). But we also should note that this advantage owes to the virtue of the extant data, not the estimation itself made by them. If an evolutionary model is wrong, the estimation will be always wrong. The point here is that there is no way to prove the authenticity of the evolutionary model (Huang et al. 2007; Majoros and Ohler 2010). In other words, a molecular evolutionary model is, after all, the "open-loop" framework (Fudenberg and Levine 1988), in which we virtually cannot revise our model by feedback of direct evidence.

Many researchers acknowledge this limitation with some frustration (Schneider et al. 2000; Elena and Lenski 2003; Achaz et al. 2014). This is not an essential shortcoming of the molecular evolution, but a collateral flaw due to that our lifespan is too short to observe ongoing evolutionary processes, though.

What we are doing in the molecular evolutionary studies is somewhat indecisive extrapolation from the extant data (Brown et al. 2010; Andujar et al. 2012). This assertion might sound a bit bold, but it is certainly one side of the modern evolutionary study (Engelhardt et al. 2011).

In the natural science, strictly speaking, no model should be accepted without validation: i.e., experiments. But in several fields, this condition is somewhat relaxed due to extreme difficulty to perform effective validation: e.g., cosmology (Oreskes et al. 1994), meteorology (Baumann and Stohl 1997), and evolution itself. In this regard, computer simulation is another hope to compensate the limitation (Garcia and Velasco 2013).

Of course, we still need a "good" model. Usually, a computational model is constructed by using feasible scale experiments and/or available data. In case of the molecular evolution, we use extant data and the comparative analysis. Fortunately, we have a large amount of data owing to emerging technologies, like next generation sequencer (NGS) (Song et al. 2013). Such "big data" (Greene et al. 2014) have potential to overcome the shortcoming of the "open-loop" framework.

In case of the isochore evolution study, however, the big data approach still seems insufficient to solve the enigmatic phenomena.

Problematic part of the isochore evolution study can be summarized like this: the molecular evolution framework has been basically constructed based on coding regions in the genome: i.e., the "molecular" virtually refers to genes or transcripts. On the other hand, the isochore is the genome-wide structure of the GC content. More boldly speaking, the evolution of isochore virtually means the evolution of so-called "junk" DNAs. Unlike coding regions, it is quite tough to mathematically formalize the evolution of such intergenic regions (Moses et al. 2003). One of the reasons is that comparative analysis of "junk" DNAs is generally difficult between species.

I feel that the only way to overcome the difficulty is the extensive computer simulation, as well as the big data approach. Owing to rapid development of the information science and technology, we are beginning to have the capability to reach both of them. In the next decade, a breakthrough in this area may be made, and we may be able to reach a fundamental mechanism of the isochore evolution.

Reductionism is occasionally criticized in biology (Mayr 2004; Van Regenmortel 2004). I also admit that excessive reductionism may be harmful for healthy discussion on biology. Debate on repeatability and testability of evolution has been hot philosophical issues for long time (Popper and Eccles 1984). At least, evolution is a pathway that each organism tracked with. Even if certain rules exist, it may be difficult to predict the future of evolution due to its uncertainty and complexity. This is rather a classic debate on the evolution (Papp et al. 2011). However, still I believe that evolution is a result of reducible phenomena and follows certain "laws." Otherwise, we have to admit that evolutionary phenomena are a somewhat exceptional science.

At least, it is acceptable to claim that every life was designed through evolution. Evolutionary studies have potential to provide us a fundamental law of biological systems. Before Kimura (1968), the fundamental law used to be the Darwinian selection (Kimura 1983). But in the molecular evolution, it is too naïve to literally justify Darwin's theory. Many researchers admit that what majorly drives evolution may be something else, for example, random mutations as Nei claimed (Nei 2013). Meanwhile, as I mentioned above, the molecular evolution literally refers to evolution of coded molecules or encoded genes at most. Considering the complexity of observed data, we may need to admit that we are still standing at an entrance of the evolutionary study.

References

Achaz G, Rodriguez-Verdugo A, Gaut BS, Tenaillon O (2014) The reproducibility of adaptation in the light of experimental evolution with whole genome sequencing. Adv Exp Med Biol 781:211–231

Almasy L (2012) The role of phenotype in gene discovery in the whole genome sequencing era. Hum Genet 131:1533–1540

Andolfatto P (2005) Adaptive evolution of non-coding DNA in Drosophila. Nature 437:1149–1152

Andujar C, Serrano J, Gomez-Zurita J (2012) Winding up the molecular clock in the genus Carabus (Coleoptera: Carabidae): assessment of methodological decisions on rate and node age estimation. BMC Evol Biol 12:40

Ayala FJ (1999) Molecular clock mirages. BioEssays 21:71–75

Baier C, Katoen J-P, Hermanns H (1999) Approximative symbolic model checking of continuous-time Markov chains. In: Baeten JM, Mauw S (eds) CONCUR'99 concurrency theory. Springer, Berlin/Heidelberg, pp 146–161

Barton NH (1995) Linkage and the limits to natural selection. Genetics 140:821–841

Baumann K, Stohl A (1997) Validation of a long-range trajectory model using gas balloon tracks from the Gordon Bennett Cup 95. J Appl Meteorol 36:711–720

Belozersky AN, Spirin AS (1958) A correlation between the compositions of deoxyribonucleic and ribonucleic acids. Nature 182:111–112

Berlin S, Brandstrom M, Backstrom N, Axelsson E, Smith NG, Ellegren H (2006) Substitution rate heterogeneity and the male mutation bias. J Mol Evol 62:226–233

Bernardi G, Bernardi G (1986) Compositional constraints and genome evolution. J Mol Evol 24:1–11

Bernardi G, Ehrlich SD, Thiery JP (1973) The specificity of deoxyribonucleases and their use in nucleotide sequence studies. Nat New Biol 246:36–40

Bernardi G, Olofsson B, Filipski J, Zerial M, Salinas J, Cuny G, Meunier-Rotival M, Rodier F (1985) The mosaic genome of warm-blooded vertebrates. Science 228:953–958

Bernardi G, Mouchiroud D, Gautier C, Bernardi G (1988) Compositional patterns in vertebrate genomes: conservation and change in evolution. J Mol Evol 28:7–18

Bishop RC (2012) Fluid convection, constraint and causation. Interface Focus 2:4–12

Bromham L, Penny D (2003) The modern molecular clock. Nat Rev Genet 4:216–224

Brown TC, Jiricny J (1988) Different base/base mispairs are corrected with different efficiencies and specificities in monkey kidney cells. Cell 54:705–711

Brown CJ, Johnson AK, Daughdrill GW (2010) Comparing models of evolution for ordered and disordered proteins. Mol Biol Evol 27:609–621

Charlesworth J, Eyre-Walker A (2007) The other side of the nearly neutral theory, evidence of slightly advantageous back-mutations. Proc Natl Acad Sci U S A 104:16992–16997

Chen J-M, Cooper DN, Chuzhanova N, Ferec C, Patrinos GP (2007) Gene conversion: mechanisms, evolution and human disease. Nat Rev Genet 8:762–775

Clay O, Douady CJ, Carels N, Hughes S, Bucciarelli G, Bernardi G (2003) Using analytical ultracentrifugation to study compositional variation in vertebrate genomes. Eur Biophys J 32:418–426
Cortadas J, Macaya G, Bernardi G (1977) An analysis of the bovine genome by density gradient centrifugation: fractionation in Cs2SO4/3,6-bis(acetatomercurimethyl)dioxane density gradient. Eur J Biochem 76:13–19
Costantini M, Auletta F, Bernardi G (2007) Isochore patterns and gene distributions in fish genomes. Genomics 90:364–371
Costantini M, Cammarano R, Bernardi G (2009) The evolution of isochore patterns in vertebrate genomes. BMC Genomics 10:146
Cuny G, Soriano P, Macaya G, Bernardi G (1981) The major components of the mouse and human genomes. 1. Preparation, basic properties and compositional heterogeneity. Eur J Biochem 115:227–233
Davis BK (1998) The forces driving molecular evolution. Prog Biophys Mol Biol 69:83–150
Dean AM (2010) The future of molecular evolution. EMBO Rep 11:409
Dietrich MR (1994) The origins of the neutral theory of molecular evolution. J Hist Biol 27:21–59
Dokoumetzidis A, Iliadis A, Macheras P (2001) Nonlinear dynamics and chaos theory: concepts and applications relevant to pharmacodynamics. Pharm Res 18:415–426
Donatsch P, Gurtler J, Matter BE (1982) Critical appraisal of the 'mouse testicular DNA-synthesis inhibition test' for the detection of mutagens and carcinogens. Mutat Res 92:265–273
Duret L, Arndt PF (2008) The impact of recombination on nucleotide substitutions in the human genome. PLoS Genet 4:e1000071
Duret L, Mouchiroud D, Gautier C (1995) Statistical analysis of vertebrate sequences reveals that long genes are scarce in GC-rich isochores. J Mol Evol 40:308–317
Duret L, Semon M, Piganeau G, Mouchiroud D, Galtier N (2002) Vanishing GC-rich isochores in mammalian genomes. Genetics 162:1837–1847
Elena SF, Lenski RE (2003) Evolution experiments with microorganisms: the dynamics and genetic bases of adaptation. Nat Rev Genet 4:457–469
Engelhardt BE, Jordan MI, Srouji JR, Brenner SE (2011) Genome-scale phylogenetic function annotation of large and diverse protein families. Genome Res 21:1969–1980
Eyre-Walker A (1993) Recombination and mammalian genome evolution. Proc Biol Sci 252:237–243
Eyre-Walker A, Hurst LD (2001) The evolution of isochores. Nat Rev Genet 2:549–555
Felsenstein J (1981) Evolutionary trees from DNA sequences: a maximum likelihood approach. J Mol Evol 17:368–376
Fryxell KJ, Moon WJ (2005) CpG mutation rates in the human genome are highly dependent on local GC content. Mol Biol Evol 22:650–658
Fryxell KJ, Zuckerkandl E (2000) Cytosine deamination plays a primary role in the evolution of mammalian isochores. Mol Biol Evol 17:1371–1383
Fudenberg D, Levine DK (1988) Open-loop and closed-loop equilibria in dynamic games with many players. J Econ Theory 44:1–18
Fujita MK, Edwards SV, Ponting CP (2011) The Anolis lizard genome: an amniote genome without isochores. Genome Biol Evol 3:974–984
Galtier N, Piganeau G, Mouchiroud D, Duret L (2001) GC-content evolution in mammalian genomes: the biased gene conversion hypothesis. Genetics 159:907–911
Garcia P, Velasco M (2013) Exploratory strategies: experiments and simulations. In: Duran JM, Eckhart A (eds) Computer simulations and the changing face of scientific experimentation. Cambridge Scholars Publishing, Cambridge
Gascuel O, Bryant D, Denis F (2001) Strengths and limitations of the minimum evolution principle. Syst Biol 50:621–627
Gilbert W, Maxam A (1973) The nucleotide sequence of the lac operator. Proc Natl Acad Sci U S A 70:3581–3584

Gingerich PD (1986) Temporal scaling of molecular evolution in primates and other mammals. Mol Biol Evol 3:205–221

Giribet G (2003) Molecules, development and fossils in the study of metazoan evolution; Articulata versus Ecdysozoa revisited. Zoology (Jena) 106:303–326

Green RE, Krause J, Ptak SE, Briggs AW, Ronan MT, Simons JF, Du L, Egholm M, Rothberg JM, Paunovic M et al (2006) Analysis of one million base pairs of Neanderthal DNA. Nature 444:330–336

Greene CS, Tan J, Ung M, Moore JH, Cheng C (2014) Big data bioinformatics. J Cell Physiol 229:1896–1900

Gu X, Li WH (1998) Estimation of evolutionary distances under stationary and nonstationary models of nucleotide substitution. Proc Natl Acad Sci U S A 95:5899–5905

Gu J, Li WH (2006) Are GC-rich isochores vanishing in mammals? Gene 385:50–56

Guagliardi A, Napoli A, Rossi M, Ciaramella M (1997) Annealing of complementary DNA strands above the melting point of the duplex promoted by an archaeal protein. J Mol Biol 267:841–848

Haiminen N, Mannila H (2007) Discovering isochores by least-squares optimal segmentation. Gene 394:53–60

Hamada K, Horiike T, Ota H, Mizuno K, Shinozawa T (2003) Presence of isochore structures in reptile genomes suggested by the relationship between GC contents of intron regions and those of coding regions. Genes Genet Syst 78:195–198

Hedges SB, Dudley J, Kumar S (2006) TimeTree: a public knowledge-base of divergence times among organisms. Bioinformatics 22:2971–2972

Hipsley CA, Muller J (2014) Beyond fossil calibrations: realities of molecular clock practices in evolutionary biology. Front Genet 5:138

Holmquist GP (1989) Evolution of chromosome bands: molecular ecology of noncoding DNA. J Mol Evol 28:469–486

Huang W, Nevins JR, Ohler U (2007) Phylogenetic simulation of promoter evolution: estimation and modeling of binding site turnover events and assessment of their impact on alignment tools. Genome Biol 8:R225

Kaehler BD, Yap VB, Zhang R, Huttley GA (2014) Genetic distance for a general non-stationary Markov substitution process. Syst Biol 64:281–293

Karolchik D, Baertsch R, Diekhans M, Furey TS, Hinrichs A, Lu YT, Roskin KM, Schwartz M, Sugnet CW, Thomas DJ et al (2003) The UCSC genome browser database. Nucleic Acids Res 31:51–54

Karro JE, Peifer M, Hardison RC, Kollmann M, von Grünberg HH (2008) Exponential decay of GC content detected by strand-symmetric substitution rates influences the evolution of isochore structure. Mol Biol Evol 25:362–374

Kell DB, Lurie-Luke E (2015) The virtue of innovation: innovation through the lenses of biological evolution. J R Soc Interface 12

Kimura M (1968) Evolutionary rate at the molecular level. Nature 217:624–626

Kimura M (1983) The neutral theory of molecular evolution. Cambridge University Press, Cambridge

King MT, Wild D (1983) The mutagenic potential of hyperthermia and fever in mice. Mutat Res 111:219–226

Kirchner JW, Weil A (2000) Correlations in fossil extinction and origination rates through geological time. Proc Biol Sci 267:1301–1309

Komiya K, Kondoh T, Aotsuka T (1995) Evolution of the noncoding regions in Drosophila mitochondrial DNA: two intergenic regions. Biochem Genet 33:73–82

Koonin EV (2007) The cosmological model of eternal inflation and the transition from chance to biological evolution in the history of life. Biol Direct 2:15

Koonin EV (2009) Darwinian evolution in the light of genomics. Nucleic Acids Res 37:1011–1034

Kopp M, Matuszewski S (2014) Rapid evolution of quantitative traits: theoretical perspectives. Evol Appl 7:169–191

Kosiol C, Goldman N (2011) Markovian and non-Markovian protein sequence evolution: aggregated Markov process models. J Mol Biol 411:910–923
Kvikstad EM, Duret L (2014) Strong heterogeneity in mutation rate causes misleading hallmarks of natural selection on indel mutations in the human genome. Mol Biol Evol 31:23–36
Lesecque Y, Mouchiroud D, Duret L (2013) GC-biased gene conversion in yeast is specifically associated with crossovers: molecular mechanisms and evolutionary significance. Mol Biol Evol 30:1409–1419
Li W (2002) Are isochore sequences homogeneous? Gene 300:129–139
Liberles DA (2001) Evolution enters the genomic era. Genome Biol 2:REPORTS4026
Lio P, Goldman N (1998) Models of molecular evolution and phylogeny. Genome Res 8:1233–1244
Macaya G, Cortadas J, Bernardi G (1978) An analysis of the bovine genome by density-gradient centrifugation. Preparation of the dG+dC-rich DNA components. Eur J Biochem 84:179–188
Majoros WH, Ohler U (2010) Modeling the evolution of regulatory elements by simultaneous detection and alignment with phylogenetic pair HMMs. PLoS Comput Biol 6:e1001037
Mandel M, Marmur J, Lawrence Grossman KM (1968) [109] Use of ultraviolet absorbance-temperature profile for determining the guanine plus cytosine content of DNA. In: Nucleic acids, Part B. Academic, New York, pp 195–206
Maxam AM, Gilbert W (1977) A new method for sequencing DNA. Proc Natl Acad Sci U S A 74:560–564
Mayr E (2004) What makes biology unique?: Considerations on the autonomy of a scientific discipline. Cambridge University Press, Cambridge
Mazzocchi F (2008) Complexity in biology. Exceeding the limits of reductionism and determinism using complexity theory. EMBO Rep 9:10–14
McDonald JH (2001) Patterns of temperature adaptation in proteins from the bacteria Deinococcus radiodurans and Thermus thermophilus. Mol Biol Evol 18:741–749
Melodelima C, Gautier C (2008) The GC-heterogeneity of teleost fishes. BMC Genomics 9:632
Mikkelsen TS, Wakefield MJ, Aken B, Amemiya CT, Chang JL, Duke S, Garber M, Gentles AJ, Goodstadt L, Heger A et al (2007) Genome of the marsupial Monodelphis domestica reveals innovation in non-coding sequences. Nature 447:167–177
Mir K, Schober S (2014) Selection pressure in alternative reading frames. PLoS One 9:e108768
Montoya-Burgos JI, Boursot P, Galtier N (2003) Recombination explains isochores in mammalian genomes. Trends Genet 19:128–130
Moses A, Chiang D, Kellis M, Lander E, Eisen M (2003) Position specific variation in the rate of evolution in transcription factor binding sites. BMC Evol Biol 3:19
Nachman MW, Crowell SL (2000) Estimate of the mutation rate per nucleotide in humans. Genetics 156:297–304
Nei M (2005) Selectionism and neutralism in molecular evolution. Mol Biol Evol 22:2318–2342
Nei M (2013) Mutation-driven evolution. Oxford University Press, Oxford
Nekrutenko A, Li WH (2000) Assessment of compositional heterogeneity within and between eukaryotic genomes. Genome Res 10:1986–1995
Norell MA, Novacek MJ (1992) The fossil record and evolution: comparing cladistic and paleontologic evidence for vertebrate history. Science 255:1690–1693
Oestreicher C (2007) A history of chaos theory. Dialogues Clin Neurosci 9:279–289
O'Higgins P (2000) The study of morphological variation in the hominid fossil record: biology, landmarks and geometry. J Anat 197(1):103–120
Ohno S (1972) So much "junk" DNA in our genome. Brookhaven Symp Biol 23:366–370
Ohno S, Yomo T (1991) The grammatical rule for all DNA: junk and coding sequences. Electrophoresis 12:103–108
Oota S, Kawamura K, Kawai Y, Saitou N (2010) A new framework for studying the isochore evolution: estimation of the equilibrium GC content based on the temporal mutation rate model. Genome Biol Evol 2:558–571

Oreskes N, Shrader-Frechette K, Belitz K (1994) Verification, validation, and confirmation of numerical models in the earth sciences. Science 263:641–646
Orzack SH (2012) The philosophy of modelling or does the philosophy of biology have any use? Philos Trans R Soc Lond Ser B Biol Sci 367:170–180
Papp B, Notebaart RA, Pal C (2011) Systems-biology approaches for predicting genomic evolution. Nat Rev Genet 12:591–602
Pavlicek A, Clay O, Jabbari K, Paces J, Bernardi G (2002) Isochore conservation between MHC regions on human chromosome 6 and mouse chromosome 17. FEBS Lett 511:175–177
Pessia E, Popa A, Mousset S, Rezvoy C, Duret L, Marais GA (2012) Evidence for widespread GC-biased gene conversion in eukaryotes. Genome Biol Evol 4:675–682
Phear G, Meuth M (1989) The genetic consequences of DNA precursor pool imbalance: sequence analysis of mutations induced by excess thymidine at the hamster aprt locus. Mutat Res 214:201–206
Popper KR, Eccles JC (1984) The self and its brain, Reprint edition. Routledge
Rabinovich MI, Abarbanel HD (1998) The role of chaos in neural systems. Neuroscience 87:5–14
Rodriguez-Trelles F, Tarrio R, Ayala FJ (2002) A methodological bias toward overestimation of molecular evolutionary time scales. Proc Natl Acad Sci U S A 99:8112–8115
Rzhetsky A, Nei M (1993) Theoretical foundation of the minimum-evolution method of phylogenetic inference. Mol Biol Evol 10:1073–1095
Sanger F, Coulson AR (1975) A rapid method for determining sequences in DNA by primed synthesis with DNA polymerase. J Mol Biol 94:441–448
Sanger F, Donelson JE, Coulson AR, Kossel H, Fischer D (1973) Use of DNA polymerase I primed by a synthetic oligonucleotide to determine a nucleotide sequence in phage fl DNA. Proc Natl Acad Sci U S A 70:1209–1213
Sanger F, Nicklen S, Coulson AR (1977) DNA sequencing with chain-terminating inhibitors. Proc Natl Acad Sci U S A 74:5463–5467
Sawyer SA, Hartl DL (1992) Population genetics of polymorphism and divergence. Genetics 132:1161–1176
Schneider D, Duperchy E, Coursange E, Lenski RE, Blot M (2000) Long-term experimental evolution in Escherichia coli. IX Characterization of insertion sequence-mediated mutations and rearrangements. Genetics 156:477–488
Scientists GKCo (2009) Genome 10K: a proposal to obtain whole-genome sequence for 10 000 vertebrate species. J Hered 100:659–674
Siegel MJ (1976) The asymptotic behavior of a divergent linear birth and death process. Adv Appl Probab 8:315–338
Smith TF (1980) Occam's razor. Nature 285:620
Smith NG, Webster MT, Ellegren H (2002) Deterministic mutation rate variation in the human genome. Genome Res 12:1350–1356
Smyrlis YS, Papageorgiou DT (1991) Predicting chaos for infinite dimensional dynamical systems: the Kuramoto-Sivashinsky equation, a case study. Proc Natl Acad Sci U S A 88:11129–11132
Song S, Jarvie T, Hattori M (2013) Our second genome-human metagenome: how next-generation sequencer changes our life through microbiology. Adv Microb Physiol 62:119–144
Stamatoyannopoulos JA, Adzhubei I, Thurman RE, Kryukov GV, Mirkin SM, Sunyaev SR (2009) Human mutation rate associated with DNA replication timing. Nat Genet 41:393–395
Stephens R, Horton R, Humphray S, Rowen L, Trowsdale J, Beck S (1999) Gene organisation, sequence variation and isochore structure at the centromeric boundary of the human MHC. J Mol Biol 291:789–799
Strizhak PE, Pojman JA (1996) Infinite period and Hopf bifurcations for the pH-regulated oscillations in a semibatch reactor (H(2)O(2)-Cu(2+)-S(2)O(2-) (3)-NaOH system). Chaos 6:461–465
Subramanian S, Lambert DM (2011) Time dependency of molecular evolutionary rates? Yes and no. Genome Biol Evol 3:1324–1328

Sueoka N (1961) Correlation between base composition of deoxyribonucleic acid and amino acid composition of protein. Proc Natl Acad Sci U S A 47:1141–1149
Sueoka N (1962) On the genetic basis of variation and heterogeneity of DNA base composition. Proc Natl Acad Sci U S A 48:582–592
Sueoka N (1988) Directional mutation pressure and neutral molecular evolution. Proc Natl Acad Sci 85:2653–2657
Sueoka N, Marmur J, Doty P (1959) Heterogeneity in deoxyribonucleic acids: II. Dependence of the density of deoxyribonucleic acids on guanine-cytosine content. Nature 183:1429–1431
Takahata N (2007) Molecular clock: an anti-neo-Darwinian legacy. Genetics 176:1–6
Tikchonenko TI, Dubichev AG, Lyubchenko Yu L, Kvitko NP, Chaplygina NM, Kalinina TI, Dreizin RS, Naroditsky BS (1981) The distribution of guanine-cytosine pairs in adenovirus DNAs. J Gen Virol 54:425–429
Van Regenmortel MH (2004) Reductionism and complexity in molecular biology. Scientists now have the tools to unravel biological and overcome the limitations of reductionism. EMBO Rep 5:1016–1020
Verbyla KL, Yap VB, Pahwa A, Shao Y, Huttley GA (2013) The embedding problem for markov models of nucleotide substitution. PLoS One 8:e69187
Vinogradov AE (2003a) DNA helix: the importance of being GC-rich. Nucleic Acids Res 31:1838–1844
Vinogradov AE (2003b) Isochores and tissue-specificity. Nucleic Acids Res 31:5212–5220
Walser JC, Ponger L, Furano AV (2008) CpG dinucleotides and the mutation rate of non-CpG DNA. Genome Res 18:1403–1414
Warren WC, Hillier LW, Marshall Graves JA, Birney E, Ponting CP, Grutzner F, Belov K, Miller W, Clarke L, Chinwalla AT et al (2008) Genome analysis of the platypus reveals unique signatures of evolution. Nature 453:175–183
Webster MT, Smith NG, Ellegren H (2003) Compositional evolution of noncoding DNA in the human and chimpanzee genomes. Mol Biol Evol 20:278–286
Wolfe KH, Sharp PM, Li WH (1989) Mutation rates differ among regions of the mammalian genome. Nature 337:283–285
Wood RE, Barroso-Ruiz C, Caparros M, Jorda Pardo JF, Galvan Santos B, Higham TF (2013) Radiocarbon dating casts doubt on the late chronology of the Middle to Upper Palaeolithic transition in southern Iberia. Proc Natl Acad Sci U S A 110:2781–2786
Wu H, Zhang Z, Hu S, Yu J (2012) On the molecular mechanism of GC content variation among eubacterial genomes. Biol Direct 7:2
Yakovchuk P, Protozanova E, Frank-Kamenetskii MD (2006) Base-stacking and base-pairing contributions into thermal stability of the DNA double helix. Nucleic Acids Res 34:564–574
Zhaxybayeva O, Nesbo CL, Doolittle WF (2007) Systematic overestimation of gene gain through false diagnosis of gene absence. Genome Biol 8:402
Zuckerkandl E (1987) On the molecular evolutionary clock. J Mol Evol 26:34–46
Zuckerkandl E (2012) Fifty-year old and still ticking.... an interview with Emile Zuckerkandl on the 50th anniversary of the molecular clock. Interview by Giacomo Bernardi. J Mol Evol 74:233–236

Chapter 4
Protein-Coding and Noncoding RNA Genes

Tadashi Imanishi

Abstract During the last decade, our understanding about human genes has gradually but drastically changed. When the human genome-sequencing project announced its completion in 2004, they reported the number of human protein-coding genes to be only 20,000–25,000, which was much less than the number that many researchers expected at that time. Later, however, studies on human transcriptome have revealed existence of more transcribed regions of the genome including those without apparent open reading frames, suggesting the existence of many noncoding RNA genes. In the last few years, comprehensive surveys of human proteome on various tissues have successfully validated about 16,000 protein-coding genes at the protein level, but still many candidates of protein-coding genes remained unconfirmed. On the other hand, noncoding RNA genes have attracted more attention, and some of them have been extensively studied for their biological function. It turned out that short noncoding RNAs (ncRNAs) such as miRNAs, snRNAs, and snoRNAs work on posttranscriptional gene silencing and other essential cellular mechanisms, while longer ncRNAs (called lncRNAs) are involve in various biological functions such as chromatin modification, transcription, and splicing. In this chapter, I will describe our current view of these human protein-coding and noncoding RNA genes. In addition, I will illustrate how alternative splicing and other mechanisms diversify human proteome.

Keywords Transcriptome · Proteome · ncRNA · CAGE · Pseudogene · Alternative splicing

4.1 Human Protein-Coding Genes

In this section, I will first outline the basic concepts of human protein-coding genes and explain the structure of a typical human protein-coding gene. Then, I will explain the exceptional genes related to the immune system and pseudogenes.

T. Imanishi (✉)
Tokai University School of Medicine, Isehara, Kanagawa, Japan
e-mail: imanishi@tokai.ac.jp

N. Saitou (ed.), *Evolution of the Human Genome I*, Evolutionary Studies,
https://doi.org/10.1007/978-4-431-56603-8_4

4.1.1 Concepts of Human Protein-Coding Genes

According to a conventional textbook-like concept of human protein-coding genes, they are regions of the human genome, beginning at the transcription start sites (TSS) and ending at the transcription termination sites, sometimes including the promoter regions upstream of TSSs. Introns that are spliced out after transcription are also thought to be parts of genes. The gene regions are transcribed into RNAs, followed by modifications by splicing, 5′-capping, and polyadenylation to produce matured messenger RNAs (mRNAs). The mRNAs have open reading frames (ORFs) that carry information necessary for translation into proteins. Upstream and downstream of ORFs are 5′ and 3′ untranslated regions (5′ UTRs and 3′ UTRs), respectively.

The above classic concept of human protein-coding genes has been collapsed gradually by recent, large amount of comprehensive experimental data. For example, it has been thought that mRNA molecules are repeatedly transcribed from a particular genomic position, producing many copies of mRNAs of identical nucleotide sequences. However, transcriptome studies have revealed that the positions of TSSs are highly variable. Cap analysis of gene expression (CAGE) is a technique for precise determination of the genomic locations of TSSs at the genome-wide scale, by preparing a library of about 20 bases from the 5′ ends of full-length cDNAs, sequencing them, and then mapping the sequences onto the genome sequence. Extensive human transcriptome studies using the CAGE technique have revealed that TSS is highly variable. The FANTOM and the genome-network project used the CAGE technique and found that there are broad and narrow types of TSS in human genes, and each gene has its characteristic pattern of TSS distribution (Carninci et al. 2006). In other words, TSS is not a predetermined single nucleotide of the genome but a region of certain length. There are genes with a sharp, single dominant peak of TSS, but there are genes with multiple, broad-shaped peaks as well.

In addition, there were many important discoveries during the last decade, such as existence of many noncoding RNA genes other than rRNAs and tRNAs and the existence of many genes with alternative splicing isoforms, that made our concepts of human genes drastically changed. Some of these will be described in more detail in the following sections.

4.1.2 A Typical Human Protein-Coding Gene

Human genes have great variation in their sizes and structure. While there are many single-exon genes, there are genes with more than a hundred exons. Some of the human genes encode very small proteins of less than 100 amino acids, but others encode huge proteins with 36,000 amino acids (titin). In order to grasp a clear image of diverse human genes, a general description of an average protein-coding gene

Table 4.1 Statistics of average human genes

	Multi-exon genes		Single-exon genes	
	Mean ± s.d.	Median	Mean ± s.d.	Median
Number of genes	23,448		2938	
Length of transcript (bp)	2564.5 ± 1822.1	2156	1746.2 ± 1173.0	1438
5′ UTR (bp)	247.0 ± 432.8	120	418.2 ± 731.9	138
ORF (bp)	1520.4 ± 1529.2	1149	707.9 ± 598.7	549
3′ UTR (bp)	797.0 ± 924.7	476	620.1 ± 823.3	251
Number of exons	10.38 ± 9.23	8	–	
Number of AS isoforms	2.75 ± 2.89	2	–	

The means ± standard deviations and medians are shown. A reliable set of protein-coding genes (category = I, II, or III) was extracted from H-InvDB 9.0, and representative transcripts of each gene were used to calculate means and medians

may be of help. I thus show here some features of an average human protein-coding gene, such as length of ORFs, number of exons, and number of alternative splicing isoforms (Table 4.1). Also, I will compare them with those of single-exon genes that are the minority of human genes.

An average structure of multi-exon genes is quite different from that of single-exon genes (Table 4.1). In the case of multi-exon genes, the average length of transcripts (after splicing) is 2.6 kb, while that of single-exon genes is 1.7 kb which is much shorter than multi-exon genes. Multi-exon genes have ORFs of 1520 bps (506 amino acids), while single-exon genes have ORFs of 708 bps (235 amino acids), about half of multi-exon genes. In the upstream and downstream of these ORFs, there are short 5′ UTRs and relatively long 3′ UTRs.

Additionally, the majority of human single-exon genes encode G-protein-coupled receptors (GPCRs), including hundreds of olfactory receptors, that have important cellular functions and are frequently utilized as drug targets.

4.1.3 Number of Protein-Coding Genes

How many protein-coding genes exist in the human genome has long been a question of much attention especially among genome scientists. This is partly because solving this question was an important milestone of the human genome research. The genome scientists were so enthusiastic that while human genome-sequencing project was going on, they were betting on the number of human genes (Pennisi 2007). However, it is not very clear whether or not we have reached the final answer, even though more than 10 years have passed since the completion of the human genome-sequencing project in 2004.

The exact numbers of protein-coding genes for major eukaryotic model organisms have been revealed in the 1990s. The numbers appeared to be 6000 in yeast (Goffeau et al. 1996), 19,000 in worms (*C. elegans* Sequencing Consortium 1998), and 13,600 in fruit flies (Adams et al. 2000). On the other hand, the number of human protein-coding genes remained somewhat controversial even when the draft sequence of the human genome was published. For example, there were 39,114 predicted genes from the human genome draft sequences determined in 2000, while there were only 11,015 known human genes in RefSeq mRNA dataset, showing great discrepancy (Hogenesch et al. 2001). Around the same time, by extrapolating from the number of genes encoded on chromosome 22 that has been sequenced by then and the number of expressed sequence tags (ESTs), the number of human protein-coding genes was estimated to be around 35,000 (Ewing and Green 2000). Finally, after the completion of the human genome-sequencing project, the number of human protein-coding genes was corrected to be only 20,000–25,000 (International Human Genome Sequencing Consortium 2004). Many scientists accepted it with a big surprise, because the number was thought to be between 30,000 and 100,000 before the completion of human genome project and because the number was not much different from those of fruit flies and other animals.

Recent human proteomics studies have validated many of the human protein-coding genes. A proteomics study on 30 different human tissues (including 7 fetal tissues) has validated 17,294 human proteins by mass spectrometry (Kim et al. 2014). Another proteomics study on 32 different human tissues and organs has succeeded in identifying 17,132 human proteins by either mass spectrometry, monoclonal antibodies, or other techniques, out of 20,344 putative protein-coding genes supported by RNAseq data (Uhlén et al. 2015). These studies have provided good pieces of evidence for many human protein-coding genes at the protein level, but more studies will be needed to identify "minor" proteins that are used in a specific tissue at a particular timing. By further proteomics studies of various human tissues at various developmental stages, we will be able to better discriminate protein-coding genes from noncoding RNA genes, which will lead to a more precise number of human protein-coding genes (Imanishi et al. 2013; Gaudet et al. 2015).

4.1.4 Immunoglobulin and T-Cell Receptor Genes

The most characteristic gene complex in the human genome that is different from the majority of human protein-coding genes might be immunoglobulin (Ig) and T-cell receptor (TCR) genes. These genes that encode important proteins in the immune response experience somatic rearrangements and mutations during the maturation processes of B-cells and T-cells, which produce the greatest variation in gene sequences, resulting in the recognition of a wide variety of antigens. The mechanism underlying the hypervariability is mostly due to combinatorics; they increase the possible number of combinations by selecting one gene from many variable (V) genes, one gene from many diversity (D) genes, and one gene from many joining (J) genes during somatic rearrangements.

Ig loci are comprised of a heavy-chain locus (*IGH* locus at 14q32.33) and two light-chain loci, lambda (*IGL* at 22q11.2) and kappa (*IGK* at 2p11.2). All these loci undergo somatic V(D)J rearrangements that produce great diversity of Ig proteins.

The TCR loci are comprised of alpha/beta TCR loci (*TRA* at 14q11.2 and *TRB* at 7q34) and gamma/delta TCR loci (*TRG* at 7p14 and *TRD* at 14q11.2). In the same way as Ig loci, TCR loci undergo somatic V(D)J rearrangements to produce hypervariability in the complementarity-determining regions (CDRs) of the TCR molecules.

In fact, the immunoglobulin heavy-chain (*IGH*) locus has 167 V regions, 27 D regions, 9 J regions, and 11 constant (C) regions, according to the human gene nomenclature database (HGNC). Thus, the possible combinations of V, D, J, and C regions will be $167 \times 27 \times 9 \times 11 = 446{,}391$. The T-cell receptor beta (*TRB*) locus has 68 V regions, 2 D regions, 14 J regions, and 2 C regions if we include pseudogenes. Thus, the possible combinations of V, D, J, and C regions will be $68 \times 2 \times 14 \times 2 = 3808$. Because these molecules form heterodimers, the possible combinations can be even more. Although there are only three Ig loci and only four TCR loci, respectively, the mechanism described above produces hypervariability of these molecules.

4.1.5 Pseudogenes

Pseudogenes are genes that have structural and sequence similarity with some functional genes but lost their original function by any mechanisms. There are two most conspicuous mechanisms of producing pseudogenes. One is the destruction of a copy of functional duplicated genes. Such type of pseudogenes is quite common, because duplicated genes provide redundant copies of functional genes and there is no harm on the survival of the organism even if one of the multiple copies is lost. The second type of pseudogenes is processed pseudogenes that arise by reverse transcription of mRNAs. This type of pseudogenes lacks intronic sequences but sometimes has poly-A sequences, so they are easily recognized in the genome sequences. Also, there are two ways how pseudogenes lost their function: the fixation of nonsense mutations on ORFs and silencing of genes by mutations in gene control regions.

In general, most of functional genes are conserved during evolution. This means that functional genes are subject to functional constraint and that if they lose their function by mutations and turn into pseudogenes, they should have harmful effects on the organism. Thus, it is safe to think that pseudogenization is disadvantageous in most cases. To avoid such genetic load, redundant copies of functional genes should be produced by duplication before pseudogenization takes place. This is why many pseudogenes have functional counterpart in the genome.

There are functional genes that originate from processed pseudogenes. If we carefully examine human and mouse processed pseudogenes that can be found by transcriptome and genome sequence comparisons, we can find many transcribed pseudogenes that have intact ORFs. They are indistinguishable from functional genes, but their gene structures lack introns. Up to 1% of the processed pseudogenes seemed to have reinvigorated and became functional genes (Sakai et al. 2007). In this way, functional resurrection took place in a small fraction of processed pseudogenes, which was utilized as a way of producing new functional genes in the genome.

According to the human GENCODE database (version 24), there are 14,505 pseudogenes in the human genome, including 10,728 processed pseudogenes and 3295 unprocessed pseudogenes (Pei et al. 2012).

4.2 Noncoding RNA Genes

One of the most significant findings from human transcriptome studies that have been extensively carried out concurrently with the human genome studies is the discovery of abundant transcripts that do not contain apparent open reading frames and thus lack potential to produce proteins. This finding was not expected from the analysis of human genome sequence, because such noncoding RNA (ncRNA) genes could be by no means predicted from the genome sequences alone. Nowadays, ncRNAs became a new established category of genes, and more and more functional ncRNA genes are being discovered in the human genome.

4.2.1 Classification of Noncoding RNA Genes

In the first years of human transcriptome studies, expressed sequence tags (ESTs) have been used to identify many genes that encode human proteins (Adams et al. 1992). Later, transcribed regions of the human genome have been comprehensively surveyed using genome-wide tiling arrays (Bertone et al. 2004; Cheng et al. 2005) and cDNAs (Imanishi et al. 2004; Genome Information Integration Project and H-Invitational 2, 2008), which could gradually reveal the whole picture of human transcribed genes. These studies also identified many transcripts that have no apparent ORFs. There was a possibility that these transcripts have very short ORFs that encode functional peptides, but such hypothesis was not supported experimentally at the protein level. And later, at least some of these transcripts appeared to function as RNA molecules.

Aside from mRNAs that have genetic information for protein synthesis, classical biological textbooks introduce ribosomal RNA (rRNA) that is a component of ribosomes and transfer RNA (tRNA) that transports amino acids during protein synthesis. rRNA is an essential component of ribosomes. In eukaryotes, large ribosomal subunit contains 28S, 5.8S, and 5S rRNAs, while small ribosomal subunit contains 18S rRNA. rRNA is the most abundant RNA molecule in the cell. On the other hand, tRNA is a small RNA that functions as a transporter of specific amino acids to newly synthesized polypeptides during protein syntheses. tRNAs bind with specific amino acids to become amino acyl-tRNAs by the support of amino acyl-tRNA synthetase. Then, correct amino acids are transferred to the

Table 4.2 Classification of noncoding RNAs

Name	Full name	N[a]	HGNC[b]
rRNA	Ribosomal RNA	–	39
tRNA	Transfer RNA	610	637
lncRNA	Long noncoding RNA	76	2772
snRNA	Small nuclear RNA	–	65
snoRNA	Small nucleolar RNA	377	498
miRNA	microRNA	1881	–

[a]Number of known noncoding RNA genes registered in specialized databases. Database used are GtRNAdb (based on NCBI build 37.1) for tRNA, lncRNAdb (version 2.0) for lncRNA, snoRNA-LBME-db (version 3) for snoRNA, and miRBase (release 21) for miRNA (Chan and Lowe 2016; Lestrade and Weber 2006; Kozomara and Griffiths-Jones 2014)
[b]Number of officially approved human noncoding RNA genes by HGNC database (Gray et al. 2015)

polypeptides by referring to the codons of mRNAs in the ribosomes with the anticodons of tRNAs.

Later, many other new classes of functional ncRNAs have been discovered (Hirose et al. 2014). They can be classified into long noncoding RNAs (lncRNAs) that are typically longer than a few hundred bases and short noncoding RNAs that function in gene expression regulation. Until now, noncoding RNA genes have been roughly classified into six classes by their length, structure, and function (Table 4.2). However, noncoding RNA is one of the most enthusiastically studied subjects as of now, and it is highly probable that new members of ncRNAs as well as new classes of ncRNAs will be discovered in the future. Data in Table 4.2 should thus be regarded as a tentative snapshot of the functional ncRNAs.

4.2.2 Long Noncoding RNAs

Among RNAs other than mRNAs, rRNAs, and tRNAs, those that are longer than a few hundred bases and have poly-A sequences like mRNAs are called long noncoding RNAs (lncRNAs). This class of ncRNAs mostly binds to proteins to function in various cellular processes such as chromatin modification, transcription, and splicing. Examples of human lncRNAs include X inactive-specific transcript (XIST), H19, imprinted maternally expressed transcript (H19), and nuclear paraspeckle assembly transcript 1 (NEAT1). Among them, XIST is the most extensively studied lncRNA.

XIST is a master controller of the X chromosome inactivation in females. In female somatic cells, one of the two copies of the X chromosome is inactivated. Because the inactive X chromosome is randomly chosen, expression of X chromosomal genes will be a mosaic in the female cells. The X chromosome from which the XIST gene is expressed is inactivated. XIST is a long noncoding RNA of 17 kb that is spliced and polyadenylated. The function of XIST is to bind with inactivated X chromosomes and trigger X chromosome inactivation.

As explained above, XIST RNAs bind with genomic DNA and some proteins in the nucleus. There seem to be many other lncRNAs that function in a similar way; they bind with certain RNA-binding proteins and function as ribonucleoproteins in the nucleus. However, most of their functions remain unresolved, which are to be further studied in the future.

According to lncRNAdb version 2.0, a database of lncRNAs, there are at least 76 human lncRNA entries of possible functional significance (Quek et al. 2015). Also, there are hundreds of noncoding transcripts in a human transcriptome database H-InvDB, but only a small fraction of them have known function (Imanishi et al. 2004; Takeda et al. 2013). Such lncRNAs may include transcripts by un-induced or leaky transcription from the human genome.

4.2.3 *miRNA and Other Small ncRNAs*

Other classes of short RNAs that are apparently different from the above-mentioned RNAs have been found. These RNAs, including miRNAs, revealed to have extensive variation. This class of small ncRNAs can be classified into miRNAs, snRNAs, snoRNAs, and many other minor RNAs. Here, I will outline these small ncRNAs.

Small nuclear RNA (snRNA) is a group of small RNAs that are found in the nucleus. They are involved in various important functions such as splicing of mRNAs and maintenance of telomeres. Each of snRNAs binds to specific proteins to form small nuclear ribonuleoproteins (snRNPs). The most well-known snRNAs are five components (U1, U2, U4, U5, and U6) of spliceosomes. They bind to specific proteins to form snRNPs that function in splicing reactions. There are 65 snRNA genes registered in the HGNC database that provides official nomenclature of human genes (Table 4.2).

Small nucleolar RNA (snoRNA) is a class of small RNAs that function in chemical modifications of other RNA molecules such as rRNAs and tRNAs. There are two major classes of snoRNAs: C/D box snoRNAs function in methylation, and H/ACA box snoRNAs function in pseudouridylation. snoRNAs are associated with some proteins to form ribonucleoproteins (called snoRNPs). snoRNAs bind to target RNAs that have complementary sequences to parts of snoRNAs, and associated proteins catalyze chemical modifications. According to snoRNA-LBME-db, a database of snoRNAs, there are 269 C/D box snoRNAs and 108 H/ACA box snoRNAs. On the other hand, there are 498 entries in the HGNC database.

MicroRNA (miRNA) is yet another class of small noncoding RNAs and a key molecule of posttranscriptional regulation of gene expression. miRNAs have complementary sequences to specific protein-coding genes, and they bind with 3′ UTRs of mRNA molecules to form double-strand RNAs, which trigger degradation of target mRNAs. After transcription from the genome, miRNAs form a characteristic stem-loop structure using repetitive sequences and are then processed by Dicer proteins and RNA-induced silencing complex (RICS) to become mature miRNAs of 21–22 nucleotides. According to miRBase, a database of miRNA genes, there are 1881 miRNA genes in the human genome (Table 4.2). It has been estimated that each miRNA has about 400 target mRNAs on average (Friedman et al. 2009).

Because small noncoding RNA is an actively studied area of biology, it is highly probable that new classes of functional ncRNAs will be discovered and their classifications will be modified in the future.

4.3 Alternative Splicing

By the time when the human genome-sequencing project officially completed, it has been revealed that the total number of human genes is as low as 20,000–25,000, which is much lower than previous predictions (International Human Genome Sequencing Consortium 2004). On the other hand, some researchers predicted that the number of different human proteins is somewhere around 100,000. One of the possible mechanisms that can partly fill the gap between the two numbers is alternative splicing (AS). Now it is known that more than 90% of human genes undergo AS. Nearly 100,000 different kinds of proteins are produced by AS from some 20,000 genes, thus expanding the human proteome diversity (Nilsen and Graveley 2010). Here, one gene-one protein relationship no more holds. Furthermore, we postulate that many unidentified AS variants exist that are specific to some tissues or to some developmental stages. We thus need to investigate comprehensively the transcriptome for each of some hundred kinds of human cells, in order to reveal the whole picture of AS.

4.3.1 Mechanisms of AS

Splicing is a rigorously regulated reaction that takes place in the nucleus to cut out introns from pre-mRNA molecules that are originally transcribed from the genome. Usually, there are perceivable signal DNA sequences for splicing both in introns and exons in order to precisely determine the positions of splicing junctions. So, in principle, each gene can produce one kind of mRNA and hence one kind of protein. However, for example, when nucleotide substitutions have weakened the signals for the regulation of splicings, the mechanism of splicing does not perfectly work in a uniform manner, and splicing may occur at different positions from the original position. This is thought to be one of the possible mechanisms of AS.

Splicing reaction usually binds the 5′-end of exons with the 3′-end of preceding exons. In contrast, in the case of alternative splicing of cassette exon type, an exon will be bound not to the preceding exon but to the second or more remote exon, by

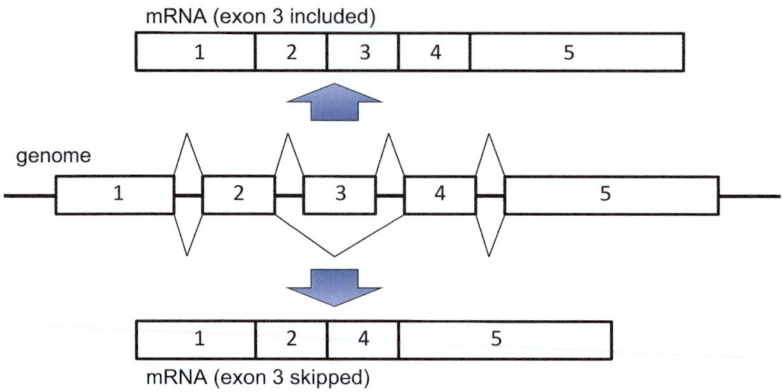

Fig. 4.1 A model of alternative splicing. Exon 3 of this gene is an alternative exon, and depending on whether exon 3 is included (upper panel) or excluded (lower panel), two kinds of mRNAs will be generated. Here, this type of alternative splicing is called cassette type

skipping the previous exon. This may produce two kinds of mRNAs: long mRNAs with the exon in question as well as short mRNAs without the exon (Fig. 4.1).

Above is the basic mechanism of alternative splicing. In this way, different kinds of mRNAs are produced by AS, which sometimes result in a diversification of protein function. A typical example is that AS, especially the exon skipping type, switches the protein sequences at the C-terminus and sometimes deletes transmembrane domains. This causes a change of protein function from membrane proteins into secreted type. There seem to be many genes that encode membrane proteins and are regulated in the same way as this example.

4.3.2 Patterns of AS

There are various forms of alternative splicing that can be classified into five major types based on the positions of exons skipped (Fig. 4.2, Table 4.3; Takeda et al. 2006). Here, I call the exons that are alternatively spliced as "AS exons." The most prevailing pattern of AS is the cassette exon in which single or multiple neighboring exons are skipped in some of the isoforms. The second most popular type of AS has variation among isoforms at the start or end positions of introns, which causes sequence length variation among AS isoforms. They are called "alternative 5′ splice site" and "alternative 3′ splice site." NAGNAG sequences, which will be introduced later in this chapter, are a kind of alternative 3′ splice sites. Mutually exclusive exons are often thought to be a typical pattern of AS, but in fact they are found in only 791 human genes. Because this type of AS can switch sequences from one exon to another, this might be the most suitable structure for substituting

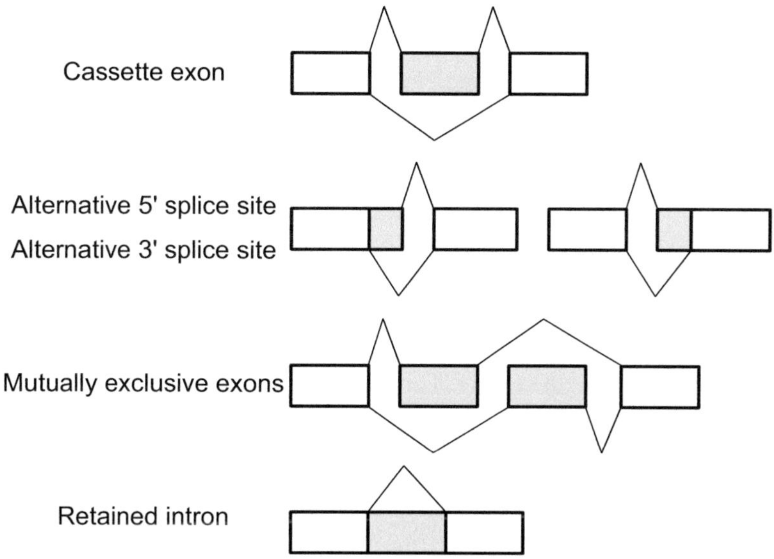

Fig. 4.2 Five most common types of alternative splicing. From upper to lower panels, cassette exon, alternative 5′ splice site, alternative 3′ splice site, mutually exclusive exons, and retained intron are shown. Alternatively spliced exons are shaded

Table 4.3 Patterns of human alternative splicings based on human full-length cDNA sequences

AS type	N
Cassette exon – including multiple cassette	23,176 (6745)
Alternative 3′ splice	14,073 (4040)
Alternative 5′ splice	13,738 (4067)
Mutually exclusive exon	2246 (791)
Retained intron	11,432 (3539)
5′-end alternative first exon	14,831 (4295)
3′-end alternative last exon	6329 (1967)

Numbers of AS exons and AS loci (in parenthesis) are shown
Modified from the H-DBAS database (Takeda et al. 2007, 2009)

functional domains of proteins. Retained intron is a type of AS in which both spliced and unspliced transcripts exist in the cell. One may think that intron-retained transcripts are caused by splicing errors, but in fact splicing reactions in these transcripts are rigorously controlled. For example, there are known cases that AS of retained intron type occurs in a tissue-specific manner.

There are other patterns of AS that do not fit to any of these five groups. They include combinations of above five AS patterns and some complicated unclassifiable patterns. Such irregular types of AS are identifiable based on the evidence of transcripts, but they are not greater in number, and functional significance of them is not yet clear.

4.3.3 Examples of Human AS

Many of human AS genes produce simple pairs of isoforms by the existence of one AS exon. On the other hand, there are human AS genes that have multiple AS exons and consequently show extensive repertoire of AS isoforms. Remarkable examples of such human AS genes include *MUC1* (mucin 1, cell surface associated), *DISC1* (disrupted in schizophrenia 1), *KCNMA1* (potassium calcium-activated channel subfamily M alpha 1), and *CALU* (calumenin) genes. Here, I will explain *KCNMA1* and *CALU* genes (Fig. 4.3).

KCNMA1 gene has many kinds of AS isoforms. Figure 4.3 shows the gene structure of *KCNMA1* gene based on the sequences of four RefSeq transcripts and six high-quality full-length cDNAs. This shows that there are isoforms with different transcription start sites (AK124355 and AK128392) as well as three isoforms with cassette exons (AK310379, NM_001161352, and NM_001161353). Because there are multiple AS exons in this locus, there is a variety of AS isoforms by combinations of these AS exons. *CALU* (calumenin) is a typical example of AS genes with mutually exclusive exons (Fig. 4.3). There are two RefSeq entries in this gene (NM_001130674 and NM_001219), and the third exons of these transcripts are selected in a mutually exclusive manner. Also, there is a cassette-type exon in 5′ UTR (AK056338). Because there are two AS exons in this gene, the number of possible AS isoforms will be four (=2^2), but only three of them have been observed so far.

The most extreme example of AS is seen in the *Dscam*gene in *Drosophila*. The gene structure of *Drosophila Dscam* is quite unique, having large numbers of alternative copies of exons 4, 6, and 9. There are 12, 48, and 33 copies of exons 4, 6, and 9, respectively, that are tandemly arranged on the genome. For each exon,

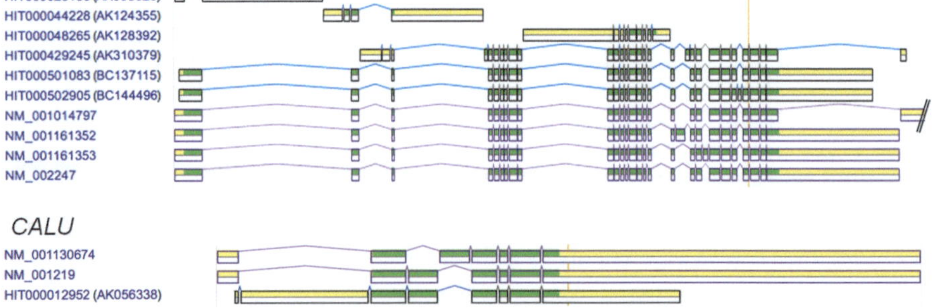

Fig. 4.3 Examples of human genes that undergo alternative splicing. Gene structures of *KCNMA1* and *CALU* genes are depicted. Exon lengths are shown in proportion to the number of nucleotides except that the last exon of NM_001014797 is shortened. Introns are shortened. Predicted ORFs are shown in green. *KCNMA1* gene has many AS isoforms. There are at least ten distinct types of transcripts that are produced by alternative splicing. *CALU* gene has at least three AS isoforms, including those with mutually exclusive exons

one of these copies will be chosen during splicing in a mutually exclusive manner. Each copy encodes different types of immunoglobulin domains. Also, there are two copies of exon 17 that encode transmembrane domains of Dscam protein. As a result, the number of possible AS isoforms will be 38,016 (Wojtowicz et al. 2004). On the other hand, Down syndrome cell adhesion molecule (*DSCAM*) gene is a human ortholog of *Dscam* in *Drosophila*. *DSCAM* belongs to the immunoglobulin superfamily and encodes cell adhesion molecule. It was identified in the Down syndrome susceptible region on chromosome 21. Human *DSCAM* gene transcribes at least two different mRNAs that encode different protein isoforms. Extremely large number of AS variants observed in *Drosophila* is not found in its human ortholog (Yamakawa et al. 1998). As is evident from this example, different species have different patterns of AS.

4.3.4 Evolutionary Conservation of AS

As has been discussed above, AS can expand the degree of human proteome variation that is translated from the human genome, which contributes to the regulation of complicated molecular mechanisms in humans. It is a very interesting problem to imagine how the precise regulatory mechanism of AS evolved.

Here, I will introduce an example of highly conserved AS in mammalian evolution. Human cysteinyl-tRNA synthetase (*CARS*) gene has two AS isoforms of cassette type (Takeda et al. 2008). One of the AS isoforms (BX647906) has exon 2 in its transcript, while the other isoform (BC002880) skips this exon. As a result, the former transcript is 249 nucleotides longer than the latter, and the protein product from the former transcript is 83 amino acids longer than that of the latter transcript. This AS exon is known to encode a functional domain, glutathione S-transferase C-terminal-like (IPR010987). What is more interesting is that the same pair of AS variants exists in mouse. This means that the common ancestor of humans and mouse about 100 million years ago might have possessed this pair of AS isoforms, and these AS isoforms have been conserved in both human and mouse lineages ever since.

As is evident from the above example, there are conserved AS isoforms in evolution, but how many of AS isoforms are conserved? To solve this problem, we conducted a comprehensive, cross-species analysis of AS between humans and mouse (Takeda et al. 2008). Because there is a large amount of transcriptome data for humans and mouse, we can comprehensively identify AS genes and AS exons and then make comparisons of AS between these two species. First, we mapped sequences of all available AS isoforms on to the genome sequence, and then we identified AS isoforms that show good correspondence between these two species, using the whole genome alignment of human and mouse. We call these isoforms as "evolutionarily conserved AS isoforms." Then, if there are two or more pairs of evolutionarily conserved AS isoforms on a particular locus between these species, we defined them as evolutionarily conserved AS. As a result of this analysis, we found only 189 genes that have multiple pairs of evolutionarily conserved AS isoforms between human and mouse. This means that only a fraction of human AS genes are completely conserved for a long time during mammalian evolution and the majority of human AS is species specific.

To sum up, the majority of the human AS arose recently in the evolutionary history. Creation of AS isoforms can be regarded as evolutionary experiments to try to generate new functional proteins from already existing genes. This is as if life is trying to make full use of all available resources during evolution. It is well established that gene duplications have created new functional genes during evolution. In a similar way, AS might have contributed to the modification and improvement of genes during evolution.

4.4 Other Mechanisms for Proteome Diversification

In addition to alternative splicing, there are other mechanisms that can extend the diversity of human genes and proteins. Although they are not observed in all human genes, they do occur in a minority of human genes. In this section I will introduce three kinds of such mechanisms: alternative open reading frames (AltORFs), NAGNAG sequences, and selenoproteins.

4.4.1 Alternative ORFs

Among geneticists, the one gene-one protein theory has been taken for granted for a very long time, but the alternative splicing has diminished the concept, and later even more exceptional phenomena appeared. If we carefully examine the mRNA sequences of human protein-coding genes, there are sometimes open reading frames that are different from the authentic, principal ones (Fig. 4.4a). We call them alternative open reading frames (AltORFs) in this chapter. AltORFs have been observed in many viral genomes, which may be due to a constraint to keep the genome size compact, but such ORFs also exist in human genes.

A systematic survey of proteins encoded by AltORFs in all human genes that are different from the primary ORF and have strong Kozak conserved sequences at the translation initiation sites was carried out by using mass spectrometry analysis of human cell lines and tissues. As a result, at least 1259 AltORF proteins were experimentally identified (Vanderperre et al. 2012, 2013). The average length of these AltORF proteins was 57 amino acids, and there were much longer AltORF proteins. In many cases, both AltORF and primary ORF are translated from one mRNA, leading to co-expression of two proteins.

There might exist many human AltORF proteins with functional importance. For example, AltORF protein from *MRV1* gene was demonstrated to bind with BRCA1 proteins in the nucleus (Vanderperre et al. 2013). Also, many AltORFs have high

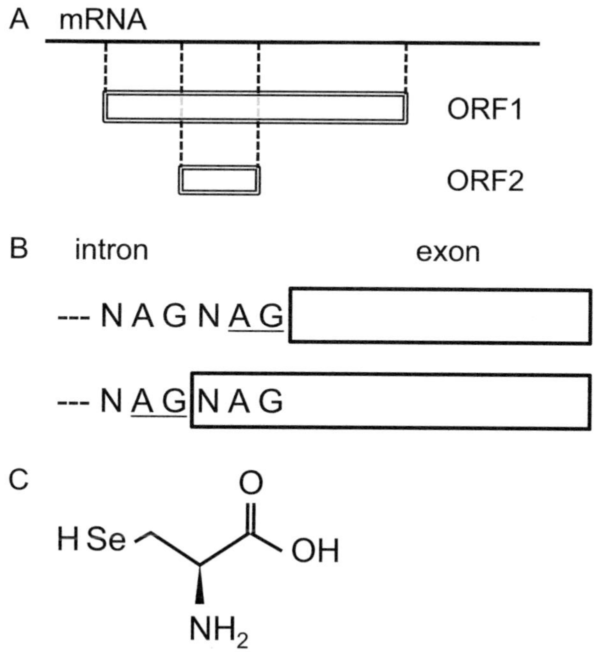

Fig. 4.4 Miscellaneous mechanisms of human gene and protein diversification. (**a**) Alternative open reading frames (AltORFs). AltORFs (ORF2) are generally overlapping with and much shorter than the authentic, principal ORFs (ORF1). (**b**) NAGNAG introns. If NAGNAG sequences are present at the 3′ ends of introns, two kinds of mRNAs can be produced by use of different splicing acceptor sites. (**c**) Molecular structure of selenocysteine (*Sec*) that is incorporated in selenoproteins. This is identical to cysteine (*Cys*), if selenium (*Se*) is replaced by sulfur (*S*)

evolutionary conservation, suggesting functional importance. However, whether or not human AltORF proteins have important function is still to be solved, and apparently further investigation is required. Experimental validation of functional significance of AltORF proteins will be hard, because proteins from AltORF could be translated in very small quantities and they could be synthesized only in specific tissues or in specific periods of developmental stages. In the future, high-resolution proteomics studies may lead to new discoveries about human AltORFs.

Furthermore, it is known that some of the human mRNAs have short upstream ORFs in their 5′ UTRs, which are called uORFs. They are typically terminated at the upstream of primary ORFs, or they are overlapping with primary ORFs. If uORFs are translated, it consumes ribosomes, and as a result it inhibits translation of primary ORFs downstream. By this mechanism, uORFs are thought to control human protein translations from mRNAs.

4.4.2 NAGNAG Introns

Nucleotide sequences at the 3′ ends of introns are called splicing acceptor sites and are known to be highly conserved. In particular, the last two nucleotides at the 3′ ends of introns attain a strong consensus sequence of "AG." However, if we examine the last six nucleotides of introns, there are some introns having "NAGNAG" sequences, where N represents one of the A, C, G, or T nucleotides

(Fig. 4.4b). In this case, splicing usually takes place immediately downstream of the fifth to sixth "AG" nucleotides, but it also occurs at the second to third "AG" nucleotides with some probability. As a result of this error, the corresponding mRNA will be only three nucleotides longer, and its protein products will be only one amino acid longer because of the extra three nucleotides. Because the reading frame of this elongated mRNA is the same as the original one, the other parts of the protein remain unaffected. So, this will not affect the protein function significantly, and both types of proteins can coexist.

As has been shown above, the presence of NAGNAG sequences at the splicing acceptor sites will cause human mRNA and protein diversification. In fact, such NAGNAG sequences are found in introns of many human genes (Hiller et al. 2004). For example, the second intron of human microfibrillar-associated protein 2 (*MFAP2*) gene has "CAGCAG" sequence at the acceptor site, and as a result of this, two types of *MFAP2* mRNAs are produced (BC015039 and AK222751 in GenBank records). Although the functional consequence of this variation has not been recognized, these mRNAs may be translated into two different proteins. According to H-InvDB, an integrated database of human genes, there are at least 5081 human protein-coding genes that have NAGNAG introns, which might have caused mRNA and protein diversification to a certain extent. They must have some functional influence on their mRNAs and protein products. Also, NAGNAG introns have been found in 31 long noncoding RNA genes (Sun et al. 2014).

How the 1-amino acid shorter or longer proteins produced by NAGNAG sequences diverge in their structure and function will be an issue to be addressed in the future. If an extra or deleted amino acid is located in a loop region of a protein, it may not change the protein structure drastically, hence not much disadvantageous. On the contrary, if the amino acid is located inside a functional domain of the protein, it will cause a serious impact on the protein function. In reality, there seems to be a weaker structural constraint of proteins on producing NAGNAG sequences, because more than half of mutations that produce NAGNAG sequences in human introns have been eliminated by negative selection (Hiller et al. 2008).

4.4.3 Selenoproteins

Selenium (Se) is the 34th element that belongs to the same group as oxygen and sulfur and is an essential trace element for humans. Selenocysteine (Sec) is an amino acid with similar molecular structure with cysteine (Cys), in which sulfur (S) is substituted by selenium (Fig. 4.4c). It is generally thought that human proteins are comprised of 20 kinds of amino acids, but in fact the Sec is the 21st amino acid that is actually incorporated in some of the human proteins. The proteins that have Sec are collectively called selenoproteins (Hatfield and Gladyshev 2002).

During their translation, selenoproteins are synthesized by incorporating Sec at the positions of UGA codons that are usually recognized as a termination signal of protein synthesis. tRNAs for Sec recognize not only the UGA codons of selenoprotein mRNAs but also the presence of a specific stem-loop structure, called Sec insertion sequence, in the 3′ UTR of mRNA. Thus, Sec is not incorporated to other UGA codons that lack the signal structure. Also, UGA codons for Sec will be recognized as premature termination codons by the regular translational machinery, and such mRNAs will be degraded by nonsense-mediated decay (NMD). However, selenoprotein mRNAs can escape from the NMD (Reeves and Hoffmann 2009).

Incorporation of Sec seems to occur only in specific proteins. It has been revealed that there are 25 selenoproteins in humans (Kryukov et al. 2003). The selenium atoms of incorporated Sec molecules are located in the reactive centers of the protein and take important roles in many of the selenoproteins. For example, selenoprotein P that is encoded by *SEPP1* gene has ten Sec residues in a protein and may function as extracellular antioxidant (Mostert 2000).

Acquisition of selenocysteine is an exceptional phenomenon that requires a special mechanism of translation. In the same way as the irregular genetic code in mitochondrial DNA, we can consider the selenoprotein as a kind of genetic code variation in human protein translation. In other words, selenoprotein can be regarded as a mechanism for diversification of human genes and proteins.

4.5 Human Gene Databases

In this section, I will introduce five major databases from which researchers can obtain information about human genes. HGNC is an official body of human gene nomenclature providing database of human gene symbols and gene names. RefSeq is a database of human reference sequences that are nonredundant and curated. GENCODE is a standard dataset of the ENCODE project for identifying all functional elements in the human genome. H-InvDB is an annotation database of human genes based on human transcriptome. lncRNAdb is a database of long noncoding RNAs. Finally, I will discuss about comparisons of these databases and future perspectives.

Because these databases are being developed independently in different research institutes, there are some discrepancies among them even now. The contradiction comes from different interpretations of experimental data and will not be cleared until more detailed and comprehensive validation studies are carried out for all human genes in the future. Nevertheless, these databases reflect our current understandings about human genes.

4.5.1 HUGO Gene Nomenclature Committee (HGNC)

HGNC is responsible for approving unique symbols and names for human genes and provides a database of human gene names (http://genenames.org/). As of August 2015, HGNC database provides information about 18,997 protein-coding genes, 2734 long noncoding RNAs, 1879 miRNAs, 65 small nuclear RNAs, and 458 small nucleolar RNAs. Furthermore, HGNC provides symbols and names for 12,444 human pseudogenes.

4.5.2 RefSeq

RefSeq (http://www.ncbi.nlm.nih.gov/refseq/) is a collection of annotated genomic, transcript, and protein sequence records developed by the National Center for Biotechnology Information (Pruitt KD et al. 2014). Human gene set of RefSeq is comprised of 26,266 genes and 47,619 transcripts. Seventy-nine percent of the transcripts are "curated," and the remaining are "models" (release 60; July 2013).

4.5.3 GENCODE

GENCODE (http://www.gencodegenes.org) is a database developed by the Wellcome Trust Sanger Institute that was used as a reference gene set in the ENCODE (Encyclopedia of DNA Elements) project (Harrow et al. 2012). GENCODE provides integrated annotation of human genes by manual curation, computational analysis, and experimental validation. The latest version M16 (Aug 2017 freeze) contains a total of 53,379 genes (21,963 protein-coding genes, 12,374 long noncoding RNA genes, 6,109 small noncoding RNA genes, 12,437 pseudogenes, and 494 immunoglobulin and T-cell receptor gene segments).

4.5.4 H-InvDB

H-InvDB (http://hinv.jp) is a human gene database based on analysis of human transcriptome (Imanishi et al. 2004). The latest release of H-InvDB (ver 9.0) contains 39,495 entries of protein-coding genes in total. Among them, 26,386 entries that belong to categories I (identical to known human proteins), II (similar to known proteins), or III (InterPro domain-containing proteins) comprise a reliable set of protein-coding genes. There are also 8591 entries of possible noncoding RNA genes.

4.5.5 lncRNAdb

lncRNAdb (http://www.lncrnadb.org) is a database of long ncRNAs of 200 bps or longer that are manually curated from literatures (Quek et al. 2015). Unlike shorter ncRNAs, long ncRNAs have diverse cellular function such as chromatin modification, transcription, and splicing. The number of entries in lncRNAdb gradually increased as research proceeds. As of August 2015, there are at least 76 entries of human long ncRNAs (lncRNAdb ver 2.0).

4.5.6 Comparisons of Databases

If we compare human gene datasets among various public databases, they will never agree to each other perfectly, because these databases have different policies of gene annotation. For example, according to a recent comparison of GENCODE and RefSeq gene annotation, there is a significant difference between the two (Frankish et al. 2015). Basic set of GENCODE is larger than the basic set of RefSeq (NM/NR only), and the comprehensive set of GENCODE is larger than the large set of RefSeq (XM/MR included). Figure 4.5 shows a comparison of three major

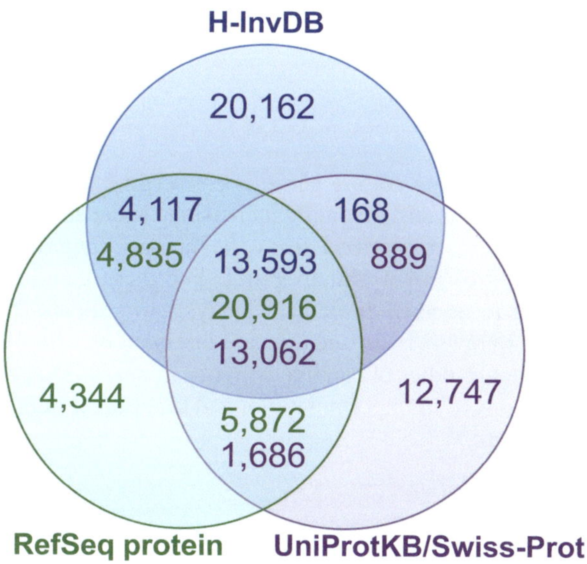

Fig. 4.5 Comparison of human gene, transcript, and protein datasets. Human protein sequences were obtained from each of RefSeq protein, H-InvDB (transcripts), and UniProtKB/Swiss-Prot (proteins; The UniProt Consortium 2015), and the protein sequences were compared to each other to examine correspondence among databases. The condition of sequence matches was set at identity >95% and coverage >80%. Values in the Venn diagram indicate numbers of protein entries in each cell (green for RefSeq proteins, blue for H-InvDB representative protein-coding transcripts, and purple for UniProtKB/Swiss-Prot proteins). Because of the AS and partial sequences, the proteins from different databases do not always show one-to-one correspondence. The following datasets were used: 35,967 protein-coding genes in RefSeq protein (release 59; dated on May 20, 2013), 28,384 human reviewed proteins in UniProtKB/Swiss-Prot (release 2013_05; dated on May 23, 2013), and 38,040 protein-coding transcripts in H-InvDB (release 9.0; dated on May 27, 2015)

databases about human genes, human transcripts, and human proteins. This also illustrates how the consensus among different sets of experimental evidence is hard to obtain.

Furthermore, each database changes its contents in periodical updates. In particular, every time when human reference genome sequence is updated, the structure and annotation of many genes and transcripts require significant modifications. Also, because of the genomic variations among individuals, we cannot expect that the DNA sequences in databases will match perfectly with the sequences of human samples that are actually used in experiments.

These problems are very hard to clear up in the future. Researchers may wish to use the most accurate database that shows the best correspondence with the actual samples used in experiments. However, none of the databases can provide perfect information after all, and we need to understand that what the databases provide is no more than an approximation of the real world. In order for us to approach the complete human gene databases, we need to produce and accumulate a larger amount of more precise data about human genes, by carrying out more comprehensive experiments than before.

4.6 Conclusion

The number of human protein-coding genes is converging to about 20,000, while that of RNA-coding genes is still gradually increasing. There are thousands of other genomic regions that are transcribed from the human genome, which can be regarded as candidates of novel human genes. For these candidate genes, it is essentially important to conduct proteomic analysis to validate the existence of protein products and to reveal their function and interactions with other molecules. Through such functional studies of candidate human genes, it is expected that many true genes that are involved in important biological function or human diseases will be discovered in the future.

References

Adams MD, Dubnick M, Kerlavage AR, Moreno R, Kelley JM, Utterback TR, Nagle JW, Fields C, Venter JC (1992) Sequence identification of 2,375 human brain genes. Nature 355 (6361):632–634

Adams MD, Celniker SE, Holt RA, Evans CA, Gocayne JD, Amanatides PG, Scherer SE, Li PW, Hoskins RA, Galle RF, George RA, Lewis SE, Richards S, Ashburner M, Henderson SN, Sutton GG, Wortman JR, Yandell MD, Zhang Q, Chen LX, Brandon RC, Rogers YH, Blazej RG, Champe M, Pfeiffer BD, Wan KH, Doyle C, Baxter EG, Helt G, Nelson CR, Gabor GL, Abril JF, Agbayani A, An HJ, Andrews-Pfannkoch C, Baldwin D, Ballew RM, Basu A, Baxendale J, Bayraktaroglu L, Beasley EM, Beeson KY, Benos PV, Berman BP, Bhandari D, Bolshakov S, Borkova D, Botchan MR, Bouck J, Brokstein P, Brottier P, Burtis

KC, Busam DA, Butler H, Cadieu E, Center A, Chandra I, Cherry JM, Cawley S, Dahlke C, Davenport LB, Davies P, de Pablos B, Delcher A, Deng Z, Mays AD, Dew I, Dietz SM, Dodson K, Doup LE, Downes M, Dugan-Rocha S, Dunkov BC, Dunn P, Durbin KJ, Evangelista CC, Ferraz C, Ferriera S, Fleischmann W, Fosler C, Gabrielian AE, Garg NS, Gelbart WM, Glasser K, Glodek A, Gong F, Gorrell JH, Gu Z, Guan P, Harris M, Harris NL, Harvey D, Heiman TJ, Hernandez JR, Houck J, Hostin D, Houston KA, Howland TJ, Wei MH, Ibegwam C, Jalali M, Kalush F, Karpen GH, Ke Z, Kennison JA, Ketchum KA, Kimmel BE, Kodira CD, Kraft C, Kravitz S, Kulp D, Lai Z, Lasko P, Lei Y, Levitsky AA, Li J, Li Z, Liang Y, Lin X, Liu X, Mattei B, McIntosh TC, McLeod MP, McPherson D, Merkulov G, Milshina NV, Mobarry C, Morris J, Moshrefi A, Mount SM, Moy M, Murphy B, Murphy L, Muzny DM, Nelson DL, Nelson DR, Nelson KA, Nixon K, Nusskern DR, Pacleb JM, Palazzolo M, Pittman GS, Pan S, Pollard J, Puri V, Reese MG, Reinert K, Remington K, Saunders RD, Scheeler F, Shen H, Shue BC, Sidén-Kiamos I, Simpson M, Skupski MP, Smith T, Spier E, Spradling AC, Stapleton M, Strong R, Sun E, Svirskas R, Tector C, Turner R, Venter E, Wang AH, Wang X, Wang ZY, Wassarman DA, Weinstock GM, Weissenbach J, Williams SM, Woodage T, Worley KC, Wu D, Yang S, Yao QA, Ye J, Yeh RF, Zaveri JS, Zhan M, Zhang G, Zhao Q, Zheng L, Zheng XH, Zhong FN, Zhong W, Zhou X, Zhu S, Zhu X, Smith HO, Gibbs RA, Myers EW, Rubin GM, Venter JC (2000) The genome sequence of *Drosophila melanogaster*. Science 287(5461):2185–2195

Bertone P, Stolc V, Royce TE, Rozowsky JS, Urban AE, Zhu X, Rinn JL, Tongprasit W, Samanta M, Weissman S, Gerstein M, Snyder M (2004) Global identification of human transcribed sequences with genome tiling arrays. Science 306(5705):2242–2246

C. elegans Sequencing Consortium (1998) Genome sequence of the nematode *C. elegans*: a platform for investigating biology. Science 282(5396):2012–2018

Carninci P, Sandelin A, Lenhard B, Katayama S, Shimokawa K, Ponjavic J, Semple CA, Taylor MS, Engström PG, Frith MC, Forrest AR, Alkema WB, Tan SL, Plessy C, Kodzius R, Ravasi T, Kasukawa T, Fukuda S, Kanamori-Katayama M, Kitazume Y, Kawaji H, Kai C, Nakamura M, Konno H, Nakano K, Mottagui-Tabar S, Arner P, Chesi A, Gustincich S, Persichetti F, Suzuki H, Grimmond SM, Wells CA, Orlando V, Wahlestedt C, Liu ET, Harbers M, Kawai J, Bajic VB, Hume DA, Hayashizaki Y (2006) Genome-wide analysis of mammalian promoter architecture and evolution. Nat Genet 38(6):626–635

Chan PP, Lowe TM (2016) GtRNAdb 2.0: an expanded database of transfer RNA genes identified in complete and draft genomes. Nucleic Acids Res 44(D1):D184–D189

Cheng J, Kapranov P, Drenkow J, Dike S, Brubaker S, Patel S, Long J, Stern D, Tammana H, Helt G, Sementchenko V, Piccolboni A, Bekiranov S, Bailey DK, Ganesh M, Ghosh S, Bell I, Gerhard DS, Gingeras TR (2005) Transcriptional maps of 10 human chromosomes at 5-nucleotide resolution. Science 308(5725):1149–1154

Ewing B, Green P (2000) Analysis of expressed sequence tags indicates 35,000 human genes. Nat Genet 25:232–234

Frankish A, Uszczynska B, Ritchie GR, Gonzalez JM, Pervouchine D, Petryszak R, Mudge JM, Fonseca N, Brazma A, Guigo R, Harrow J (2015) Comparison of GENCODE and RefSeq gene annotation and the impact of reference geneset on variant effect prediction. BMC Genomics 16 (Suppl 8):S2

Friedman RC, Farh KK, Burge CB, Bartel DP (2009) Most mammalian mRNAs are conserved targets of microRNAs. Genome Res 19(1):92–105

Gaudet P, Michel PA, Zahn-Zabal M, Cusin I, Duek PD, Evalet O, Gateau A, Gleizes A, Pereira M, Teixeira D, Zhang Y, Lane L, Bairoch A (2015) The neXtProt knowledgebase on human proteins: current status. Nucleic Acids Res 43(D1):D764–D770

Genome Information Integration Project and H-Invitational 2 (2008) The H-Invitational Database (H-InvDB), a comprehensive annotation resource for human genes and transcripts. Nucleic Acids Res 36(Database Issue):D793–D799

Goffeau A, Barrell BG, Bussey H, Davis RW, Dujon B, Feldmann H, Galibert F, Hoheisel JD, Jacq C, Johnston M, Louis EJ, Mewes HW, Murakami Y, Philippsen P, Tettelin H, Oliver SG (1996) Life with 6000 genes. Science 274(5287):563–567

Gray KA, Yates B, Seal RL, Wright MW, Bruford EA (2015) Genenames.org: the HGNC resources in 2015. Nucleic Acids Res 43(Database issue):D1079–D1085

Harrow J, Frankish A, Gonzalez JM, Tapanari E, Diekhans M, Kokocinski F, Aken BL, Barrell D, Zadissa A, Searle S, Barnes I, Bignell A, Boychenko V, Hunt T, Kay M, Mukherjee G, Rajan J, Despacio-Reyes G, Saunders G, Steward C, Harte R, Lin M, Howald C, Tanzer A, Derrien T, Chrast J, Walters N, Balasubramanian S, Pei B, Tress M, Rodriguez JM, Ezkurdia I, van Baren J, Brent M, Haussler D, Kellis M, Valencia A, Reymond A, Gerstein M, Guigó R, Hubbard TJ (2012) GENCODE: the reference human genome annotation for The ENCODE Project. Genome Res 22(9):1760–1774

Hatfield DL, Gladyshev VN (2002) How selenium has altered our understanding of the genetic code. Mol Cell Biol 22(11):3565–3576

Hiller M, Huse K, Szafranski K, Jahn N, Hampe J, Schreiber S, Backofen R, Platzer M (2004) Widespread occurrence of alternative splicing at NAGNAG acceptors contributes to proteome plasticity. Nat Genet 36(12):1255–1257

Hiller M, Szafranski K, Huse K, Backofen R, Platzer M (2008) Selection against tandem splice sites affecting structured protein regions. BMC Evol Biol 8:89

Hirose T, Mishima Y, Tomari Y (2014) Elements and machinery of non-coding RNAs: toward their taxonomy. EMBO Rep 15(5):489–507

Hogenesch JB, Ching KA, Batalov S, Su AI, Walker JR, Zhou Y, Kay SA, Schultz PG, Cooke MP (2001) A comparison of the Celera and Ensembl predicted gene sets reveals little overlap in novel genes. Cell 106(4):413–415

Imanishi T, Itoh T, Suzuki Y, O'Donovan C, Fukuchi S, Koyanagi KO, Barrero RA, Tamura T, Yamaguchi-Kabata Y, Tanino M, Yura K, Miyazaki S, Ikeo K, Homma K, Kasprzyk A, Nishikawa T, Hirakawa M, Thierry-Mieg J, Thierry-Mieg D, Ashurst J, Jia L, Nakao M, Thomas MA, Mulder N, Karavidopoulou Y, Jin L, Kim S, Yasuda T, Lenhard B, Eveno E, Suzuki Y, Yamasaki C, Takeda J, Gough C, Hilton P, Fujii Y, Sakai H, Tanaka S, Amid C, Bellgard M, Bonaldo Mde F, Bono H, Bromberg SK, Brookes AJ, Bruford E, Carninci P, Chelala C, Couillault C, de Souza SJ, Debily MA, Devignes MD, Dubchak I, Endo T, Estreicher A, Eyras E, Fukami-Kobayashi K, Gopinath GR, Graudens E, Hahn Y, Han M, Han ZG, Hanada K, Hanaoka H, Harada E, Hashimoto K, Hinz U, Hirai M, Hishiki T, Hopkinson I, Imbeaud S, Inoko H, Kanapin A, Kaneko Y, Kasukawa T, Kelso J, Kersey P, Kikuno R, Kimura K, Korn B, Kuryshev V, Makalowska I, Makino T, Mano S, Mariage-Samson R, Mashima J, Matsuda H, Mewes HW, Minoshima S, Nagai K, Nagasaki H, Nagata N, Nigam R, Ogasawara O, Ohara O, Ohtsubo M, Okada N, Okido T, Oota S, Ota M, Ota T, Otsuki T, Piatier-Tonneau D, Poustka A, Ren SX, Saitou N, Sakai K, Sakamoto S, Sakate R, Schupp I, Servant F, Sherry S, Shiba R, Shimizu N, Shimoyama M, Simpson AJ, Soares B, Steward C, Suwa M, Suzuki M, Takahashi A, Tamiya G, Tanaka H, Taylor T, Terwilliger JD, Unneberg P, Veeramachaneni V, Watanabe S, Wilming L, Yasuda N, Yoo HS, Stodolsky M, Makalowski W, Go M, Nakai K, Takagi T, Kanehisa M, Sakaki Y, Quackenbush J, Okazaki Y, Hayashizaki Y, Hide W, Chakraborty R, Nishikawa K, Sugawara H, Tateno Y, Chen Z, Oishi M, Tonellato P, Apweiler R, Okubo K, Wagner L, Wiemann S, Strausberg RL, Isogai T, Auffray C, Nomura N, Gojobori T, Sugano S (2004) Integrative annotation of 21,037 human genes validated by full-length cDNA clones. PLoS Biol 2:856–875

Imanishi T, Nagai Y, Habara T, Yamasaki C, Takeda J, Mikami S, Bando Y, Tojo H, Nishimura T (2013) Full-length transcriptome-based H-InvDB throws a new light on chromosome-centric proteomics. J Proteome Res 12(1):62–66

International Human Genome Sequencing Consortium (2004) Finishing the euchromatic sequence of the human genome. Nature 431:931–945

Kim MS, Pinto SM, Getnet D, Nirujogi RS, Manda SS, Chaerkady R, Madugundu AK, Kelkar DS, Isserlin R, Jain S, Thomas JK, Muthusamy B, Leal-Rojas P, Kumar P, Sahasrabuddhe NA, Balakrishnan L, Advani J, George B, Renuse S, Selvan LD, Patil AH, Nanjappa V, Radhakrishnan A, Prasad S, Subbannayya T, Raju R, Kumar M, Sreenivasamurthy SK,

Marimuthu A, Sathe GJ, Chavan S, Datta KK, Subbannayya Y, Sahu A, Yelamanchi SD, Jayaram S, Rajagopalan P, Sharma J, Murthy KR, Syed N, Goel R, Khan AA, Ahmad S, Dey G, Mudgal K, Chatterjee A, Huang TC, Zhong J, Wu X, Shaw PG, Freed D, Zahari MS, Mukherjee KK, Shankar S, Mahadevan A, Lam H, Mitchell CJ, Shankar SK, Satishchandra P, Schroeder JT, Sirdeshmukh R, Maitra A, Leach SD, Drake CG, Halushka MK, Prasad TS, Hruban RH, Kerr CL, Bader GD, Iacobuzio-Donahue CA, Gowda H, Pandey A (2014) A draft map of the human proteome. Nature 509(7502):575–581

Kozomara A, Griffiths-Jones S (2014) miRBase: annotating high confidence microRNAs using deep sequencing data. Nucleic Acids Res 42(Database issue):D68–D73

Kryukov GV, Castellano S, Novoselov SV, Lobanov AV, Zehtab O, Guigó R, Gladyshev VN (2003) Characterization of mammalian selenoproteomes. Science 300(5624):1439–1443

Lestrade L, Weber MJ (2006) snoRNA-LBME-db, a comprehensive database of human H/ACA and C/D box snoRNAs. Nucleic Acids Res 34(Database issue):D158–D162

Mostert V (2000) Selenoprotein P: properties, functions, and regulation. Arch Biochem Biophys 376(2):433–438

Nilsen TW, Graveley BR (2010) Expansion of the eukaryotic proteome by alternative splicing. Nature 463(7280):457–463

Pei B, Sisu C, Frankish A, Howald C, Habegger L, Mu XJ, Harte R, Balasubramanian S, Tanzer A, Diekhans M, Reymond A, Hubbard TJ, Harrow J, Gerstein MB (2012) The GENCODE pseudogene resource. Genome Biol 13(9):R51

Pennisi E (2007) Working the (gene count) numbers: finally, a firm answer? Science 316:1113

Pruitt KD, Brown GR, Hiatt SM, Thibaud-Nissen F, Astashyn A, Ermolaeva O, Farrell CM, Hart J, Landrum MJ, McGarvey KM, Murphy MR, O'Leary NA, Pujar S, Rajput B, Rangwala SH, Riddick LD, Shkeda A, Sun H, Tamez P, Tully RE, Wallin C, Webb D, Weber J, Wu W, DiCuccio M, Kitts P, Maglott DR, Murphy TD, Ostell JM (2014) RefSeq: an update on mammalian reference sequences. Nucleic Acids Res 42(D1):D756–D763

Quek XC, Thomson DW, Maag JL, Bartonicek N, Signal B, Clark MB, Gloss BS, Dinger ME (2015) lncRNAdb v2.0: expanding the reference database for functional long noncoding RNAs. Nucleic Acids Res 43(D1):D168–D173

Reeves MA, Hoffmann PR (2009) The human selenoproteome: recent insights into functions and regulation. Cell Mol Life Sci 66(15):2457–2478

Sakai H, Koyanagi KO, Imanishi T, Itoh T, Gojobori T (2007) Frequent emergence and functional resurrection of processed pseudogenes in the human and mouse genomes. Gene 389 (2):196–203

Sun X, Lin SM, Yan X (2014) Computational evidence of NAGNAG alternative splicing in human large intergenic noncoding RNA. Biomed Res Int 2014:736798

Takeda J, Suzuki Y, Nakao M, Barrero RA, Koyanagi KO, Jin L, Motono C, Hata H, Isogai T, Nagai K, Otsuki T, Kuryshev V, Shionyu M, Yura K, Go M, Thierry-Mieg J, Thierry-Mieg D, Wiemann S, Nomura N, Sugano S, Gojobori T, Imanishi T (2006) Large-scale identification and characterization of alternative splicing variants of human gene transcripts using 56,419 completely sequenced and manually annotated full-length cDNAs. Nucleic Acids Res 34 (14):3917–3928

Takeda J, Suzuki Y, Nakao M, Kuroda T, Sugano S, Gojobori T, Imanishi T (2007) H-DBAS: alternative splicing database of completely sequenced and manually annotated full-length cDNAs based on H-invitational. Nucleic Acids Res 35(Database issue):D104–D109

Takeda J, Suzuki Y, Sakate R, Sato Y, Seki M, Irie T, Takeuchi N, Ueda T, Nakao M, Sugano S, Gojobori T, Imanishi T (2008) Low conservation and species-specific evolution of alternative splicing in humans and mice: comparative genomics analysis using well-annotated full-length cDNAs. Nucleic Acids Res 36(20):6386–6395

Takeda J, Suzuki Y, Sakate R, Sato Y, Gojobori T, Imanishi T, Sugano S (2009) H-DBAS: human-transcriptome database for alternative splicing: update 2010. Nucleic Acids Res 38(Database issue):D86–D90

Takeda J, Yamasaki C, Murakami K, Nagai Y, Sera M, Hara Y, Obi N, Habara T, Gojobori T, Imanishi T (2013) H-InvDB in 2013: an omics study platform for human functional gene and transcript discovery. Nucleic Acids Res 41(D1):D915–D919

The UniProt Consortium (2015) UniProt: a hub for protein information. Nucleic Acids Res 43 (D1):D204–D212

Uhlén M, Fagerberg L, Hallström BM, Lindskog C, Oksvold P, Mardinoglu A, Sivertsson Å, Kampf C, Sjöstedt E, Asplund A, Olsson I, Edlund K, Lundberg E, Navani S, Szigyarto CA, Odeberg J, Djureinovic D, Takanen JO, Hober S, Alm T, Edqvist PH, Berling H, Tegel H, Mulder J, Rockberg J, Nilsson P, Schwenk JM, Hamsten M, von Feilitzen K, Forsberg M, Persson L, Johansson F, Zwahlen M, von Heijne G, Nielsen J, Pontén F (2015) Proteomics. Tissue-based map of the human proteome. Science 347(6220):1260419

Vanderperre B, Lucier JF, Roucou X (2012) HAltORF: a database of predicted out-of-frame alternative open reading frames in human. Database 2012:bas025

Vanderperre B, Lucier JF, Bissonnette C, Motard J, Tremblay G, Vanderperre S, Wisztorski M, Salzet M, Boisvert FM, Roucou X (2013) Direct detection of alternative open reading frames translation products in human significantly expands the proteome. PLoS One 8(8):e70698

Wojtowicz WM, Flanagan JJ, Millard SS, Zipursky SL, Clemens JC (2004) Alternative splicing of Drosophila Dscam generates axon guidance receptors that exhibit isoform-specific homophilic binding. Cell 118(5):619–633

Yamakawa K, Huot YK, Haendelt MA, Hubert R, Chen XN, Lyons GE, Korenberg JR (1998) DSCAM: a novel member of the immunoglobulin superfamily maps in a Down syndrome region and is involved in the development of the nervous system. Hum Mol Genet 7 (2):227–237

Chapter 5
Duplicated Genes

Takashi Kitano

Abstract Gene duplication can be categorized into four types: tandem, genome, transposition-based, and mRNA-based. In humans, 11 HOXA genes are clustered on the short arm of chromosome 7 as tandemly duplicated genes. In the common ancestor of vertebrates, the four HOX gene clusters (HOXA, HOXB, HOXC, and HOXD), which are located on different chromosomes, were produced by two-round whole genome duplications. Thus, HOXA genes were produced by two types of gene duplications – tandem and genome duplication. As it is likely that most genes arose by gene duplications, the "gene duplication" phenomenon appears to be a fundamental component of the evolution of organisms. Nowadays, genomics research conducted on several organisms has revealed that many duplicated genes exist in the genome of multicellular organisms, and research into gene duplication events and duplicated genes is ongoing.

Keywords Duplication · Evolution · Gene · Genome · Orthologous · Paralogous · Pseudogene · Synteny · Tandem

5.1 Fate of Duplicated Genes: Pseudogenization or Gain of New Function

Let us assume that a functional gene (A) has duplicated, forming A1 and A2 (Fig. 5.1). Shortly after duplication, the two genes (A1 and A2) are identical or share homologies with each other. However, the two genes become different nucleotide sequences by accumulating mutations over time. As is often the case, since one gene (A1) is sufficient in a genome, the other gene (A2) becomes free from functional constraints. If this occurs, A2 can accumulate a greater number of mutations compared with A1. Ultimately, the A2 gene may become a pseudogene

T. Kitano (✉)
Department of Biomolecular Functional Engineering, College of Engineering, Ibaraki University, Hitachi, Japan
e-mail: takashi.kitano.evolution@vc.ibaraki.ac.jp

N. Saitou (ed.), *Evolution of the Human Genome I*, Evolutionary Studies,
https://doi.org/10.1007/978-4-431-56603-8_5

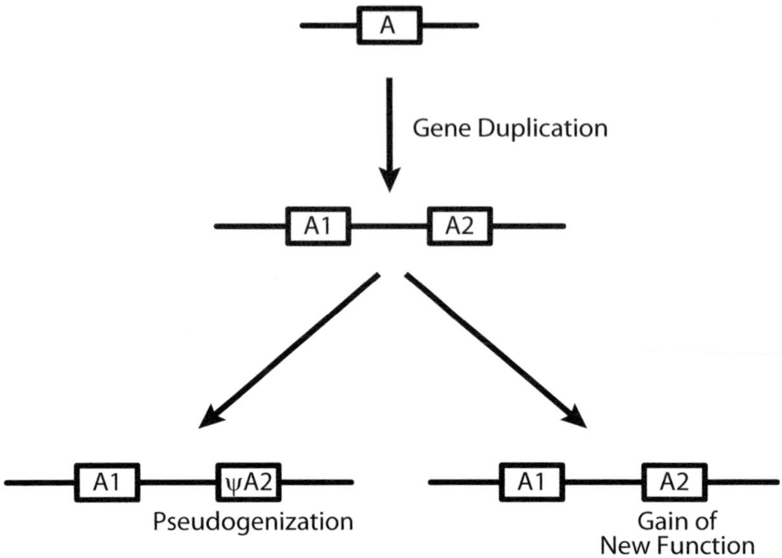

Fig. 5.1 A scheme for evolution of duplicated genes. Each box indicates a gene. For details, see text

(ψA2) by obtaining a nonsense mutation (a mutation that leads to the appearance of a stop codon where previously a codon specified an amino acid) and/or a frameshift mutation (a mutation that shifts the codon reading frames by indels (insertions and deletions) in non-multiples of three). Alternatively, the A2 gene may evolve into a new functional gene by accumulating a higher number of nonsynonymous mutations, or it may obtain a new expression pattern through mutations in its promoter region.

Genes encoding α-chains of hemoglobin are duplicated genes. Two proteins of α-chains and two proteins of β-chains constitute the heterotetramer molecule hemoglobin. Genes encoding α-chains are located on the short arm of chromosome 16 (Fig. 5.2). These genes were formed by tandem gene duplications. Genes encoding β-chains are clustered on the short arm of chromosome 11. Genes with "ψ" indicate pseudogenes. The two loci of ζ and ψζ were also formed by tandem gene duplication. Interestingly, the hemoglobin of the early embryonic stages consists of ζ- and ε-chains, although α- and β-chains form hemoglobin in adults.

It is likely that most genes arose by gene duplications. Therefore, it can be started that the "gene duplication" phenomenon is a fundamental component of the evolution of organisms.

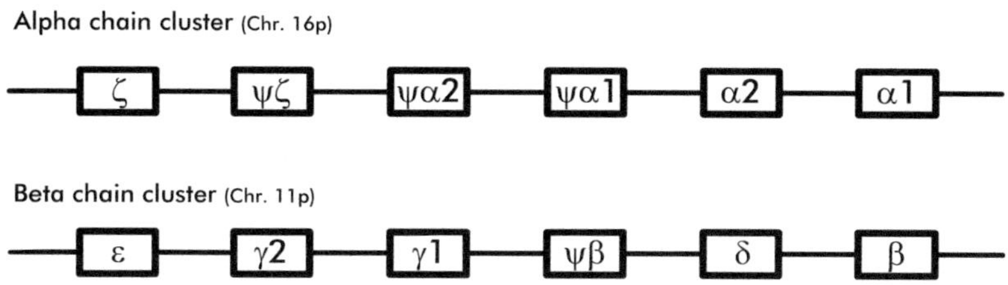

Fig. 5.2 A scheme of the human hemoglobin α-chain gene cluster. Each box indicates a gene. This gene cluster is located on chromosome 16 (16p13.3)

5.2 Short History of Gene Duplication Studies

Gene duplication is a type of mutation event in which the genome of a cell was doubled or a part of the region of the genome was copied. The former is referred to as polyploidization, which has been reported in some plants since olden days. For example, in 1911, Kuwada reported numerous variations in the number of chromosomes in maize (*Zea mays*) (Kuwada 1911). He proposed that the variations arose due to chromosomal duplication events. It is now known that the maize genome has undergone paleopolyploidization (Paterson et al. 2004) followed by an additional whole genome duplication event (Blanc and Wolfe 2004; Swigonova et al. 2004; see also Schnable et al. 2009). In 1919, Bridges reported the earliest observations of a segmental duplication event as a chromosomal aberration in *Drosophila* species. Ohno (1970) further emphasized the importance of gene duplication in evolution, stating "Natural selection merely modified, while redundancy created". Gene duplication is increasingly being considered as important based on the current accumulation of genome data, and the study of Ohno (1970) is still cited today. In the 1970s and 1980s, Ohta (1980) performed a series of studies on multigene families and proposed the concept of concerted evolution of tandemly duplicated genes. Concerted evolution is caused by mechanisms of unequal crossing-over and/or gene conversion (Ohta 1980). In contrast, Nei and Rooney (2005) proposed birth-and-death evolution of multigene families. In their model, new genes are created by gene duplication, and some duplicated genes remain in the genome for a long time, whereas others are deactivated or deleted from the genome. Nowadays, genomics research conducted on several organisms has revealed that many duplicated genes exist in the genome of multicellular organisms, and research into gene duplication events and duplicated genes is ongoing.

5.3 Types of Gene Duplication

Gene duplication can be categorized into four types: tandem, genome, transposition-based, and mRNA-based. In tandem duplication, more than two copies of genes are tandemly duplicated on a chromosome. Tandemly duplicated genes occupy 14–17% of genes in mammalian genomes (Shoja and Zhang 2006). More than two tandem copies are referred to as a multigene family. Tandemly duplicated genes can be categorized into three types of gene orientations: (1) two genes are oriented in the same direction, that is, 5′ end of one gene faces the 3′ end of the other gene (head to tail, 5′−3′ 5′-3′), (2) the 5′ ends of the two genes face each other (head to head, 3′−5′ 5′−3′), and (3) the 3′ ends of the two genes face each other (tail to tail, 5′−3′ 3′−5′). In rodent genomes, the head-to-tail (5′−3′ 5′−3′) orientation constitutes approximately 70% of tandemly duplicated genes (Ezawa et al. 2006). The RH blood group genes are an example of tandemly duplicated genes. In human RH blood group genes, two homologous (approximately 92% of amino acid identity) loci, RHD and RHCE, are located on 1p36.11 with tail-to-tail orientation, and a small gene termed TMEM50A (transmembrane protein 50A or SMP1 (small membrane protein 1)) is located on an approximately 30-kb region between these loci (Suto et al. 2000; Wagner and Flegel 2000) (Fig. 5.3). It is assumed that gene conversion occurs frequently at both loci (Kitano and Saitou 1999; Innan 2003). A so-called RH (−) person does not have RHD proteins because they lack the RHD locus. Since orangutans and Old World and New World monkeys all have a single RH locus, it has been suggested that gene duplication occurred in the lineage of the common ancestor of humans, chimpanzees, and gorillas (Blancher et al. 1992). Kitano et al. (2007) reported that following gene duplication, the rate of nonsynonymous substitution increased on exon 7, which is located on an RHD-specific motif. In addition, Kitano et al. (2016) determined the genome structure of the gene cluster of chimpanzee RH blood group genes and characterized three complete loci in chimpanzees (Fig. 5.3). Their sequence comparisons with the human RH blood group genes suggested that rearrangements and gene conversions frequently occurred between these genes and that the classic orthology/paralogy dichotomy no longer holds between human and chimpanzee RH blood group genes.

When speciation occurs after a tandem duplication event, duplicated genes are inherited in each species. Figure 5.4a shows a tandem duplication event followed by two speciation events. In this scheme, a relationship between, for example, the A1 gene of species x (xA1) and the A1 gene of species y (yA1) is referred to as orthologous, indicating that the genes were inherited directly from an ancestor. In contrast, a relationship of the two genes from one species, such as xA1 and xA2, is referred to as paralogous, indicating that the genes evolved parallel in one species. A relationship between xA1 and yA2 is also paralogous. A phylogenetic tree based on these genes is shown in Fig. 5.4b. When constructing a species tree, it is vital to use orthologous genes only. If a paralogous gene is included in a data, incorrect phylogenies may be constructed. For example, when constructing a species tree using the A1 genes of species x, y, and z, if zA2 is mistakenly used for zA1, an incorrect phylogenetic relationship could be obtained, i.e., species x (xA1) and y (yA1) can form a cluster as closely related species and z (zA2) may be located outside of this cluster (Fig. 5.4c).

Recently, Ezawa et al. (2011) proposed that physically relatively distant duplicated genes on a chromosome were generated in a different manner from the usual

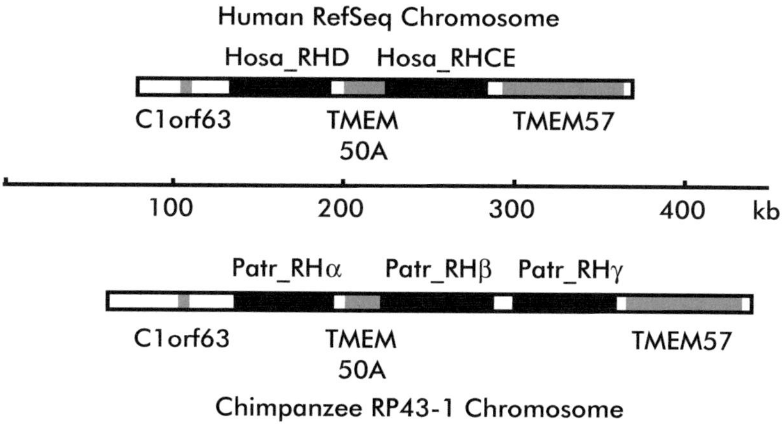

Fig. 5.3 Genome structures of the gene cluster of human (top) and chimpanzee (bottom) RH blood group genes. Black boxes denote RH genes, and gray boxes represent C1orf63, TMEM50A, and TMEM57 genes (Modified from Kitano et al. 2016)

tandem duplication event. They suggested that intrachromosomal duplogs (duplog is a synonym of intraspecies paralogs) with fairly long physical distances (and sometimes inverted) were generated at once, rather than resulting from tandem duplications and subsequent genomic rearrangements, and termed as "drift duplication." They also showed that the drift duplication has been producing duplicate copies at paces comparable with tandem duplications since the common ancestor of vertebrates.

Genome duplication is a type of gene duplication where the entire genome in a cell is copied at one time. In plants, this well-known phenomenon is referred to as polyploidization. For example, polyploidization is used to produce seedless watermelon (*Citrullus lanatus*). Here, a tetraploidy (4*n*) watermelon is developed by colchicine treatment of wild-type (2*n*) watermelon, which is then pollinated with wild-type (2*n*) watermelon to produce triploidy (3*n*). Then, triploid watermelon is pollinated with wild-type watermelon; however, it cannot produce seeds because triploid watermelon cannot undergo normal meiosis. As a result, this watermelon does not contain seeds (Kihara and Nishiyama 1947).

In animals, genome duplication events are known from several fish species. It is also well known that two rounds of whole genome duplications have occurred in the common ancestor of vertebrates (Ohno 1970). When two-round whole genome duplications occur, four homologous genes arise. Members of the RH gene family are examples of this process. The main members of the RH gene family are RH, RHAG (RH-associated glycoprotein, also known as the RH50 glycoprotein), RHBG (RH type B glycoprotein), and RHCG (RH type C glycoprotein). The RH gene is the original blood group gene; its nucleotide sequence was determined independently by Cherif-Zahar et al. (1990) and Avent et al. (1990). In humans, the RH genes are located on chromosome 1p34-p36 (Ruddle et al. 1972; Cherif-Zahar et al. 1991), which is referred to as 1p36.11. The RHAG protein was obtained

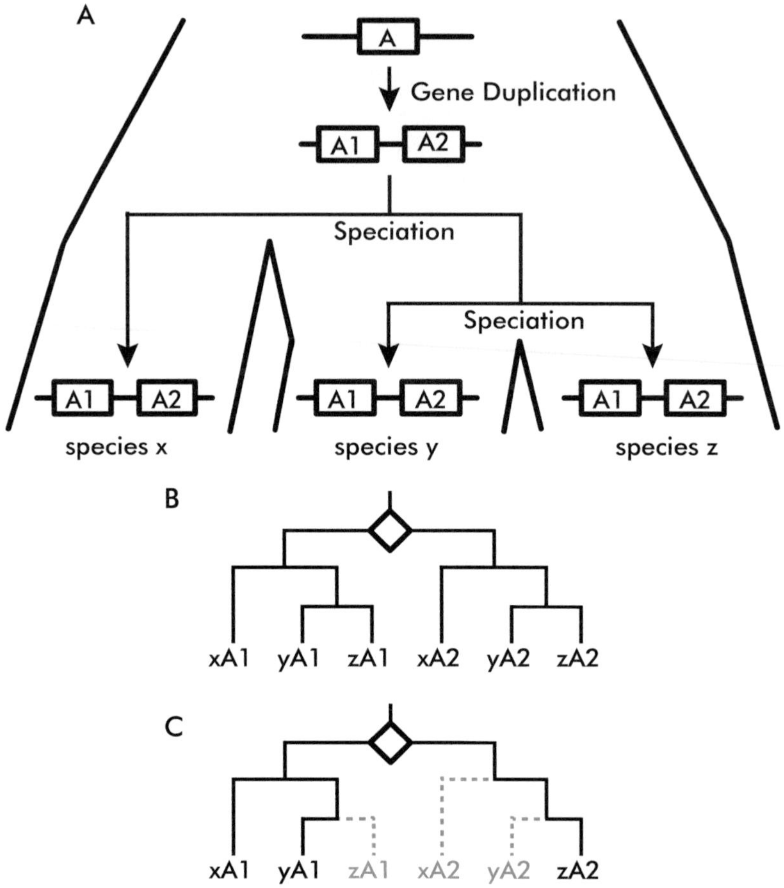

Fig. 5.4 Orthologous and paralogous genes. (**a**) Scheme for a single gene duplication, producing the two duplicated genes A1 and A2 and two subsequent speciations. The divergence between species x and a common ancestor of species y and z and the divergence between species y and z are indicated. Each box indicates a gene. (**b**) A phylogenetic tree among these genes. For example, xA1 indicates the A1 gene of species x. The gene duplication is shown by a white diamond. (**c**) A phylogenetic relationship among xA1, yA1, and zA2. Genes zA1, xA2, and yA2, which were not used for the phylogenetic tree construction, are shown in gray letters with broken gray lines

together with the RH proteins on immunoprecipitation with anti-RH antibodies from humans (Moore and Green 1987). The nucleotide sequence of the human RHAG (also known as RH50) gene was determined by Ridgwell et al. (1992) using a degenerate PCR-based assay on the N-terminal amino acid sequence of the RH-associated glycoproteins. Liu et al. (2000) reported the RHCG genes for humans and mouse. They blast-searched against the mouse EST database using mouse RHAG cDNA as a query and found a partial sequence of the gene, which was used as a beachhead to sequence the entire mouse and human RHCG cDNAs. Further, Liu et al. (2001) detailed the RHBG genes for humans and mouse using the

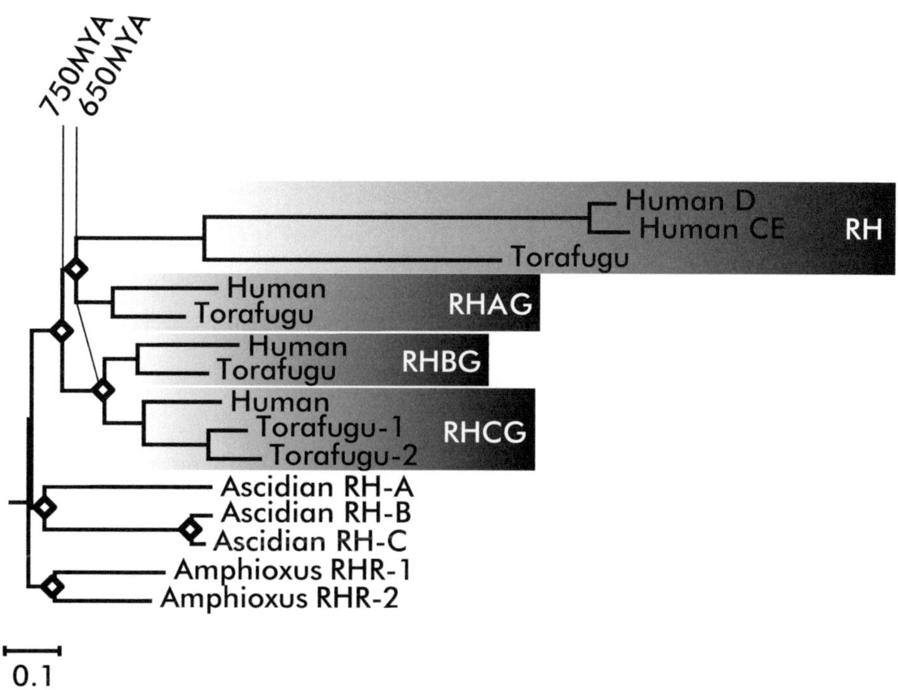

Fig. 5.5 A phylogenetic tree of chordates RH family genes. The root was determined using purple sea urchin as an outgroup. Human and torafugu genes were used as representatives of vertebrates. Gene duplications are shown by open diamonds. *MYA* Million years ago (Modified from Kitano et al. 2010)

same strategy. In humans, RHAG, RHBG, and RHCG genes are located on 6p11-21.1 (Cherif-Zahar et al. 1996; Ridgwell et al. 1992), 1q21.3 (Liu et al. 2001), and 15q25 (Liu et al. 2000), respectively. Gene duplication event times among the four members of the RH gene family are estimated as approximately 650–750 million years ago (Kitano et al. 2010) (Fig. 5.5). This period roughly corresponds to the two-round whole genome duplications in the common ancestor of vertebrates.

Because the whole genome is duplicated, the physical co-localization of some genetic loci around the four homologous genes is also preserved, which is referred to as synteny. The two-round whole genome duplications in the common ancestor of vertebrates are exemplified by the fact that 2–4 synteny regions are frequently observed in vertebrate genomes. The four HOX gene cluster regions (Carroll 1995) are examples of this. Hox genes encode transcription factors with a DNA-binding domain termed a homeobox. In the human genome, the four clusters of HOXA, HOXB, HOXC, and HOXD are located on different chromosomes (Fig. 5.6). While some of these have been lost in the course of vertebrate evolution, such as HOXA8 and HOXD7, the physical co-localizations have been preserved. Genes in a cluster may have been duplicated by tandem gene duplications before the two-round whole

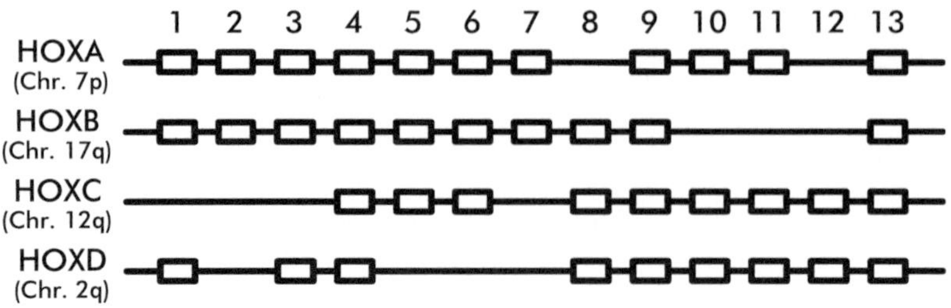

Fig. 5.6 Scheme of the HOX gene clusters in humans. Each box indicates a gene. Chromosomal location of each cluster is shown in parentheses

genome duplication. For example, 11 HOXA genes (A1–A7, A9–A11, and A13) in the HOXA cluster region may have been duplicated by repeated tandem gene duplications in the common ancestor of Animalia. Furthermore, conserved noncoding sequences within the HOX gene clusters have been duplicated at the same time (Matsunami et al. 2010).

The third type of gene duplication is transposition-based duplication. There are two types of transposable elements: the DNA transposon and the retrotransposon (retroposon). DNA transposons encode transposase that catalyzes the movement of DNA transposon to another region of a genome – effectively a "cut and paste" type of transfer. In contrast, a retrotransposon's DNA region is transcribed into RNA and then reverse transcribed into complementary DNA (cDNA) by reverse transcriptase, followed by the transfer of the cDNA fragment to another region of the genome. This may be termed a "copy and paste" type of transfer. From the perspective of self-duplication, retrotransposon could be considered as a type of gene duplication. Some retrotransposons encode reverse transcriptase, but others do not encode the enzyme. The former is represented by LTR (long terminal repeat) elements and LINEs (long interspersed elements), the latter by SINEs (short interspersed elements). The Alu sequence is one of the most famous SINEs. Alu is comprised of approximately 300 bp of short interspersed elements, and there are approximately one million copies (occupying about 10.6% of a genome) in the human genome (Table 5.1). It is assumed that Alu sequences evolved from 7SL RNA (Ullu and Tschudi 1984), and there are some subfamilies such as AluJo, AluSc, AluSg, AluSp, AluSq, AluSx, AluY, and AluYa (Price et al. 2004).

The fourth type of gene duplication is mRNA-based duplication, which is related to transposition-based duplication. Typically, genes are transcribed into RNA with a cap structure on the 5′ end and a poly(A) tail on the 3′ end and then spliced to yield mature mRNA. This mature mRNA is reverse transcribed into complementary DNA (cDNA) by reverse transcriptase, and the cDNA fragment is transferred to another region of the genome. This newly inserted gene exists as an intronless gene compared with the original genes and frequently becomes a pseudogene. If some transcription regulatory DNA sequences are located on the 5′ side of the inserted gene, the inserted intronless gene might have a new mode of expression. There are an estimated 4000 of these duplicated genes in the human genome (Marques et al.

Table 5.1 Number of copies and fraction of the draft genome sequence for classes of interspersed repeats

	Number of copies (×1000)	Total number of bases in the draft genome (Mb)	Fraction of the draft genome sequence (%)
SINE	1558	359.6	13.14
(Alu)	1090	290.1	10.60
LINE	868	558.8	20.42
LTR elements	443	227.0	8.29
DNA elements	294	77.6	2.84
Unclassified	3	3.8	0.14

Cited from Lander et al. (2001)

2005). Two genes, GLUD1 and GLUD2, encode L-glutamate dehydrogenase in humans (Fig. 5.7). The GLUD1 gene is located on 10q23.3 and consists of 13 exons. In contrast, the GLUD2 gene is an intronless (one exon) gene and is located on Xq24-q25. The expression patterns of these genes differ. Although GLUD1 expressed in various organs as a housekeeping gene, expression of GLUD2 gene is restricted to organs such as the brain, retina, and testes (Shashidharan et al. 1994). It is believed that GLUD2 arose by mRNA-based duplication from GLUD1 following the divergence between the common ancestor of hominoids and the common ancestor of Old World monkeys (Burki and Kaessmann 2004). It could be assumed that a transcription-regulatory DNA sequence may be located on the 5′ side of the inserted location of the GLUD2 gene fragment.

5.4 The Number of Tandemly Duplicated Genes in Humans

Zhang (2003) estimated the number of tandemly duplicated genes in the human lineage to be approximately 1800 genes since the divergence between humans and chimpanzees. This was estimated by assuming a tandem gene duplication rate of 0.01 per gene per million years (Lynch and Conery 2000), a total of 30,000 genes in mammals (Waterston et al. 2002), and the divergence time between humans and chimpanzees to be 6 million years ago ($0.01/1,000,000 \times 30,000 \times 6,000,000$). The tandem gene duplication rate, i.e., the gene birth rate, varies between species, ranging from approximately 0.002 to 0.02 per gene per million years among *Homo sapiens*, *Mus musculus*, *Gallus gallus*, *Danio rerio*, *Caenorhabditis elegans*, *Drosophila melanogaster*, *Arabidopsis thaliana*, *Oryza sativa*, and Saccharomyces cerevisiae. This rate is 5–50 times larger than the de novo nucleotide substitution rate of 0.00004 per site per million years in the human lineage. This was estimated using the average de novo mutation rate of 1.2×10^{-8} per site per generation (Kong

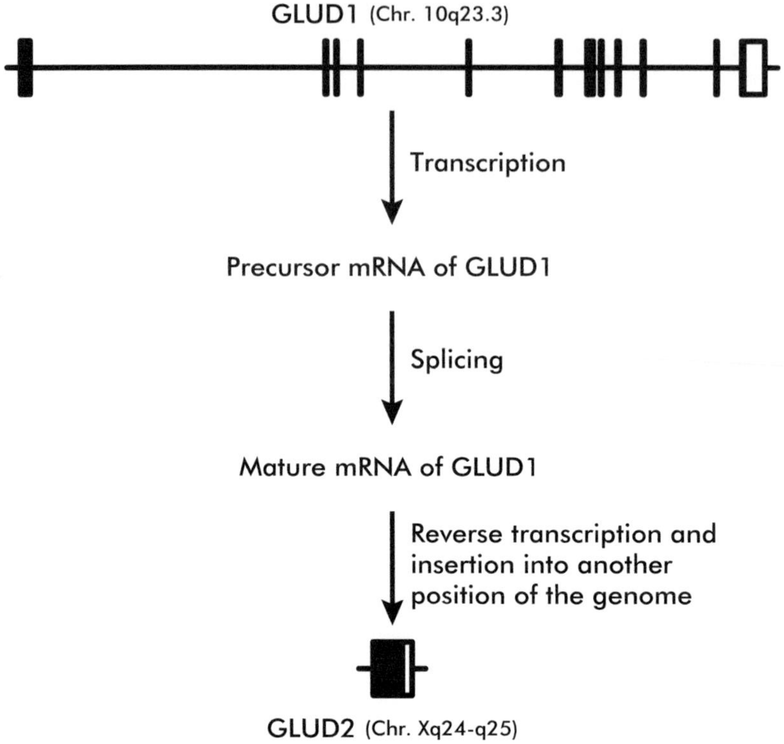

Fig. 5.7 Scheme for the gene duplication to yield the GLUD2 gene from the GLUD1 gene as an example of mRNA-based duplication. The chromosomal location of each gene is shown in parentheses

et al. 2012) and a 30-year generation time ($1.2 \times 10^{-8}/30 \times 1,000,000$). Nevertheless, it is thought that a number of duplicated genes disappeared through pseudogenization.

One of the best-known tandem gene duplication events, followed by neofunctionalization, is the ECP (eosinophil cationic protein, also termed RNASE3 (ribonuclease, RNase A family, 3)) gene for hominoids and Old World monkeys (Zhang et al. 1998). The ECP gene has been duplicated from the EDN (eosinophil-derived neurotoxin, also termed RNASE2 (ribonuclease, RNase A family, 2)) gene, both of which are located on 14q11.2 in the human genome. As New World monkeys only possess EDN genes, this tandem gene duplication must have occurred in the common ancestral lineage of hominoids and Old World monkeys after the divergence between the common ancestor of hominoids and Old World monkeys and the common ancestor of New World monkeys. The EDN protein has ribonuclease activity in eosinophil granulocytes (Rosenberg et al. 1995), whereas ECP has very low ribonuclease activity but contains toxins to various pathogenic bacteria and parasites (Slifman et al. 1986; Rosenberg and Dyer 1995). It is likely that this functional shift from EDN to ECP occurred in the common ancestral lineage of hominoids and Old World monkeys after the tandem gene duplication, based on positive Darwinian selection accumulating amino acid replacements (Zhang et al. 1998). In particular, a number of amino acid changes to arginine residues have occurred in the ECP lineage. The accumulation of arginine residues may have acquired toxic activity through the production of pores in pathogen cell membranes (Young et al. 1986).

References

Avent ND, Ridgwell K, Tanner MJ, Anstee DJ (1990) cDNA cloning of a 30 kDa erythrocyte membrane protein associated with Rh (Rhesus)-blood-group-antigen expression. Biochem J 271:821–825

Blanc G, Wolfe KH (2004) Widespread paleopolyploidy in model plant species inferred from age distributions of duplicate genes. Plant Cell 16:1667–1678

Blancher A, Calvas P, Ruffie J (1992) Etude des equivalents des antigens Rhesus chez les primates non hominiens. CR Soc Biol 186:682–695

Bridges CB (1919) Duplication. Anat Rec 15:357–358

Burki F, Kaessmann H (2004) Birth and adaptive evolution of a hominoid gene that supports high neurotransmitter flux. Nat Genet 36:1061–1063

Carroll SB (1995) Homeotic genes and the evolution of arthropods and chordates. Nature 376:479–485

Cherif-Zahar B, Bloy C, Le Van Kim C, Blanchard D, Bailly P, Hermand P, Salmon C, Cartron JP, Colin Y (1990) Molecular cloning and protein structure of a human blood group Rh polypeptide. Proc Natl Acad Sci U S A 87:6243–6247

Cherif-Zahar B, Mattei MG, Le Van Kim C, Bailly P, Cartron JP, Colin Y (1991) Localization of the human Rh blood group gene structure to chromosome region 1p34.3–1p36.1 by in situ hybridization. Hum Genet 86:398–400

Cherif-Zahar B, Raynal V, Gane P, Mattei MG, Bailly P, Gibbs B, Colin Y, Cartron JP (1996) Candidate gene acting as a suppressor of the RH locus in most cases of Rh-deficiency. Nat Genet 12:168–173

Ezawa K, OOta S, Saitou N (2006) Proceedings of the SMBE Tri-National Young Investigators' Workshop 2005. Genome-wide search of gene conversions in duplicated genes of mouse and rat. Mol Biol Evol 23:927–940

Ezawa K, Ikeo K, Gojobori T, Saitou N (2011) Evolutionary patterns of recently emerged animal duplogs. Genome Biol Evol 3:1119–1135

Innan H (2003) A two-locus gene conversion model with selection and its application to the human RHCE and RHD genes. Proc Natl Acad Sci U S A 100:8793–8798

Kihara H, Nishiyama I (1947) An application of sterility of autotriploids to the breeding of seedless watermelons. Seiken Ziho 3:5–15

Kitano T, Saitou N (1999) Evolution of Rh blood group genes have experienced gene conversions and positive selection. J Mol Evol 49:615–626

Kitano T, Umetsu K, Tian W, Yamazaki K, Saitou N (2007) Tempo and mode of evolution of the Rh blood group genes before and after gene duplication. Immunogenetics 59:427–431

Kitano T, Satou M, Saitou N (2010) Evolution of two Rh blood group-related genes of the amphioxus species Branchiostoma floridae. Genes Genet Syst 85:121–127

Kitano T, Kim CG, Blancher A, Saitou N (2016) No distinction of orthology/paralogy between human and chimpanzee Rh blood group genes. Genome Biol Evol 8:519–527

Kong A, Frigge ML, Masson G, Besenbacher S, Sulem P, Magnusson G, Gudjonsson SA, Sigurdsson A, Jonasdottir A, Jonasdottir A, Wong WS, Sigurdsson G, Walters GB,

Steinberg S, Helgason H, Thorleifsson G, Gudbjartsson DF, Helgason A, Magnusson OT, Thorsteinsdottir U, Stefansson K (2012) Rate of de novo mutations and the importance of father's age to disease risk. Nature 488:471–475

Kuwada Y (1911) Meiosis in the pollen mother cells of Zea mays L. Bot Mag 25:163–181

Lander ES, Linton LM, Birren B, Nusbaum C, Zody MC, Baldwin J, Devon K, Dewar K, Doyle M, FitzHugh W, Funke R, Gage D, Harris K, Heaford A, Howland J, Kann L, Lehoczky J, LeVine R, McEwan P, McKernan K, Meldrim J, Mesirov JP, Miranda C, Morris W, Naylor J, Raymond C, Rosetti M, Santos R, Sheridan A, Sougnez C, Stange-Thomann N, Stojanovic N, Subramanian A, Wyman D, Rogers J, Sulston J, Ainscough R, Beck S, Bentley D, Burton J, Clee C, Carter N, Coulson A, Deadman R, Deloukas P, Dunham A, Dunham I, Durbin R, French L, Grafham D, Gregory S, Hubbard T, Humphray S, Hunt A, Jones M, Lloyd C, McMurray A, Matthews L, Mercer S, Milne S, Mullikin JC, Mungall A, Plumb R, Ross M, Shownkeen R, Sims S, Waterston RH, Wilson RK, Hillier LW, McPherson JD, Marra MA, Mardis ER, Fulton LA, Chinwalla AT, Pepin KH, Gish WR, Chissoe SL, Wendl MC, Delehaunty KD, Miner TL, Delehaunty A, Kramer JB, Cook LL, Fulton RS, Johnson DL, Minx PJ, Clifton SW, Hawkins T, Branscomb E, Predki P, Richardson P, Wenning S, Slezak T, Doggett N, Cheng JF, Olsen A, Lucas S, Elkin C, Uberbacher E, Frazier M, Gibbs RA, Muzny DM, Scherer SE, Bouck JB, Sodergren EJ, Worley KC, Rives CM, Gorrell JH, Metzker ML, Naylor SL, Kucherlapati RS, Nelson DL, Weinstock GM, Sakaki Y, Fujiyama A, Hattori M, Yada T, Toyoda A, Itoh T, Kawagoe C, Watanabe H, Totoki Y, Taylor T, Weissenbach J, Heilig R, Saurin W, Artiguenave F, Brottier P, Bruls T, Pelletier E, Robert C, Wincker P, Smith DR, Doucette-Stamm L, Rubenfield M, Weinstock K, Lee HM, Dubois J, Rosenthal A, Platzer M, Nyakatura G, Taudien S, Rump A, Yang H, Yu J, Wang J, Huang G, Gu J, Hood L, Rowen L, Madan A, Qin S, Davis RW, Federspiel NA, Abola AP, Proctor MJ, Myers RM, Schmutz J, Dickson M, Grimwood J, Cox DR, Olson MV, Kaul R, Raymond C, Shimizu N, Kawasaki K, Minoshima S, Evans GA, Athanasiou M, Schultz R, Roe BA, Chen F, Pan H, Ramser J, Lehrach H, Reinhardt R, McCombie WR, de la Bastide M, Dedhia N, Blöcker H, Hornischer K, Nordsiek G, Agarwala R, Aravind L, Bailey JA, Bateman A, Batzoglou S, Birney E, Bork P, Brown DG, Burge CB, Cerutti L, Chen HC, Church D, Clamp M, Copley RR, Doerks T, Eddy SR, Eichler EE, Furey TS, Galagan J, Gilbert JG, Harmon C, Hayashizaki Y, Haussler D, Hermjakob H, Hokamp K, Jang W, Johnson LS, Jones TA, Kasif S, Kaspryzk A, Kennedy S, Kent WJ, Kitts P, Koonin EV, Korf I, Kulp D, Lancet D, Lowe TM, McLysaght A, Mikkelsen T, Moran JV, Mulder N, Pollara VJ, Ponting CP, Schuler G, Schultz J, Slater G, Smit AF, Stupka E, Szustakowski J, Thierry-Mieg D, Thierry-Mieg J, Wagner L, Wallis J, Wheeler R, Williams A, Wolf YI, Wolfe KH, Yang SP, Yeh RF, Collins F, Guyer MS, Peterson J, Felsenfeld A, Wetterstrand KA, Patrinos A, Morgan MJ, de Jong P, Catanese JJ, Osoegawa K, Shizuya H, Choi S, Chen YJ (2001) Initial sequencing and analysis of the human genome. Nature 409:860–921

Liu Z, Chen Y, Mo R, Hui C, Cheng JF, Mohandas N, Huang CH (2000) Characterization of human RhCG and mouse Rhcg as novel nonerythroid Rh glycoprotein homologues predominantly expressed in kidney and testis. J Biol Chem 275:25641–25651

Liu Z, Peng J, Mo R, Hui C, Huang CH (2001) Rh type B glycoprotein is a new member of the Rh superfamily and a putative ammonia transporter in mammals. J Biol Chem 276:1424–1433

Lynch M, Conery JS (2000) The evolutionary fate and consequences of duplicate genes. Science 290:1151–1155

Marques AC, Dupanloup I, Vinckenbosch N, Reymond A, Kaessmann H (2005) Emergence of young human genes after a burst of retroposition in primates. PLoS Biol 3:1970–1979

Matsunami M, Sumiyama K, Saitou N (2010) Evolution of conserved non-coding sequences within the vertebrate Hox clusters through the two-round whole genome duplications revealed by phylogenetic footprinting analysis. J Mol Evol 71:427–436

Moore S, Green C (1987) The identification of specific Rhesus-polypeptide-blood-group-ABH-active-glycoprotein complexes in the human red-cell membrane. Biochem J 244:735–741

Nei M, Rooney AP (2005) Concerted and birth-and-death evolution of multigene families. Annu Rev Genet 39:121–152

Ohno S (1970) Evolution by Gene Duplication. Springer, Berlin/Heidelberg/New York

Ohta T (1980) Evolution and variation of multigene families. Springer, Berlin/Heidelberg/New York

Paterson AH, Bowers JE, Chapman BA (2004) Ancient polyploidization predating divergence of the cereals, and its consequences for comparative genomics. Proc Natl Acad Sci U S A 101:9903–9908

Price AL, Eskin E, Pevzner PA (2004) Whole-genome analysis of Alu repeat elements reveals complex evolutionary history. Genome Res 14:2245–2252

Ridgwell K, Spurr NK, Laguda B, MacGeoch C, Avent ND, Tanner MJ (1992) Isolation of cDNA clones for a 50 kDa glycoprotein of the human erythrocyte membrane associated with Rh (rhesus) blood-group antigen expression. Biochem J 287:223–228

Rosenberg HF, Dyer KD (1995) Eosinophil cationic protein and eosinophil-derived neurotoxin. Evolution of novel function in a primate ribonuclease gene family. J Biol Chem 270:21539–21544

Rosenberg HF, Dyer KD, Tiffany HL, Gonzalez M (1995) Rapid evolution of a unique family of primate ribonuclease genes. Nat Genet 10:219–223

Ruddle F, Ricciuti F, McMorris FA, Tischfield J, Creagan R, Darlington G, Chen T (1972) Somatic cell genetic assignment of peptidase C and the Rh linkage group to chromosome A-1 in man. Science 176:1429–1431

Schnable PS, Ware D, Fulton RS, Stein JC, Wei F, Pasternak S, Liang C, Zhang J, Fulton L, Graves TA, Minx P, Reily AD, Courtney L, Kruchowski SS, Tomlinson C, Strong C, Delehaunty K, Fronick C, Courtney B, Rock SM, Belter E, Du F, Kim K, Abbott RM, Cotton M, Levy A, Marchetto P, Ochoa K, Jackson SM, Gillam B, Chen W, Yan L, Higginbotham J, Cardenas M, Waligorski J, Applebaum E, Phelps L, Falcone J, Kanchi K, Thane T, Scimone A, Thane N, Henke J, Wang T, Ruppert J, Shah N, Rotter K, Hodges J, Ingenthron E, Cordes M, Kohlberg S, Sgro J, Delgado B, Mead K, Chinwalla A, Leonard S, Crouse K, Collura K, Kudrna D, Currie J, He R, Angelova A, Rajasekar S, Mueller T, Lomeli R, Scara G, Ko A, Delaney K, Wissotski M, Lopez G, Campos D, Braidotti M, Ashley E, Golser W, Kim H, Lee S, Lin J, Dujmic Z, Kim W, Talag J, Zuccolo A, Fan C, Sebastian A, Kramer M, Spiegel L, Nascimento L, Zutavern T, Miller B, Ambroise C, Muller S, Spooner W, Narechania A, Ren L, Wei S, Kumari S, Faga B, Levy MJ, McMahan L, Van Buren P, Vaughn MW, Ying K, Yeh CT, Emrich SJ, Jia Y, Kalyanaraman A, Hsia AP, Barbazuk WB, Baucom RS, Brutnell TP, Carpita NC, Chaparro C, Chia JM, Deragon JM, Estill JC, Fu Y, Jeddeloh JA, Han Y, Lee H, Li P, Lisch DR, Liu S, Liu Z, Nagel DH, McCann MC, San Miguel P, Myers AM, Nettleton D, Nguyen J, Penning BW, Ponnala L, Schneider KL, Schwartz DC, Sharma A, Soderlund C, Springer NM, Sun Q, Wang H, Waterman M, Westerman R, Wolfgruber TK, Yang L, Yu Y, Zhang L, Zhou S, Zhu Q, Bennetzen JL, Dawe RK, Jiang J, Jiang N, Presting GG, Wessler SR, Aluru S, Martienssen RA, Clifton SW, McCombie WR, Wing RA, Wilson RK (2009) The B73 maize genome: complexity, diversity, and dynamics. Science 326:1112–1115

Shashidharan P, Michaelidis TM, Robakis NK, Kresovali A, Papamatheakis J, Plaitakis A (1994) Novel human glutamate dehydrogenase expressed in neural and testicular tissues and encoded by an X-linked intronless gene. J Biol Chem 269:16971–16976

Shoja V, Zhang L (2006) A roadmap of tandemly arrayed genes in the genomes of human, mouse, and rat. Mol Biol Evol 23:2134–2141

Slifman NR, Loegering DA, McKean DJ, Gleich GJ (1986) Ribonuclease activity associated with human eosinophil-derived neurotoxin and eosinophil cationic protein. J Immunol 137:2913–2917

Suto Y, Ishikawa Y, Hyodo H, Uchikawa M, Juji T (2000) Gene organization and rearrangements at the human Rhesus blood group locus revealed by fiber-FISH analysis. Hum Genet 106:164–171

Swigonová Z, Lai J, Ma J, Ramakrishna W, Llaca V, Bennetzen JL, Messing J (2004) Close split of sorghum and maize genome progenitors. Genome Res 14:1916–1923

Ullu E, Tschudi C (1984) Alu sequences are processed 7SL RNA genes. Nature 312:171–172

Wagner FF, Flegel WA (2000) RHD gene deletion occurred in the Rhesus box. Blood 95:3662–3668

Waterston RH, Lindblad-Toh K, Birney E, Rogers J, Abril JF, Agarwal P, Agarwala R, Ainscough R, Alexandersson M, An P, Antonarakis SE, Attwood J, Baertsch R, Bailey J, Barlow K, Beck S, Berry E, Birren B, Bloom T, Bork P, Botcherby M, Bray N, Brent MR, Brown DG, Brown SD, Bult C, Burton J, Butler J, Campbell RD, Carninci P, Cawley S, Chiaromonte F, Chinwalla AT, Church DM, Clamp M, Clee C, Collins FS, Cook LL, Copley RR, Coulson A, Couronne O, Cuff J, Curwen V, Cutts T, Daly M, David R, Davies J, Delehaunty KD, Deri J, Dermitzakis ET, Dewey C, Dickens NJ, Diekhans M, Dodge S, Dubchak I, Dunn DM, Eddy SR, Elnitski L, Emes RD, Eswara P, Eyras E, Felsenfeld A, Fewell GA, Flicek P, Foley K, Frankel WN, Fulton LA, Fulton RS, Furey TS, Gage D, Gibbs RA, Glusman G, Gnerre S, Goldman N, Goodstadt L, Grafham D, Graves TA, Green ED, Gregory S, Guigó R, Guyer M, Hardison RC, Haussler D, Hayashizaki Y, Hillier LW, Hinrichs A, Hlavina W, Holzer T, Hsu F, Hua A, Hubbard T, Hunt A, Jackson I, Jaffe DB, Johnson LS, Jones M, Jones TA, Joy A, Kamal M, Karlsson EK, Karolchik D, Kasprzyk A, Kawai J, Keibler E, Kells C, Kent WJ, Kirby A, Kolbe DL, Korf I, Kucherlapati RS, Kulbokas EJ, Kulp D, Landers T, Leger JP, Leonard S, Letunic I, Levine R, Li J, Li M, Lloyd C, Lucas S, Ma B, Maglott DR, Mardis ER, Matthews L, Mauceli E, Mayer JH, McCarthy M, McCombie WR, McLaren S, McLay K, McPherson JD, Meldrim J, Meredith B, Mesirov JP, Miller W, Miner TL, Mongin E, Montgomery KT, Morgan M, Mott R, Mullikin JC, Muzny DM, Nash WE, Nelson JO, Nhan MN, Nicol R, Ning Z, Nusbaum C, O'Connor MJ, Okazaki Y, Oliver K, Overton-Larty E, Pachter L, Parra G, Pepin KH, Peterson J, Pevzner P, Plumb R, Pohl CS, Poliakov A, Ponce TC, Ponting CP, Potter S, Quail M, Reymond A, Roe BA, Roskin KM, Rubin EM, Rust AG, Santos R, Sapojnikov V, Schultz B, Schultz J, Schwartz MS, Schwartz S, Scott C, Seaman S, Searle S, Sharpe T, Sheridan A, Shownkeen R, Sims S, Singer JB, Slater G, Smit A, Smith DR, Spencer B, Stabenau A, Stange-Thomann N, Sugnet C, Suyama M, Tesler G, Thompson J, Torrents D, Trevaskis E, Tromp J, Ucla C, Ureta-Vidal A, Vinson JP, Von Niederhausern AC, Wade CM, Wall M, Weber RJ, Weiss RB, Wendl MC, West AP, Wetterstrand K, Wheeler R, Whelan S, Wierzbowski J, Willey D, Williams S, Wilson RK, Winter E, Worley KC, Wyman D, Yang S, Yang SP, Zdobnov EM, Zody MC, Lander ES (2002) Initial sequencing and comparative analysis of the mouse genome. Nature 420: 520–562

Young JD, Peterson CG, Venge P, Cohn ZA (1986) Mechanism of membrane damage mediated by human eosinophil cationic protein. Nature 321:613–616

Zhang J (2003) Evolution by gene duplication: an update. Trends Ecol Evol 18:292–298

Zhang J, Rosenberg HF, Nei M (1998) Positive Darwinian selection after gene duplication in primate ribonuclease genes. Proc Natl Acad Sci U S A 95:3708–3713

Chapter 6
Recombination

Ludovica Montanucci and Jaume Bertranpetit

Abstract Recombination is a key molecular process of chromosome reshuffling that takes place producing the gametes during sexual reproduction. The frequency of recombination varies greatly along the chromosomes, and its variation along the chromosome can be quantified through the measure of linkage disequilibrium (LD). Thus, recombination, through the measure of linkage disequilibrium, is at the base of all progress in mapping traits in the genome, is critical for GWAS studies, and has therefore fostered advancements in medical genetics. A major limitation that hampers a more extended use of LD is due to the fact that most current experimental technologies do not solve the phasing, along with a low accuracy of computational phasing algorithms.

Besides LD, recombination events by themselves, that is, the presence or absence of specific recombination events, can be used as genetic markers to study human population diversity as well as the dynamics of the recombination process. Differences in the recombination rates are found not only along the genome but also between populations and individuals. When LD has been applied to reconstructing the demographic history of modern human populations, the complementation of genetic diversity measures with LD has proved to be critical in solving the problem of defining ancestral populations. Despite the wealth of information contained in the recombination footprint, the full use of recombination and linkage disequilibrium data in population genetics is still in its infancy, and many more possibilities are waiting to be uncovered.

Keywords Linkage disequilibrium · Recombination · Population genetics · Genetic variation · Modern human origins

---

L. Montanucci (✉) · J. Bertranpetit (✉)
Institut de Biologia Evolutiva (UPF-CSIC), CEXS-Universitat Pompeu Fabra, Barcelona, Catalonia, Spain
e-mail: ludovica.montanucci@upf.edu; jaume.bertranpetit@upf.edu

N. Saitou (ed.), *Evolution of the Human Genome I*, Evolutionary Studies,
https://doi.org/10.1007/978-4-431-56603-8_6

6.1 Recombination and Its Distribution Along the Genome, Linkage Disequilibrium

Sexual reproduction requires diploid number of chromosomes, which reduces to haploid by meiosis to generate the gametes (eggs or spermatozoa). During this process, homologous chromosomes align and switch over segments. As a consequence, the haploid content of a gamete does not have any full chromosomes from the maternal or paternal genome, so each gamete is made up of pieces or a complex mosaic of their parental chromosomes.

Recombination happens all over the genome except in a few specific locations: the mitochondrial DNA (mtDNA) and most of the Y chromosomes (the non-recombinant part or NRY). Thus, genes will only be transmitted in a group if recombination has not occurred between them. From classical genetics principles, it is possible to calculate the number of cases in which two genes (or any genetic fragment or locus) which are neighbors on a chromosome were set apart by recombination while producing gametes. This ascertains a distance between the two genes (or two loci) defined as the percentage of cases in which recombination has occurred between them. This is the base of genetic distances, measured in centimorgans (one centimorgan means that 1% of the gametes will have a recombination event between the two specific points for which the distance is measured). A general average for the genome is that 1% of recombination events happen at a distance of one megabase (1 cM = 1 MB).

The frequency of recombination varies greatly along the chromosomes featuring regions with high recombination, while others are depleted or even totally absent (Paigen and Petkov 2010). The short regions (1–2 Kb) where most of the recombinations happen are called recombination hotspots and have a very strong concentration of recombination events. One of the most interesting advances in genomics in the last few years has been the discovery of the molecular bases for the formation of recombination hotspots in the genome and their complex evolutionary dynamics (Baudat et al. 2013).

Differences in the recombination landscape are found not only along the genome or a chromosome but also between sexes (women recombine more than men), between populations (slight differences among human populations have been observed, with some hotspots being population specific), and among species (the recombination map of humans and chimpanzees differs greatly). Kong et al. (2010) were able to construct the first recombination map based on directly observed recombinations with a resolution that is effective down to 10 kilobases (Kb) from the observation of thousands of parent-offspring pairs. They were able to pinpoint differences among sexes, populations, and even individuals. Thus, recombination is a fast-evolving trait with a birth and death process of recombination hotspots.

The existence of recombination breaks indicates that the genome content along a single arm of a chromosome (or along the DNA that constitutes it) may be different from one to another. This is why we talk about haplotypes to refer to the genetic content along a single DNA molecule; if several variants (or alleles) for the consecutive genome regions exist, the specific combination in each chromosome constitutes a haplotype. With n number of variants in a given fragment of a chromosome, the number of possible haplotypes is 2^{n-1}. In fact, given the information of diploid variants (like a DNA sequence), it may be impossible to know the phase in which the variants are found along the two chromosomes, but we explore below methods to solve it, based on the information in pedigrees or using statistical methods.

Given the irregular pattern of recombination endowed with strong differences among genome regions, we can wonder to which extent the information at a specific genome position is independent of the information at a position nearby: if we have two close genomic positions, the genomic information of the two may be independent if it is a recombination hotspot between them, or they may be fully correlated if there is no recombination, and the information in one tells us what the state is of the other. It is similar to the concept of pants and jackets—they can be separate objects, yet linked completely like in the case of a suit. This idea is unified by a concept: linkage disequilibrium (LD), which tells the degree of interdependence between two positions in the genome, from independence (total lack of LD) to full dependence (total LD) with all possible states in between. It is a statistical concept and the most usual measures vary from 0 (no LD) to 1 (total LD).

When looking at the LD pattern in a given genome region (Fig. 6.1), information for a population of individuals is needed, because LD is a population-level statistic. Linkage disequilibrium is at the base of all progress in mapping traits in the genome. The advancement of medical genetics owes most of its achievements to the existence of LD, as it has aided in the mapping of traits in the genome to the linkage between the gene of interest and a specific genetic marker, usually a SNP (single nucleotide polymorphism). All the GWAS (genome-wide association studies) that analyze hundreds of thousands or millions of SNPs in a group of affected people and unaffected controls will be able to find the genetic bases of the trait if LD is present between one (or several) of the SNPs and the causative variant(s) in the genome.

The haplotype is of interest in many areas of genome studies, as variants in the genome are not independent among them. But, most of the present technologies give information of both chromosomes together; in fact many technologies allow working on a single DNA strand, but difficulties remain in separating chromosomes. Some basic techniques based on long-range PCR or on the separation of single chromosomes may be used to solve the issue, but both are difficult and with limitations. The technology of second-generation sequencing does not solve the issue, but third-generation technology will be able to sequence long single DNA molecules with long reads and haplotype information.

Computationally, there are interesting statistical methods to solve the phasing of a set of linked variants. These methods use coalescent theory to reconstruct the phase of several variants, and their power is strongly dependent on the amount of

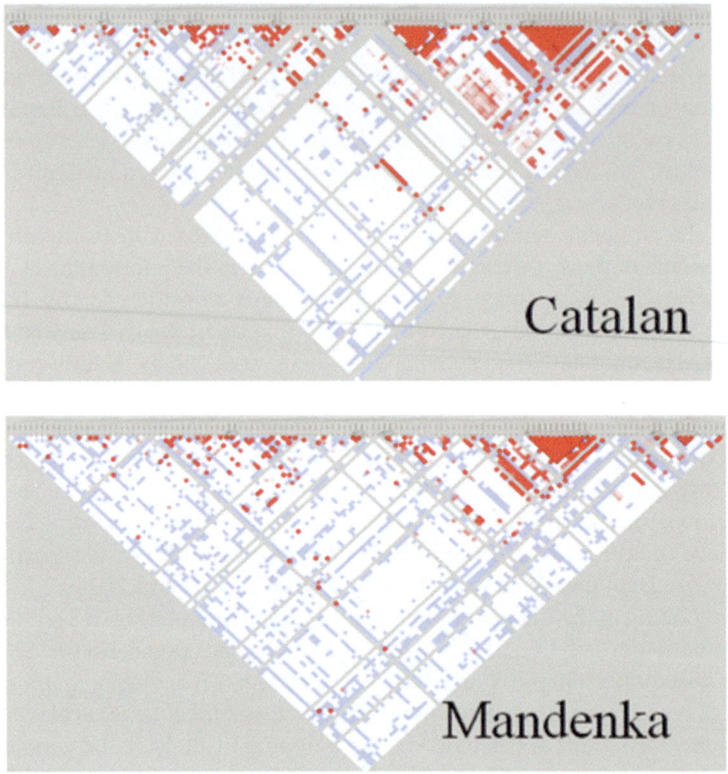

Fig. 6.1 Structure of linkage disequilibrium in a region in the human genome. Red squares indicate high linkage disequilibrium between the two SNPs. It is possible to see long stretches of strong linkage disequilibrium, which usually are flanked by recombination hotspots that are the cause of breakage of linkage disequilibrium blocks. Data for European (Catalans) and African (Mandenka) populations. African populations show much less linkage disequilibrium than in non-African

LD: the more LD, the better the reconstruction will be. One important consideration is that the accuracy of the phasing of genotyped data depends on LD, and thus no algorithm can make accurate phasing estimations in the absence of LD. The low accuracy of these phasing algorithms, especially in populations with very low linkage disequilibrium, is often not adequately taken into account, like in the data provided in the 1000 Genomes project.

The detection of specific recombination events in a set of sequences allows the study of "junctions" in the genome (as in the classical approach by Fisher) as genetic markers and opens the door to the study of recent population events (Melé et al. 2010). In fact, at the beginning of modern genetics, Ronald Fisher postulated that beyond the diversity given by the allelic state of genetic variation (a change of one nucleotide in the case of a SNP), the way in which the successive pieces of DNA had been assembled by recombinations could be used as genetic characteristics of a sequence. Nonetheless, this approach has not been perused due to the difficulties of recognizing these junctions. To test this hypothesis, it is necessary to reconstruct the ARG (ancestral recombination graph), which is a graph that tries to reconstruct the gene genealogy of a given amount of linked variants not only by mutation but also by recombination.

The inference of past recombination events and the reconstruction of the ARG have been addressed, for example, through the IRIS software (Melé et al. 2010). This software uses the patterns of adjacent SNPs created by linkage disequilibrium to infer past recombination events by means of a combinatory as well as statistical algorithm based on pattern-switch recognition. Through this and other methods, it is possible to infer recombination events with the specific position and the parental haplotypes. Recombination events inferred though these methods strongly correlate with recombination rates inferred through methods based on linkage disequilibrium and on sperm typing (Fig. 6.2). The presence or absence of a specific recombination event can be taken as a genetic marker, opening the way to consider the whole set of recombinations in a chromosome as a set of consecutive markers, which have been named "recotypes" that can be analyzed with the same toolkit available for SNPs and haplotypes. The utilization of recombination events as genetic markers is not only useful for human population genetics but also for achieving a deeper understanding of how recombination shapes genomes. Figure 6.3 shows the coalescence with recombination of the extant genetic variation for Africa (in blue) and Out of Africa (Europe and East Asia). The small gray spheres represent coalescent nodes and blue spheres indicate recombinant nodes. Red spheres are the recombinant nodes that have been inferred. In all, it is a quite complete reconstruction of the history of a genome fragment, including its demographic history.

6.2 Linkage Disequilibrium and a Population View

We have seen that LD is a population concept that enriches the most used concepts of genome variation. In fact, haplotypes contain more information than SNPs; nonetheless, their use is hampered by the difficulties of defining their length in terms of the number of variant positions they encompass.

Much effort has been concentrated on describing genome variation among human populations based on single nucleotide variants, but very few studies are based on the information that the whole process of recombination can provide. A single genetic distance could be much more informative than just the independent information of each SNP. It is interesting to mention a very efficient study in the analysis of a worldwide survey of high-density SNPs (Jakobsson et al. 2008), in

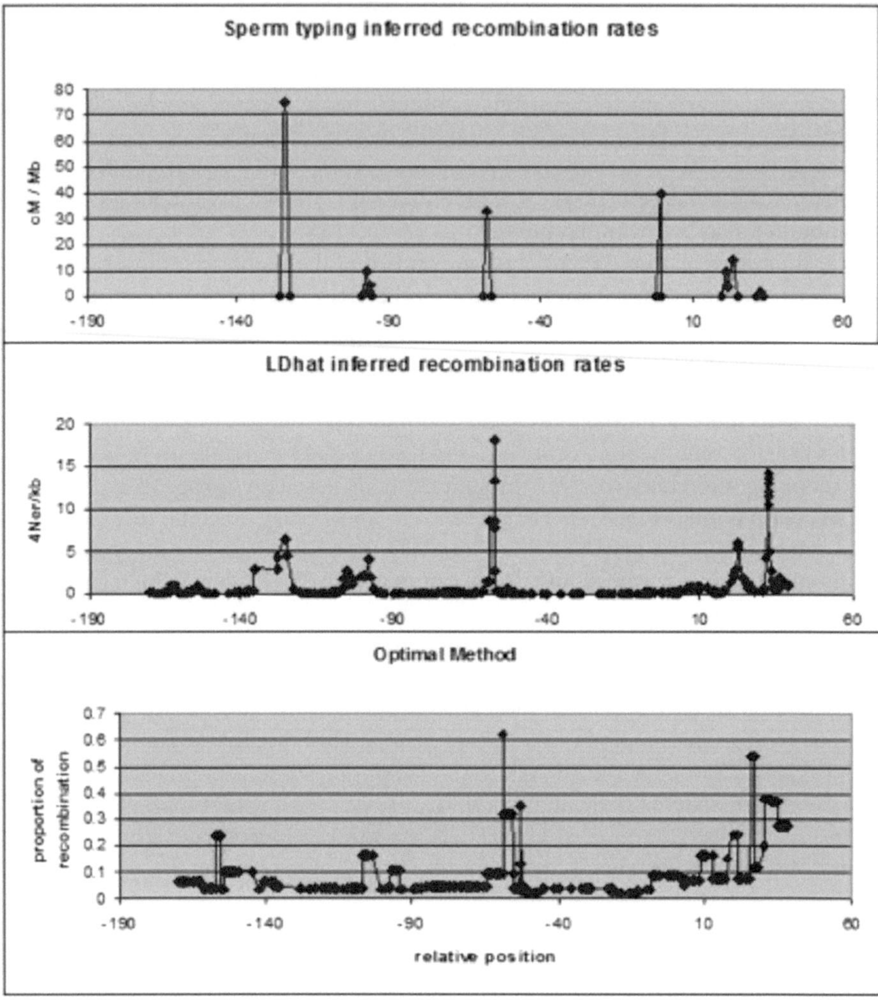

Fig. 6.2 Recombination rates obtained by three different methods: direct observation by sperm typing (top), inferred through linkage disequilibrium using the program LDhat (center), and reconstructing the ancestral recombination graph and, from it, counting the recombinations that have taken place (bottom). Produced for a region of chromosome 1, near MS32 minisatellite, with 365 SNPs and using 120 sequences from HapMap CEU population. A very strong correlation is observed among the methods

which the haplotype information proved to be more reliable than the independent information given by SNPs. This approach should be further developed and used in population genetics studies.

It is becoming clearer that recombination rates vary among populations, and in fine detail differences (like the presence/absence of recombination hotspots in specific populations) may become a powerful tool for population genetics. As an

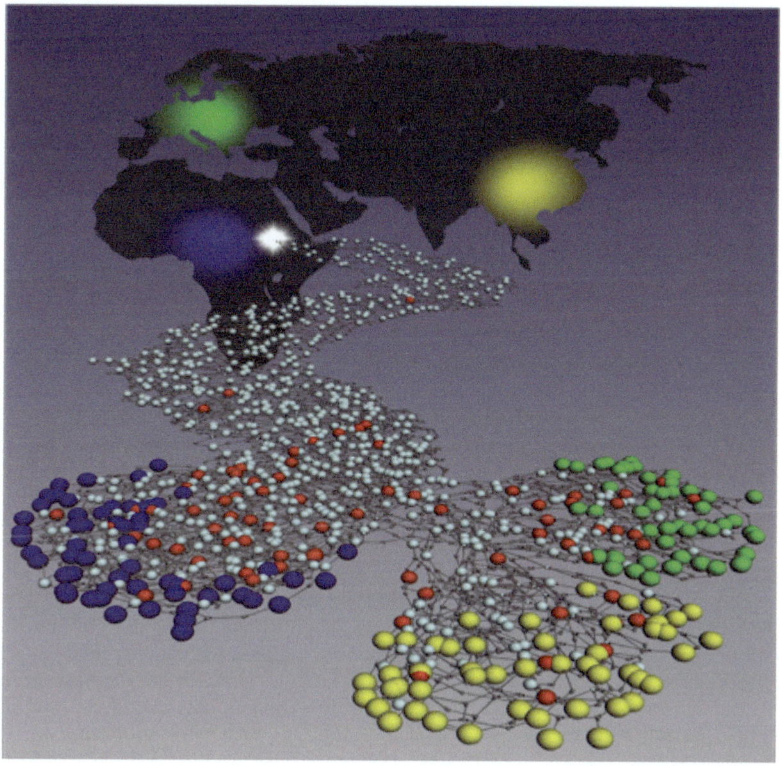

Fig. 6.3 A representation of the evolutionary diversification of the human genome since the origin of modern humans. The Out of Africa is clear and both mutations and recombinations have been considered

example, Fig. 6.4, with data extracted from Li et al. (2008), showed that the recombination landscapes vary among populations and continental groups: some hotspots are population specific and others follow a continental pattern, in agreement with the pace of the expansion of modern humans. The wealth of information contained in the recombination footprint in the genome has hardly been used in human population genetics and deserves further attention.

Differences in recombination rates among human populations provide a useful temporal framework to analyze the evolution of the recombination landscape, which is recent enough to capture fast evolutionary changes. The basal branches of the genetic diversification of human populations happened some 150,000 years ago, a much shorter time than the split between humans and chimpanzees (around 6 million years). The comparison of the recombination patterns among human populations provides a means to verify whether recombination landscapes evolve over time. To address this issue, Laayouni et al. (2011) analyzed whether differences in recombination rates among human populations are correlated with

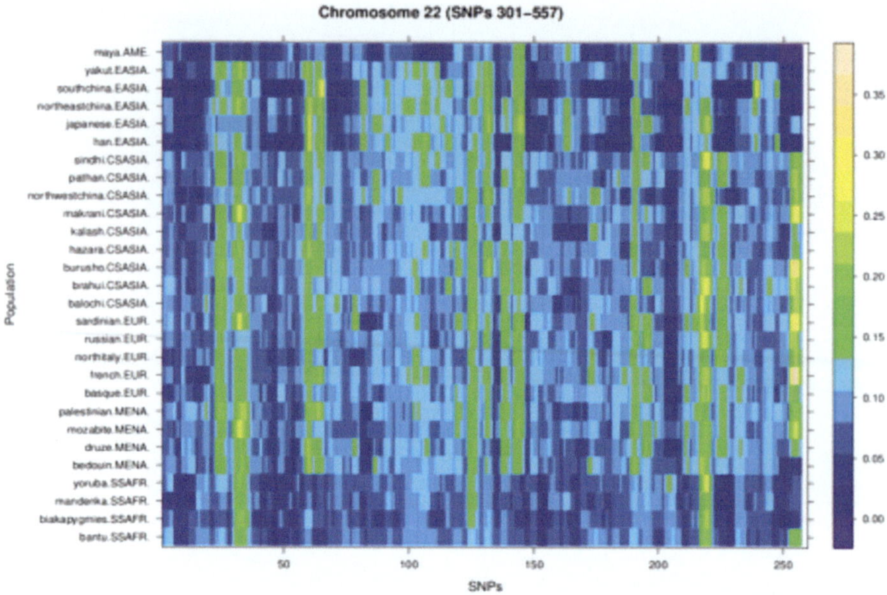

Fig. 6.4 Heatmap showing recombination rates, including recombination hotspots, observed for 258 SNPs of chromosome 22 for 28 human populations, grouped according to their geographical region (*AME*, Americas; *EASIA*, Eastern Asia; *CSASIA*, Central and South Asia; *EUR*, Europeans; *MENA*, Middle East and North Africa; *SSAFR*, sub-Saharan Africa). The 258 SNPs of chromosome 22 are reported on the X axis. In color, for each population, the value of the recombination estimate ($4N_er$/kb) (corrected for effective population size) for that SNP is shown by a gradient from blue (low recombination values) to green (high recombination values). This figure shows that recombination rates vary not only along the genome (with clear hotspots) but also among continental groups and, to a lesser extent, among populations. Some recombination hotspots are strong in some populations and do not exist in others. This plot indicates the complexity of the distribution of recombination rates and subsequently of linkage disequilibrium and haplotype distribution which are key factors for interpreting the genetic diversity data in rich and robust terms of population history

their genetic differences computed as genetic distances. Results show a positive and strong correlation of 0.894 ($p < 0.0001$) indicating that differences in recombination rates among populations increase with their genetic distance. It is a clear indicator of the pace of evolution of recombination, which in longer time periods (like that of separation of humans and chimpanzees) would give a totally different recombination landscape along the genome.

Recombination rate appears to be a rapidly changing parameter, indicating that the underlying factors shaping the likelihood of a recombination event, such as DNA sequences controlling recombination rate variation, also change. This shows that recombination is not a fixed feature of the genome of a species but a phenotype with broad genetic variation.

6.3 The Role of Linkage Disequilibrium in Reconstructing Our Past and the Origins of Modern Humans

Traditionally, most inferences in human population genetics have been based on the non-recombining mtDNA and NRY, while the action of recombination in autosomal and X-chromosome data has been considered a hurdle. However, it is possible to study human population diversity using recombination events as genetic markers. In fact, the fragmentation produced by recombinations in the genome produces a mosaic structure of haplotypes rich in information that has yet to be fully used. This analysis focuses on reconstructing the recombination events, using haplotypes instead of SNPs to encompass the genetic diversity, and the direct use of recombination rates and measures of LD.

A paramount accomplishment of genetics has been to provide an incredible source of data from which the demographic history of modern human populations can be deeply investigated. A landmark achievement in this field has been the discovery of the African origin of modern humankind, which was followed by the "Out of Africa" migration, leading to the spread of modern human populations over the whole world. A major piece of evidence of the African origin is that the genetic variability of all non-African populations is reduced and comprised of African populations, seen clearly in mtDNA (Behar et al. 2008) and Y-chromosome (Underhill and Kivisild 2007) studies. Reconstructing the demographic history of African populations before the migration, however, remains a difficult task. Indeed genetic studies of African populations, carried out with different methods (uniparental markers (mtDNA and Y chromosome), microsatellites, SNP genotyping arrays, or whole genome sequences), have revealed a high level of complexity in the structure of African populations, suggesting a complicated intertwining of human groups, some of which have been recognized in their present form, thanks to distinctive forms of leaving and languages. In fact, the complex history of human groups before the "Out of Africa" migration (Tishkoff et al. 2009; Sikora et al. 2011) makes it arduous (or even nearly impossible) to ultimately identify the ancestral genetic pool of modern humans and its region of origin. Genetic data analyzed so far have generally supported the hypothesis that the most ancient population of modern humans originated in eastern Africa, mainly because populations outside of Africa carry a subset of the genetic diversity found in eastern Africa, along with tentative support from the fossil record. The issue of finding the ancestral genetic pool of modern humans and thus localizing their place of origin within the African continent has been readdressed (Henn et al. 2011) and postulates an origin of modern humans in South Africa, with a shared common ancestry of all hunter-gatherer populations distinct from agriculturalists.

Since the pioneering work of Cavalli-Sforza, genetic data have provided formidable evidence to elucidate relationships between extant populations and reconstructed past events of their ancestors, making it possible to infer many aspects of ancient population history such as expansion and migration events, admixture, gene flow, and dramatic changes in population size. Among the tools that have been used, recombination has an interesting role. The key question is to which extent genetic footprints of past historical and demographic events allow us to disentangle the "source" population from those stemming from it at different times during the history of the species. In situations where the evolutionary process is not simply a linear dynamic process of growth, dispersion, and fusion and fission in which the initial characteristics would be much conserved, a founder situation in genetic terms has to reshape the original characteristics, a kind of "zero point" of the process. This can only be created by an offshoot of a small number of individuals and would have two main characteristics: reduction of genetic diversity and creation of linkage disequilibrium. Thus, we should look for, as a proxy for ancestral human populations, those groups having more genetic diversity and less LD. High levels of genetic diversity for a given population may be simply found as a high number of nucleotide variants or a higher heterozygosity for a genome region, though it is something difficult to assess if done through the genotyping of already known variant positions or SNPs and not through retrieving all the variants via DNA sequencing. In the case of genotypes, instead of using direct measures of gene diversity, it is more useful to use the information of genetic distances among populations, as those populations less-derived (and thus, closer to the ancestral one) will have higher differences (measured by the standard Fst statistic, easily understood if the differentiation process is mainly produced by drift) than the rest. In this case it is assumed that the amount of variation in the pre-existing African populations is higher than that produced in non-Africans due to the differentiation process of time and drift, and the data for worldwide surveys confirms this.

But the use of LD enhances our power to understand the past. No doubt the LD differences between Africans and non-Africans (descendants of the Out of Africa) are dramatic, with strong LD landscape outside Africa and low levels in South Saharan Africa. The African analysis done by Henn et al. (2011) makes an interesting use of LD data: low values of LD reflect higher effective population sizes and, therefore, denote a more ancient population because it is closer to the equilibrium than any derived population. By regressing LD estimates for the different African populations with geographical coordinates, they are able to localize a hypothetical point of origin of modern humans as the region for which this correlation is highest; and this region has been identified as South Africa.

There have been other attempts to use recombination to reconstruct the human past. As the number of past recombination events in a population sample is a function of its effective population size (Ne), it has been possible to detect specific past recombination events in Old World populations to infer their Ne. Results show that sub-Saharan African populations have an Ne that is ~4 times greater than those of non-African populations, and outside of Africa, the South Asian populations had the largest Ne. It has also been possible to use this information to reconstruct the "Out of Africa" route into Eurasia. Observing how the patterns of recombination diversity of the Eurasian populations correlates with distance from Africa, which shows a clearly significant correlation with that distance measured along a path crossing South Arabia, while no such correlation is found through the Sinai route (usually taken as the route of the Out of Africa), suggesting that anatomically modern humans first left Africa through the Bab-el-Mandeb Strait rather than through present Egypt (Melé et al. 2012).

The full use of recombination and linkage disequilibrium data in population genetics is still in its infancy, and many more possibilities are waiting to be uncovered and fully used. Its full development requires good algorithms for detection of recombination points, a close approach to the ancestral recombination graph, and a high-resolution description of the genome, something that is being achieved with the third generation of sequencing technologies that allow obtaining long DNA sequences (and haploid sequences) of single molecules. Thus, not only the genetic variation is of interest but also how that variation is organized along the chromosomes, the departure point of making diploid individuals and population genome pools.

Acknowledgments Ongoing work on evolutionary genetics is supported by grant BFU2016-77961-P (AEI/FEDER, UE) awarded by the Ministerio de Economía y Competitividad (Spain) and with the support of the Secretaria d'Universitats i Recerca del Departament d'Economia i Coneixement de la Generalitat de Catalunya (GRC 2014 SGR 866).

References

Baudat F, Imai Y, de Massy B (2013) Meiotic recombination in mammals: localization and regulation. Nat Rev Genet 14(11):794–806

Behar DM, Villems R, Soodyall H, Blue-Smith J, Pereira L, Metspalu E, Scozzari R, Makkan H, Tzur S, Comas D, Bertranpetit J, Quintana-Murci L, Tyler-Smith C, Wells RS, Rosset S, Genographic Consortium (2008) The dawn of human matrilineal diversity. Am J Hum Genet 82(5):1130–1140

Henn BM, Gignoux CR, Jobin M, Granka JM, Macpherson JM, Kidd JM, Rodríguez-Botigué L, Ramachandran S, Hon L, Brisbin A, Lin AA, Underhill PA, Comas D, Kidd KK, Norman PJ, Parham P, Bustamante CD, Mountain JL, Feldman MW (2011) Hunter-gatherer genomic diversity suggests a Southern African origin for modern humans. Proc Natl Acad Sci U S A 108(13):5154–5162

Jakobsson M et al (2008) Genotype, haplotype and copy-number variation in worldwide human populations. Nature 451(7181):998–1003

Kong A, Thorleifsson G, Gudbjartsson DF, Masson G, Sigurdsson A, Jonasdottir A, Walters GB, Jonasdottir A, Gylfason A, Kristinsson KT, Gudjonsson SA, Frigge ML, Helgason A, Thorsteinsdottir U, Stefansson K (2010) Fine-scale recombination rate differences between sexes, populations and individuals. Nature 467(7319):1099–1103

Laayouni H, Montanucci L, Sikora M, Melé M, Dall'Olio GM, Lorente-Galdos B, McGee KM, Graffelman J, Awadalla P, Bosch E, Comas D, Navarro A, Calafell F, Casals F, Bertranpetit J (2011) Similarity in recombination rate estimates highly correlates with genetic differentiation in humans. PlosOne 6(3):e17913

Li JZ, Absher DM, Tang H, Southwick AM, Casto AM, Ramachandran S, Cann HM, Barsh GS, Feldman M, Cavalli-Sforza LL, Myers RM (2008) Worldwide human relationships inferred from genome-wide patterns of variation. Science 319:1100–1104

Melé M, Javed A, Pybus M, Calafell F, Parida L, Bertranpetit J, Genographic Consortium (2010) A new method to reconstruct recombination events at a genomic scale. PLoS Comput Biol 6 (11):e1001010

Melé M, Javed A, Pybus M, Zalloua P, Haber M, Comas D, Netea MG, Balanovsky O, Balanovska E, Jin L, Yang Y, Pitchappan RM, Arunkumar G, Parida L, Calafell F, Bertranpetit J, Genographic Consortium (2012) Recombination gives a new insight in the effective population size and the history of the old world human populations. Mol Biol Evol 29 (1):25–30

Paigen K, Petkov P (2010) Mammalian recombination hot spots: properties, control and evolution. Nat Rev Genet 11(3):221–233

Sikora M et al (2011) A genomic analysis identifies a novel component in the genetic structure of sub-Saharan African populations. Eur J Hum Genet 19(1):84–88

Tishkoff SA, Reed FA, Friedlaender FR, Ehret C, Ranciaro A, Froment A, Hirbo JB, Awomoyi AA, Bodo JM, Doumbo O, Ibrahim M, Juma AT, Kotze MJ, Lema G, Moore JH, Mortensen H, Nyambo TB, Omar SA, Powell K, Pretorius GS, Smith MW, Thera MA, Wambebe C, Weber JL, Williams SM (2009) The genetic structure and history of Africans and African Americans. Science 324(5930):1035–1044

Underhill PA, Kivisild T (2007) Use of y chromosome and mitochondrial DNA population structure in tracing human migrations. Annu Rev Genet 41:539–536

Chapter 7
CNVs and Microsatellite DNA Polymorphism

Naoko Takezaki

Abstract Microsatellite DNAs are short tandem repeats (often abbreviated as STR) of nucleotides (1–6 bp) and abundant in eukaryotic genomes. Microsatellites are used as genetic markers for population studies and forensics because of the high mutation rate. Recently, copy number variations (CNVs) of DNA segments of a kilobase to a few megabases are found to be prevalent and cover about 10% of the human genome. This chapter describes (1) the mutational pattern of microsatellites and the application for evolutionary studies of populations, (2) the distribution of CNVs in human genome and its relationship with multigene families of chemosensory receptor genes, and (3) the evolutionary pattern of microsatellites and CNVs and their effects on genome size variation in human populations by taking microsatellites as a neutral model of CNV evolution.

Keywords Mutational pattern · Genetic marker · Genetic distance · Genetic differentiation · Chemosensory receptor gene · Genome size · STR

7.1 Microsatellite DNA in Human Genome

7.1.1 Genomic Distribution

Microsatellite DNAs are tandem repeats of short nucleotide sequence (1–6 bp). Microsatellites consist of 3% of the human genome (International Human Genome Sequencing Consortium 2001) and are widely used as genetic markers for population studies and forensics because of the high mutation rates (10^{-2}–10^{-6} per locus per generation). Microsatellites are also known to be a cause of more than 40 hereditary diseases (Pearson et al. 2005). A high extent of mutation in microsatellites (microsatellite instability) is observed in certain types of cancer (Sideris and Papagrigoriadis 2014).

N. Takezaki (✉)
Life Science Research Center, Kagawa University, Takamatsu, Japan
e-mail: takezaki@med.kagawa-u.ac.jp

N. Saitou (ed.), *Evolution of the Human Genome I*, Evolutionary Studies,
https://doi.org/10.1007/978-4-431-56603-8_7

Most of microsatellites are located in intergenic and intronic regions because exonic regions are only up to 3% of the whole human genome (Alexander et al. 2010). In these regions densities of microsatellites with small repeat unit size tend to be higher than those of microsatellites with large unit size (Toth et al. 2000). In the human genome, di- and mononucleotide repeats occupy the largest and the second largest regions. However, the region that trinucleotide repeats occupy is smaller than those of tetra-, penta-, and hexanucleotide repeats. In exonic regions density of microsatellites is generally lower than those in intergenic and intronic regions because in protein-coding region, repeat number change of microsatellites often results in frameshift that causes truncation or a loss of function of the coded protein. However, density of trinucleotide repeats is much higher than those of repeats of the other unit sizes in exonic region as expected, because repeat number change of trinucleotides does not shift the reading frame.

7.1.2 Mutational Pattern

Expansion or contraction of microsatellite DNA by change of the repeat number often occurs at mutation through mechanism called replication slippage (Ellegren 2004). During DNA replication pausing of DNA polymerase takes place momentarily, and the replication complex dissociates from the DNA. On the dissociation, the newly synthesized strand can separate from the template strand and mispairing of repeats in the two strands is likely to occur. Mispairing that involves in the nonpairing region of the daughter strand or the template strand will cause insertion or deletion of repeats, respectively.

Because of the frequent repeat number changes, mutational pattern of microsatellite DNA roughly follows the stepwise mutation model (SMM) (Ohta and Kimura 1973), which was originally developed for electrophoretically identified isozymes (Di Rienzo et al. 1994; Ellegren 2000, 2004; Schlötterer 2000). In the SMM, expansion and contraction of microsatellites by one repeat occur with equal probability, and the mutation rate is assumed constant for microsatellites with different repeat numbers. However, the mutational pattern of microsatellites is actually complex and it deviates from the SMM. These deviations include multistep changes of repeat number, an upper limit of repeat numbers, unequal rate of expansion and contraction, and dependency of mutation rate on repeat number, as will be described in the following. Furthermore, the mutation rate also varies with size and compositional motif of the repeat unit (Chakraborty et al. 1997; Lai and Sun 2003; Kelkar et al. 2008). Microsatellites with shorter repeat unit sizes tend to have higher mutation rate (Lai and Sun 2003; Kelkar et al. 2008). Previously this tendency was found in a study with human population data (Chakraborty et al. 1997). Mononucleotide repeats with A/T motif (grouping complementary motifs together) are much more abundant and longer than G/C motif and have higher mutation rate than those with G/C motif unless the number of G/C repeat is >18. Dinucleotide repeats with AT/TA, AC/CA/GT/TG, and AG/GA/CT/TC motifs (grouping motifs that appear by shifting the starting position of repeats together in addition to complementary ones) are more abundant and longer than those of CG/GC motif, and the mutation rates of the former kinds are higher than that of the latter (Kelkar et al. 2008).

There are some multistep changes of repeat number. Frequency of multistep change is <15% in most experimental studies of human data (Ellegren 2004). To take into account multistep changes, the two-phase model that allows multistep change with a certain probability (Di Rienzo et al. 1994) and the general multistep mutation model in which any number of repeat change from $-m$ to m can occur with mean 0 and variance $m/2$ (Chakrabory and Nei 1982; Kimmel et al. 1996) were applied. Because of the multistep change, the mutational pattern of microsatellites deviates from the SMM to the direction of the infinite allele model (IAM) (Kimura and Crow 1964) in which a new allele always appears at mutation (Shriver et al. 1993; Valdes et al. 1993). The probability that multistep change occurs estimated by assuming the two-phase model for human population data (0–0.2) (Di Rienzo et al. 1994) is in agreement with those in the experimental studies.

In the SMM, there is no stationary distribution for repeat numbers. However, in microsatellites there seems to be an upper limit for the repeat numbers (Garza et al. 1995), and mutation rate is higher for microsatellites with a larger repeat number (Ellegren 2000, 2004; Kelkar et al. 2008). In order to take into account the equilibrium distribution of repeat number of microsatellites and higher mutation rate for a larger repeat number, the Markov chain model was developed. In this model the equilibrium distribution is obtained by the balance of repeat number change by replication slippage and disruption of repeats by point mutation (Kruglyak et al. 1998). This model assumes one-step change of repeat number, equal mutation rate for expansion and contraction, and linear rate for the repeat number (constant per a repeat).

The mutational rates of expansion and contraction vary for different repeat numbers. In an early human germline study of dinucleotide repeats, expansion exceeded contraction (Ellegren 2000). In a large-scale human pedigree study, the rate of contraction increased with the number of repeats of loci, whereas the rate of expansion was independent of the repeat number (Xu et al. 2000). However, in other studies the rate of contraction was higher than that of expansion for alleles with a large number of repeats, whereas the rate of expansion was higher than that of contraction in alleles with a small number of repeats (Huang et al. 2002; Sun et al. 2012). These observations support the idea that the rate of expansion is higher than that of contraction when the repeat number is lower than a certain target value and the rate of contraction is higher than that of expansion when the repeat number is higher than the target value (Garza et al. 1995).

Modifications of the Markov chain model (Kruglyak et al. 1998) were developed to incorporate different expansion and contraction rates for repeat numbers. In these models slippage rate can have a linear, exponential, quadratic, or piecewise relationship with the repeat number of microsatellites and can be asymmetric to expansion and contraction (Calabrese and Durrett 2003; Whittaker et al. 2003; Sainudiin et al. 2004) or no particular relationship with the repeat number (Lai and

Sun 2003). For dinucleotide repeats in human genomic data, asymmetric quadratic rates or piecewise biased rates best fitted, and contraction rate exceeded expansion rate at repeat number 24 and 14 for repeat unit motifs AC/CA/GT/TG and AG/GA/CT/TC, respectively, and at repeat number $\geq$10 for repeat unit motif AT/TA (Calabrese and Durrett 2003). Another study using genomic sequences of human and chimpanzee estimated this focal repeat number at which contraction rate exceeds expansion rate for dinucleotide repeats mostly consisting of repeat unit motifs AC/CA/GT/TG as 14–18 (Sainudiin et al. 2004). An experimental study using human cell lines showed the focal point for motifs CA/GT and AG/TC as 20 repeats (Baptiste et al. 2013) within a similar range to those in the computational studies. Lai and Sun (2003) estimated for human genome data that the slippage rate exponentially increases for longer microsatellites and that expansion rates are larger than contraction for shorter microsatellites and smaller for longer microsatellites.

The slippage rate at microsatellites becomes higher as the repeat number increases and exceeds that of non-repetitive genomic region at a certain repeat number (e.g., Messier et al. 1996; Rose and Falush 1998). Lai and Sun (2003) estimated the threshold values of repeat number at which replication slippage starts to occur more often than the genome average as 9 for mononucleotide repeats and 4 for repeats with the other unit sizes. Ananda et al. (2013) estimated the threshold values as 8, 5, 4, and 4 for mono-, di-, tri-, and tetranucleotide repeats, respectively, using human polymorphism data.

7.1.3 Microsatellite DNA as Genetic Markers

Microsatellites are widely used as genetic markers to study relationships among populations. The extent of genetic differentiation between populations can be measured by genetic distances (Nei and Kumar 2000). The genetic distance $(\delta\mu)^2$ was developed for microsatellites by assuming the SMM (Goldstein et al. 1995). $(\delta\mu)^2$ is defined as follows.

$$(\delta\mu)^2 = \sum_{j}^{r} \frac{\left(\mu_{X_j} - \mu_{Y_j}\right)^2}{r}$$

where r is the number of loci examined, $\mu_{X_j}\left(= \sum_i ix_{ij}\right)$ and $\mu_{Y_j}\left(= \sum_i iy_{ij}\right)$ are the average number of repeats of alleles at the jth locus, and x_{ij} and y_{ij} are the frequencies of the ith allele at the jth locus in populations X and Y, respectively. Under the SMM with mutation-drift balance where the population size is constant and there is no effect of gene flow between them and from other populations for a long time (or an inverse of mutation rate), $\mathrm{E}[(\delta\mu)^2] = 2vt$ where v and t are the mutation rate per locus per generation and time after the two populations diverged in the unit of generation, respectively. It indicates that under this condition, the expected value of $(\delta\mu)^2$ increases linearly with time after the populations diverge.

The $(\delta\mu)^2$ distances calculated for 30 microsatellite loci of human populations (Bowcock et al. 1994) fit the linear relationship quite well with the divergence time known from archaeological records such as 43,000 years ago for European versus East Asian and Amerind $[(\delta\mu)^2 = 2.07]$ and 100,000 years ago for African and non-African $[(\delta\mu)^2 = 6.47]$. The mutation rate for this data was estimated as $7.96 \times {}^{-4}$ per locus per generation by assuming that generation time is 27 years (Goldstein et al. 1995).

The $(\delta\mu)^2$ distance calculated for 145 loci taken out of genotypic data of 783 loci generated by Ramachandran et al. (2005) was 1.24 between African and non-African and 0.60 between European and Asian (Takezaki and Nei 2009). These values indicate that the mutation rate of these loci is one-fourth of Bowcock et al.'s (1994) data. Approximately 80% of the 145 loci were tetranucleotide repeat loci, whereas 30 loci of Bowcock et al. (1994) mostly consist of dinucleotide repeat loci. Thus, the mutation rate of tetranucleotide repeat loci is smaller than that of dinucleotide repeat loci in agreement with the results of a population study (Chakraborty et al. 1997) and studies using human genome sequence (Lai and Sun 2003; Kelkar et al. 2008).

Standard genetic distance (D_S) was originally developed for classical markers such as blood type groups and isozymes, by assuming the IAM (Nei 1972). It is given by

$$D_S = -\ln \frac{J_{XY}}{\sqrt{J_X J_Y}},$$

where $J_X = \sum_j^r \sum_i^{m_j} x_{ij}^2/r$, $J_Y = \sum_j^r \sum_i^{m_j} y_{ij}^2/r$, $J_{XY} = \sum_j^r \sum_i^{m_j} x_{ij}y_{ij}/r$. x_{ij} and y_{ij} are frequencies of the ith allele at the jth locus of populations X and Y, respectively. m_j is the number of alleles at the jth locus, and r is the number of loci examined.

Under the IAM with mutation-drift balance, the expected value of D_S has a linear relationship with the time after the population divergence $[\mathrm{E}(D_S) = 2vt]$. For the 145 microsatellite loci of Ramachandran et al. (2005), the values of D_S were 0.24 between African and non-African and 0.10 between European and Asian (Takezaki and Nei 2009). The values of D_S also appear to be proportional to the divergence time of populations. It agrees with that the mutational pattern of microsatellites deviates from the SMM to the direction of the IAM (Shriver et al. 1993; Valdes et al. 1993).

It is informative to construct a phylogenetic tree to infer the evolutionary relationship of populations. For this purpose one can estimate genetic distance measures from allele frequency data and construct phylogenetic trees with the distance values by the tree-making method such as the neighbor-joining method (Saitou and Nei 1987) or UPGMA (Sneath and Sokal 1973). The expected values of $(\delta\mu)^2$ and D_S increase linearly with time after the population divergence under mutation-drift equilibrium in the SMM and the IAM, respectively. However, because of the large sampling error and the limited number of loci and samples examined in actual data, the probability of obtaining a correct tree topology is not high in the case where these distance measures are used (Takezaki and Nei 1996, 2008). Chord distance (Cavalli-Sforza and Edwards 1967)

$$D_C = \left[\frac{1}{\pi r}\sum_j^r \left[2\left(1 - \sum_i^{m_j} \sqrt{x_{ij}y_{ij}}\right]^{1/2}\right]$$

and D_A distance (Nei et al. 1983)

$$D_A \left[= 1 - \frac{1}{r}\sum_j^r \sum_i^{m_j} \sqrt{x_{ij}y_{ij}}\right],$$

which are based on the angular transformation between two populations located on the multidimensional hypersphere, are more efficient in constructing population trees because of the small sampling error.

The extent of genetic differentiation among subdivided populations can be measured by G_{ST}, which can be defined as

$$G_{ST} = \frac{H_T - H_S}{H_T}$$

where H_T is the heterozygosity for the entire population and H_S is the average heterozygosity within subpopulations (Nei 1977). However, G_{ST} can have a small value even when there are no shared alleles between populations if mutation rate is high (Nei and Kumar 2000). Standardized G_{ST} (Hedrick 2005), G_{ST} divided by its maximum value ($1 - H_S$), and Jost's (2008) D, which is approximately $G_{ST} \times H_T/(1 - H_S)$ in the case where there are a large number of subpopulations, were proposed to alleviate the problem. These measures may be more appropriate for high variability data such as microsatellite DNA (Meirmans and Hedrick 2011).

7.2 CNV on Human Genome

7.2.1 What Is CNV

Genome-scanning technologies uncovered a substantial amount of copy number variants (CNV) of DNA segments of a kilobase to a few megabases caused by duplication, insertion, deletion, etc. (e.g., Tuzun et al. 2005; Feuk et al. 2006; Redon et al. 2006). They cover about 10% of the human genome (Redon et al. 2006; Wong et al. 2007). CNVs are known to generate some complex medical disorders such as cancer and autoimmune diseases (e.g., Fanciulli et al. 2007). In CNV regions genes involved in extracellular processes such as cell adhesion, recognition, and communication are overrepresented, whereas genes involved in intracellular processes such as biosynthetic and metabolic pathways are underrepresented (e.g., Conrad et al. 2010). CNVs often occur in regions that contain multigene families such as olfactory receptor genes, major histocompatibility complex class III genes, and β-defensin antimicrobial genes (Freeman et al. 2006).

7.2.2 *Variation of CNVs in Human Population*

Redon et al. (2006) screened for CNV regions in 270 healthy individuals in HapMap2 Project (The International HapMap Consortium 2007), which consist of 90 Africans (30 parent-child trios from Yoruba, Nigeria), 90 European descent (30 parent-child trios from Utah, USA), and 90 Asians (45 Japanese from Tokyo, Japan, and 45 Han Chinese from Beijing, China). They identified 1447 CNV regions in total of 360 megabases covering 12% of the human genome.

Within these CNV regions, Nozawa et al. (2007) identified 3144 genes out of 22,218 human protein-coding genes annotated in the Ensembl database (Hubbard et al. 2007). It suggests that about 14% of human protein-coding genes are polymorphic with respect to the copy number. The frequency distribution of the relative gene copy number from the reference individuals of the 270 individuals revealed a large gene copy number difference between individuals. The standard deviation (SD) of this distribution was 54.4, indicating that it is not rare that there is more than 100 gene copy number difference between two individuals. The average of the gene copy number difference between two individuals relative to all gene number in the human reference genome was 0.28% (61.5/22,218). This is much higher than the difference of single nucleotide polymorphisms between two randomly chosen haploid genomes (about 0.1%) (e.g., Levy et al. 2007).

The average relative gene copy number of Africans (61.3) was significantly higher than those of Europeans (16.3) and Asians (14.8). The SD of Africans (52.4) was also larger than those of Europeans (46.0) and Asians (48.3). This indicates that the genetic variation in Africans is larger than those in non-Africans, consistent with African origin hypothesis of modern humans (Cann et al. 1987).

7.2.3 *CNVs and Chemosensory Receptor Genes*

CNV regions often contain genes involved in sensory perceptions. Chemosensory receptors for sensing taste and odors are encoded by multigene families in vertebrate genomes. In the human genome, there are about 35 bitter taste receptor genes (T2R), though one-third are pseudogenes. Particularly olfactory receptor (OR) genes compose a large gene family with over 800 genes including about 400 pseudogenes (Nei et al. 2008).

In the CNV of HapMap samples, the proportion of polymorphic genes for functional genes and pseudogenes are 30% and 35% for OR genes and 50% and

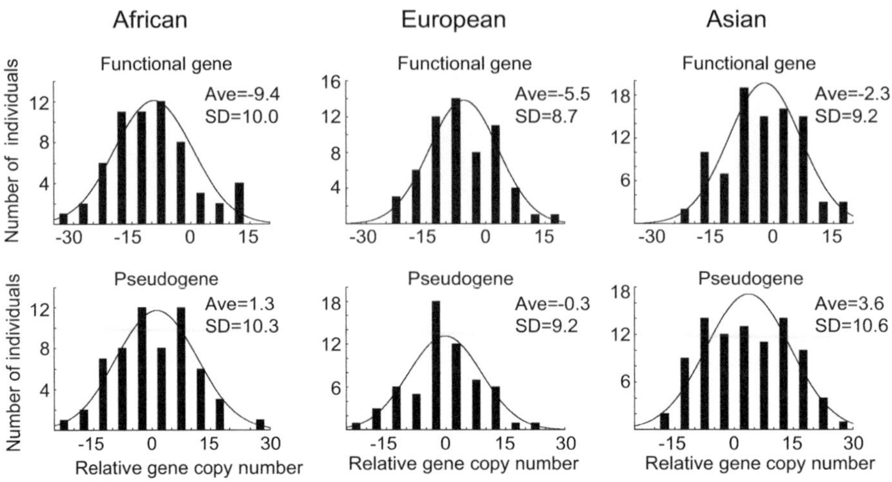

Fig. 7.1 Frequency distribution of copy number for OR genes in African, European, and Asian populations (From Nozawa et al. 2007)

80% for T2R genes, respectively (Nozawa et al. 2007). The higher proportions of polymorphic genes in pseudogenes than in functional genes suggest that purifying selection operates on functional genes of OR and T2R genes, because pseudogenes are expected to have no function and evolve neutrally. However, the difference is not statistically significant, and the average copy number differences between two individuals for both functional genes and the pseudogenes are both 11 for OR genes and 1–2 for T2R genes. In addition, the average differences of the copy number between two individuals for the functional OR genes and T2R genes (2.8–9.2%) are much higher than that for all the protein-coding genes (0.24–0.28%). These may suggest a possibility that even functional OR and T2R genes evolve neutrally rather than under purifying selection.

Figure 7.1 shows the frequency distribution of copy numbers of OR genes for Africans, Europeans, and Asians. The copy numbers of both functional genes and pseudogenes for the three populations are normally distributed and have the similar values of SD. This also suggests that the functional OR genes may evolve neutrally similarly to the pseudogenes and the copy number of functional genes varies randomly (genomic drift). The normal distribution is generated by the birth-death process in probability theory. Therefore, it is possible that the copy number change of both functional OR genes and the pseudogenes is subjected to a random process due to duplication, deletion, insertion, and inactivation.

It should be noted that sign of positive selection was also identified for a few OR genes (e.g., Moreno-Estrada et al. 2008; International HapMap 3 Consortium 2010). Therefore, although many of OR genes may evolve neutrally, it is possible that some OR genes are under positive or purifying selection.

7.3 Microsatellite DNA and CNV

7.3.1 Variation in Human Populations of Microsatellite DNA

Because most of microsatellite DNA are located in intergenic and intronic regions and have no function and the repeat number of microsatellite loci mostly change by one in a random fashion at mutation, we can consider microsatellite loci as a model of genomic drift in which the copy number change occurs more or less randomly.

Figure 7.2 shows the frequency distribution of a total number of repeats (TNR) in an individual at the 145 microsatellite loci for Africans, Europeans, and Asians (data from Ramachandran et al. 2005, which are different from HapMap samples) (Takezaki and Nei 2009). They approximately follow the normal distribution, as expected from the random change of repeat number. The average of TNR (relative value by taking the minimum value as zero) is smaller for Africans (75.1) than for Europeans (95.3) and Asians (81.9) with statistical significance. Because the repeat number change at microsatellite loci can be considered neutral, the differences in

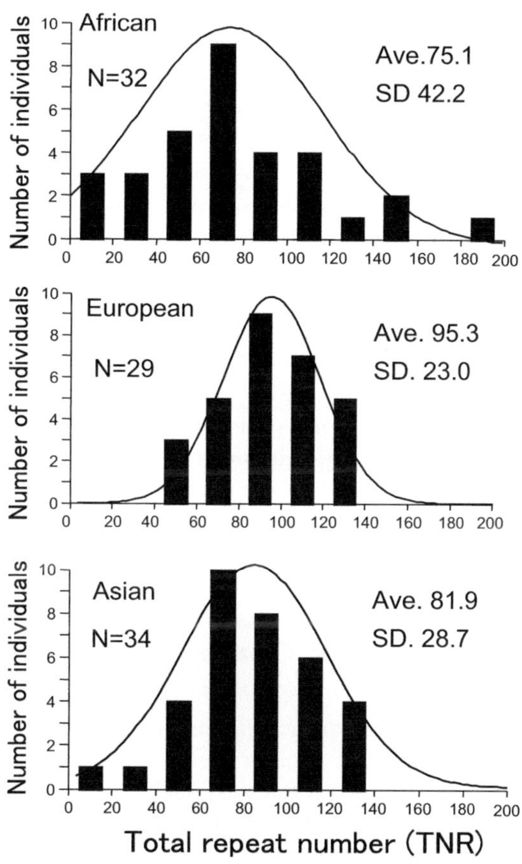

Fig. 7.2 Frequency distribution of total repeat number of 145 microsatellite loci for individuals of African, European, and Asian populations (From Takezaki and Nei 2009)

Table 7.1 MAD of TNR between two randomly chosen individuals within and between populations

Population	African	European	Asian
African	47.0		
European	41.1	30.4	
Asian	40.0	26.8	32.3

From Takezaki and Nei (2009)

Table 7.2 MAD of copy number of OR genes

Category	Between human populations	Between humans and chimpanzee
Functional genes	4.7	16
Pseudogenes	2.4	11

Modified from Nozawa et al. (2007)

the TNRs among populations are caused by the stochastic errors, not due to natural selection. The SD of Africans (42.2) is 50–80% larger than those of Europeans (23.0) and Asians (28.7). This is consistent with African origin of modern humans as in the case of CNVs of all protein-coding genes (Nozawa et al. 2007).

Table 7.1 shows the mean average difference (MAD) of TNR between two randomly chosen individuals within and between the three populations. MAD in Africans (47.0) is about 50% higher than those in Europeans (30.4) and Asians (32.3). This is consistent with the large SD value of the frequency distribution of TNRs for Africans. The MADs between Africans and non-Africans (41.1 for Europeans and 40.0 for Asians) are 1.5 times higher than that between Europeans and Asians (26.8). If we assume that Africans and the common ancestor of Europeans and Asians diverged 100,000 years ago and Europeans and Asians diverged 50,000 years ago, this indicates that the MADs between populations increase with time after the divergence, though the increase is not linearly proportional to time.

Table 7.2 shows the MADs of OR genes. The average MADs among populations are higher for functional genes than for pseudogenes among human populations (4.7 and 2.4) as well as between human and chimpanzee (16 and 11). This suggests that CNV of functional OR genes is under positive selection. Thus, the pattern of CNV in OR genes is complex, and it may not necessarily evolve neutrally, as mentioned earlier.

7.3.2 CNV and Genome Size

The SD of the frequency distribution of TNR of OR genes in Africans is 10 (Nozawa et al. 2007). OR gene consists of about 930 nucleotides (310 amino acids per protein). The SD of the total nucleotide number (TNN) of OR genes [SD(TNN)] can be obtained by multiplying the SD of TNR by 930. Therefore, SD (TNN) $= 930 \times 10 = 9300$ for OR genes.

For the 145 microsatellite loci, the SD of TNR in Africans was 42.2. The variance of TNR for r loci can be written as rV. $(42.2)^2 = 145 \times (\mathrm{SD}_1)^2$ where $\mathrm{SD}_1 = \sqrt{V}$. Therefore, $\mathrm{SD}_1 = 42.2/\sqrt{145} = 3.5$. $[\mathrm{SD(TNN)}]^2 = r\,b^2\,(\mathrm{SD}_1)^2$ where b is the nucleotide length of the repeat unit of microsatellite loci. In the search of microsatellite loci in chromosome 21 using orthologous regions between human and chimpanzee genomes, the numbers of variable loci found are 16,767, 1974, 118, 217, and 51 for mono-, di-, tri-, tetra-, and pentanucleotide repeats, respectively (Takezaki and Nei 2009). By assuming that chromosome 21 consists of about 1% of human genome, the variance of TNN for microsatellites in the whole human genome can be estimated as

$$\begin{aligned}[\mathrm{SD(TNN)}]^2 &= \left(1^2 \times 16,767 + 2^2 \times 1974 + 3^2 \times 118 + 4^2 \times 217 + 5^2 \times 21\right)\\ &\quad \times 100 \times (3.5)^2\\ &= (6034)^2\end{aligned}$$

If we consider the expected difference ($4 \times \mathrm{SD} = 4 \times 6034 = 24{,}136$) between the upper and lower 5% levels of individuals, contribution of the repeat number variation in microsatellite loci to genome size variation is quite large.

References

Alexander RP, Fang G, Rozowsky J, Snyder M, Gerstein M (2010) Annotating non-coding regions of the genome. Nat Rev Genet 11:559–571

Ananda G, Walsh E, Jacob KD, Krasilnikova M, Eckert KA, Chiaromonte F, Makova KD (2013) Distinct mutational behaviors differentiate short tandem repeats from microsatellites in the human genome. Genome Biol Evol 5:606–620

Baptiste BA, Guruprasad A, Strubczewski N, Lutzkanin A, Khoo SJ et al (2013) Mature microsatellites: mechanisms underlying dinucleotide microsatellite mutational biases in human cells. G3 3:451–463

Bowcock AM, Ruiz-Linares A, Tomfohrde J, Minch E, Kidd JR, Cavalli-Sforza LL (1994) High resolution of human evolutionary trees with polymorphic microsatellites. Nature 368:455–457

Calabrese P, Durrett R (2003) Dinucleotide repeats in the Drosophila and human genomes have complex, length-dependent mutation processes. Mol Biol Evol 20:715–725

Cann RL, Stoneking M, Wilson AC (1987) Mitochondrial DNA and human evolution. Nature 325:31–36

Cavalli-Sforza LL, Edwards AWF (1967) Phylogenetic analysis: models and estimation procedures. Am J Hum Genet 19:233–257

Chakraborty R, Kimmel M, Stivers DN, Davidson LJ, Deka R (1997) Relative mutation rates at di-, tri-, and tetranucleotide microsatellite loci. Proc Natl Acad Sci U S A 94:1041–1046

Chakrabory R, Nei M (1982) Genetic differentiation of quantitative characters between populations or species. Genet Res 39:303–314

Conrad DF, Bird C, Blackburne B, Lindsay S, Mamanova L, Lee C et al (2010) Mutation spectrum revealed by breakpoint sequencing of human germline CNVs. Nat Genet 42:385–393

Di Rienzo A, Peterson AC, Garza JC, Valdes AM, Slatkin M, Freimer NB (1994) Mutational processes of simple-sequence repeat loci in human populations. Proc Natl Acad Sci U S A 91:3166–3170

Ellegren H (2000) Heterogeneous mutation processes in human microsatellite DNA sequences. Nat Genet 24:400–402
Ellegren H (2004) Microsatellites: simple sequences with complex evolution. Nat Rev Genet 5:435–445
Fanciulli M, Norsworthy PJ, Petretto E, Dong R, Harper L, Kamesh L et al (2007) FCGR3B copy number variation is associated with susceptibility to systemic, but not organ-specific, autoimmunity. Nat Genet 39:721–723
Feuk L, Carson AR, Schere SW (2006) Structural variation in the human genome. Nat Rev Genet 7:85–97
Freeman JL, Perry GH, Feuk L, Redon R, McCarroll SA, Altshuler DM, Aburatani H, Jones KW et al (2006) Copy number variation: new insights in genome diversity. Genome Res 16:949–961
Garza JC, Slatkin M, Freimer NB (1995) Microsatellite allele frequencies in humans and chimpanzees, with implications for constraints on allele size. Mol Biol Evol 12:594–603
Goldstein DB, Linares AR, Cavalli-Sforza LL, Feldman MW (1995) Genetic absolute dating based on microsatellites and the origin of modern humans. Proc Natl Acad Sci U S A 92:6723–6727
Hedrick PW (2005) A standardized genetic differentiation measure. Evolution 59:1633–1638
Huang Q, Xu FH, Shen H, Deng Y, Liu J et al (2002) Mutation patterns at dinucleotide microsatellite loci in humans. Am J Hum Genet 70:625–635
Hubbard TJP, Aken BL, Beal K, Ballester B, Caccamo M, Chen Y, Clarke L, Coates G et al (2007) Ensembl 2007. Nucleic Acids Res 35:D610–D617
International Hap Map Consortium (2007) A second generation human haplotype map of over 3.1 million SNPs. Nature 449:851–862
International HapMap 3 Consortium (2010) Integrating common and rare genetic variation in diverse human populations. Nature 467:52–58
International Human Genome Sequencing Consortium (2001) Initial sequencing and analysis of the human genome. Nature 409:860–921
Jost L (2008) G_{ST} and its relatives do not measure differentiation. Mol Ecol 17:4015–4026
Kelkar YD, Tyekucheva S, Chiaromonte F, Makova KD (2008) The genome-wide determinants of human and chimpanzee microsatellite evolution. Genome Res 18:30–38
Kimmel M, Chakraborty R, Silvers DN, Deka R (1996) Dynamics of repeat polymorphisms under a forward-backward mutation model: within- and between-population variability at microsatellite loci. Genetics 143:549–555
Kimura M, Crow JF (1964) The numbers of alleles that can be maintained in a finite population. Genetics 49:523–538
Kruglyak S, Durrett RT, Schug MD, Aquadro C (1998) Equilibrium distributions of microsatellite repeat length resulting from a balance slippage events and point mutations. Proc Natl Acad Sci U S A 95:10774–10778
Lai Y, Sun F (2003) The relationship between microsatellite slippage mutation rate and the number of repeat units. Mol Biol Evol 20:2123–2131
Levy S, Sutton G, Ng PC, Feuk L, Halpern AL, Walenz BP, Axelrod N et al (2007) The diploid genome sequence of an individual human. PLoS One 5:e254
Meirmans PG, Hedrick PW (2011) Assessing population structure: FST and related measures. Mol Ecol Resour 11:5–18
Messier W, Li SH, Stewart CB (1996) The birth of microsatellites. Nature 381:481–483
Moreno-Estrada A, Casals F, Ramirez-Soriano A, Oliva V, Calafell F, Bertranpetit J, Bosch E (2008) Signatures of selection in human olfactory receptor OR511 gene. Mol Biol Evol 25:144–154
Nei M (1972) Genetic distance between populations. Am Nat 106:283–292
Nei M (1977) F-statistics and analysis of gene diversity in subdivided populations. Ann Hum Genet 41:225–233
Nei M, Kumar S (2000) Molecular evolution and phylogenetics. Oxford University Press, Oxford

Nei M, Tajima F, Tateno Y (1983) Accuracy of estimated phylogenetic trees from molecular data. II. Gene frequency data. J Mol Evol 19:153–170
Nei M, Niimura Y, Nozawa M (2008) The evolution of animal chemosensory receptor gene repertoires: roles of chance and necessity. Nat Rev Genet 9:951–963
Nozawa M, Kawahara Y, Nei M (2007) Genomic drift and copy number variation of sensory receptor genes in humans. Proc Natl Acad Sci U S A 104:20421–20426
Ohta T, Kimura M (1973) A model of mutation appropriate to estimate the number of electrophoretically detectable alleles in a finite population. Genet Res 22:201–204
Pearson CE, Edamura KN, Cleary JD (2005) Repeat instability; mechanisms of dynamic mutations. Nat Rev Genet 6:729–742
Ramachandran S, Deshpande O, Roseman CC, Rosenberg NA, Feldman MW, Cavalli-Sforza LL (2005) Support from the relationship of genetic and geographic distance in human populations for a serial founder effect originating in Africa. Proc Natl Acad Sci U S A 102:15942–15947
Redon R, Ishikawa S, Fitch KR, Feuk L, Perry GH, Andrews TD et al (2006) Global variation in copy number in the human genome. Nature 444:444–454
Rose O, Falush D (1998) A threshold size for microsatellite expansion. Mol Biol Evol 15:613–615
Sainudiin R, Durrett RT, Aquadro CF, Nielsen R (2004) Microsatellite mutation models: insights from a comparison of humans and chimpanzees. Genetics 168:383–395
Saitou N, Nei M (1987) The neighbor-joining method: a new method for reconstructing phylogenetic trees. Mol Biol Evol 4:406–425
Schlötterer C (2000) Evolutionary dynamics of microsatellite DNA. Chromosoma 109:365–371
Shriver MD, Jin L, Chakraborty R, Boerwinkle E (1993) VNTR allele frequency distributions under the stepwise mutation model: a computer simulation approach. Genetics 134:983–993
Sideris M, Papagrigoriadis S (2014) Molecular biomarkers and classification models in the evaluation of the prognosis of colorectal cancer. Anticancer Res 34:2061–2068
Sneath PHA, Sokal RR (1973) Numerical taxonomy. W. H. Freeman, San Francisco
Sun JX, Helgason A, Masson G, Ebenesersdottir LH, Mallick S, Gnerre S, Patterson N, Kong A, Reich D, Stefansson (2012) A direct characterization of human mutation based on microsatellites. Nat Genet 44:1161–1167
Takezaki N, Nei M (1996) Genetic distance and reconstruction of phylogenetic trees from microsatellite DNA. Genetics 144:385–392
Takezaki N, Nei M (2008) Empirical tests of the reliability of phylogenetic trees constructed with microsatellite DNA. Genetics 178:1835–1840
Takezaki N, Nei M (2009) Genomic drift and evolution of microsatellite DNAs in human populations. Mol Biol Evol 26:1835–1840
Toth G, Gaspari Z, Jurka J (2000) Microsatellites in different eukaryotic genomes: survey and analysis. Genome Res 10:967–981
Tuzun E, Sharp AJ, Bailey JA, Kaul R, Morrison VA, Pertz LM et al (2005) Fine-scale structural variation of the human genome. Nat Genet 37:727–732
Valdes AM, Slatkin M, Freimer N (1993) Allele frequencies at microsatellite loci: the stepwise mutation model revisited. Genetics 133:737–749
Whittaker JC, Harbord RM, Boxall N, Mackay I, Dawson et al (2003) Likelihood-based estimation of microsatellite mutation rates. Genetics 164:781–787
Wong KK, deLeeuw RJ, Dosanjh NS, Kimm LR, Cheng Z, Horsman DE et al (2007) A comprehensive analysis of common copy-number variations in the human genome. Am J Hum Genet 80:91–104
Xu X, Peng M, Fang Z, Xu X (2000) The direction of microsatellite mutation is dependent upon allele length. Nat Genet 24:396–399

Part II
The Human Genome Viewed Through Genes

Chapter 8
Genes on X and Y Chromosomes

Yoko Satta, Yukako Katsura, and Mineyo Iwase

Abstract It is widely accepted that therian (placental and marsupial mammal) X and Y chromosomes were differentiated from a pair of autosomes by means of either chromosomal inversions or accumulation of linked sex-determining genes or both. This evolutionary process has been dynamic and involved stepwise differentiation of proto-sex chromosomes with leaving footprints of ancient pseudo-autosomal boundaries on the X chromosome and deterioration of gametologous genes on the Y chromosome without recombination. Genes or gene families on the X and Y chromosomes have been originated in three different ways: allelic pairs on the proto-sex chromosomes, retropositions, or translocations of autosomal genes. Gene duplication and conversion have also played important roles in the evolution of these genes particularly when they form palindromic structures. Because of the rapid evolution and unique structure of the therian X and Y chromosomes, there exist quite a few genes that are specific to humans. In this chapter, we overview our current knowledge about the origin and evolution of the therian sex chromosomes and genes and discuss future perspectives.

Y. Satta (✉)
Department of Evolutionary Studies of Biosystems, The Graduate University for Advanced Studies (Sokendai), Hayama, Kanagawa, Japan

Center for Integration and Promotion of Sciences, The Graduate University for Advanced Studies (Sokendai), Hayama, Kanagawa, Japan
e-mail: satta@soken.ac.jp

Y. Katsura
Department of Biology, Pennsylvania State University, University Park, PA, USA

Department of Biology, Temple University, Philadelphia, PA, USA
e-mail: kyuca00@gmail.com

M. Iwase
Center for Educational Research and Development, Shimane University, Matsue, Shimane, Japan
e-mail: miwase@soc.shimane-u.ac.jp

N. Saitou (ed.), *Evolution of the Human Genome I*, Evolutionary Studies,
https://doi.org/10.1007/978-4-431-56603-8_8

Keywords Recombination suppression · Evolutionary strata · Gametologs · Gene conversion · Pseudo-autosomal region · Palindromic regions · Human-specific genes

8.1 The Origin of Sex Chromosomes

Muller (1914) and Ohno (1967) suggested that the sex chromosomes have originated from a pair of autosomes. In animals the sex-determining system varies even when sex chromosomes play definite roles. Phylogenetically unrelated *Drosophila* and mammals share superficially the same X/Y sex-determining system, but birds have evolved Z/W, and other vertebrates such as fish and frogs promiscuously use either X/Y or Z/W male-determining system.

As expected, these sex chromosomes have different origins and evolved independently in different phyla or even within a family. In mammals, Metatheria (marsupials) and Eutheria (placental mammals) have a single pair of X and Y chromosomes. Heterogametic individuals with respect to the Y chromosome are males, because the male-determining gene, *sex-determining region Y* (*SRY*), encoding a transcription factor bound to *SRY-box 9* (*SOX9*) is located on the Y chromosome (Sinclair et al. 1990). On the other hand, Prototheria (monotremes) such as platypus and echidna is known to have five pairs of X and Y chromosomes, but curiously part of these X and Y are rather similar to ZW chromosomes in birds (Grützner et al. 2004; Veyrunes et al. 2008). Any homolog to *SRY* has not been confirmed in Prototheria, although it is proposed that *anti-müllerian hormone* (*AMH*) is the primary male-determining gene in Prototheria that is located on a Y chromosome of monotremes (Cortez et al. 2014).

SRY in Eutheria and Metatheria is originated from an allele at the *SRY-box 3* (*SOX3*) locus on the proto-sex chromosome, because proto-X and proto-Y chromosomes originated from autosomes. In an early ancestral state, the allele had accumulated characteristic amino acid substitutions and acquired a novel function for testis determination. Surely, the emergence of such *SRY* was one critical step for the proto-X and proto-Y chromosomes to differentiate into the present-day sex chromosomes. It is speculated that the differentiation toward authentic sex chromosomes has proceeded in a stepwise manner by suppression of recombination (Ohno 1967; Nei 1969; Charlesworth and Charlesworth 2000; Iwase et al. 2003; Charlesworth et al. 2005; Graves 2006). Such suppression may have resulted from chromosomal inversions on one of the proto-sex chromosomes (Ohno 1967; Nei 1969; Lahn and Page 1999). Chromosomal inversions are, in fact, frequently observed in the early stage of sex chromosomal differentiation in plants and animals (Matsuranaga 2006; Ross and Peichel 2008). Alternatively, though not mutually exclusively, requirement of tight linkage among male-determining genes might have favored suppression of recombination (Nei 1969). In either event, such recombination suppression appears to have occurred in stepwise fashion, and the stepwise evolution of the mammalian sex chromosomes has left footprints of so-called strata between the extant mammalian X and Y chromosomes (Lahn and Page 1999; Iwase et al. 2003).

As recombination becomes infrequent, the hemizygous Y chromosome might easily accumulate deleterious mutations by Muller's ratchet mechanism (Engelstädter 2008). Some of these mutations could be harmful, and if evolutionarily accepted, they should deteriorate functional genes on the Y chromosome: Unless otherwise, most genes on the Y chromosome might be nonessential under the presence of functional gametologous genes on the X chromosome. If the male-determining gene translocates onto another autosome or a new male-determining gene emerges on a different chromosome, the Y chromosome may disappear from the genome. An example of such loss is indeed reported for Ryukyu spiny rats (Kuroiwa et al. 2010), although any gene that has replaced the SRY function has not been identified yet.

The X chromosome, on the other hand, shows a relatively stable mode of evolution because paring of chromosomes in female germ cells makes genetic information exchange possible and therefore facilitates removal of detrimental mutations. Because of this recombinational advantage, the gene content has been well conserved among different mammalian orders (Katsura et al. 2012). However, even in this conserved mode, several novel genes and unique genomic structures such as palindromes have evolved on the X chromosome. In what follows, we review our study about evolutionary strata, the origin of sex chromosome-linked genes and the evolution of palindromic structures on the human X and Y chromosomes. We would also like to ask questions concerning evolution of genes specific to the human X and Y chromosomes.

8.2 History of Human X-Y Chromosome Differentiation

Human sex chromosomes have originated from a pair of autosomes in their stem lineage of Theria (Eutheria and Metatheria) approximately 180 million years ago (mya). Even today, several coding sequences (CDS) on both sex chromosomes can be easily aligned, indicating their relatively recent descents from common ancestral genes. These homologous genes on sex chromosomes are what we call gametologs, although their original chromosomal locations may no longer be maintained, particularly in the case of Y-linked genes (Fig. 8.1).

Based on the synonymous sequence divergences (K_S) between 19 gametologous CDS (gene) pairs together with information about their locations on the X chromosome, the hypothesis of evolutionary strata was proposed by Lahn and Page (1999). As a result, all genes could be classified into four classes: class 1 contains three genes with $K_S > 1.0$, class 2 has two genes with $K_S \approx 0.5$, and both classes 3 and 4 have seven genes with $K_S = 0.2$ and 0.1, respectively. Thus, genes of classes 1 to 4 are arranged in this order on the X chromosome from the tip of the long arm to the tip of the short arm. Based on these observations, a region that genes of each class belong to was named as a stratum. These four evolutionary strata reflect the timing of recombination suppression between the sex chromosomes. The earliest suppression occurred on the long arm of the proto-sex chromosomes in an ancestor of Theria, and subsequently in the same stem linage, the second one is supposed to have occurred in a region covering from the centromere to the middle of the short

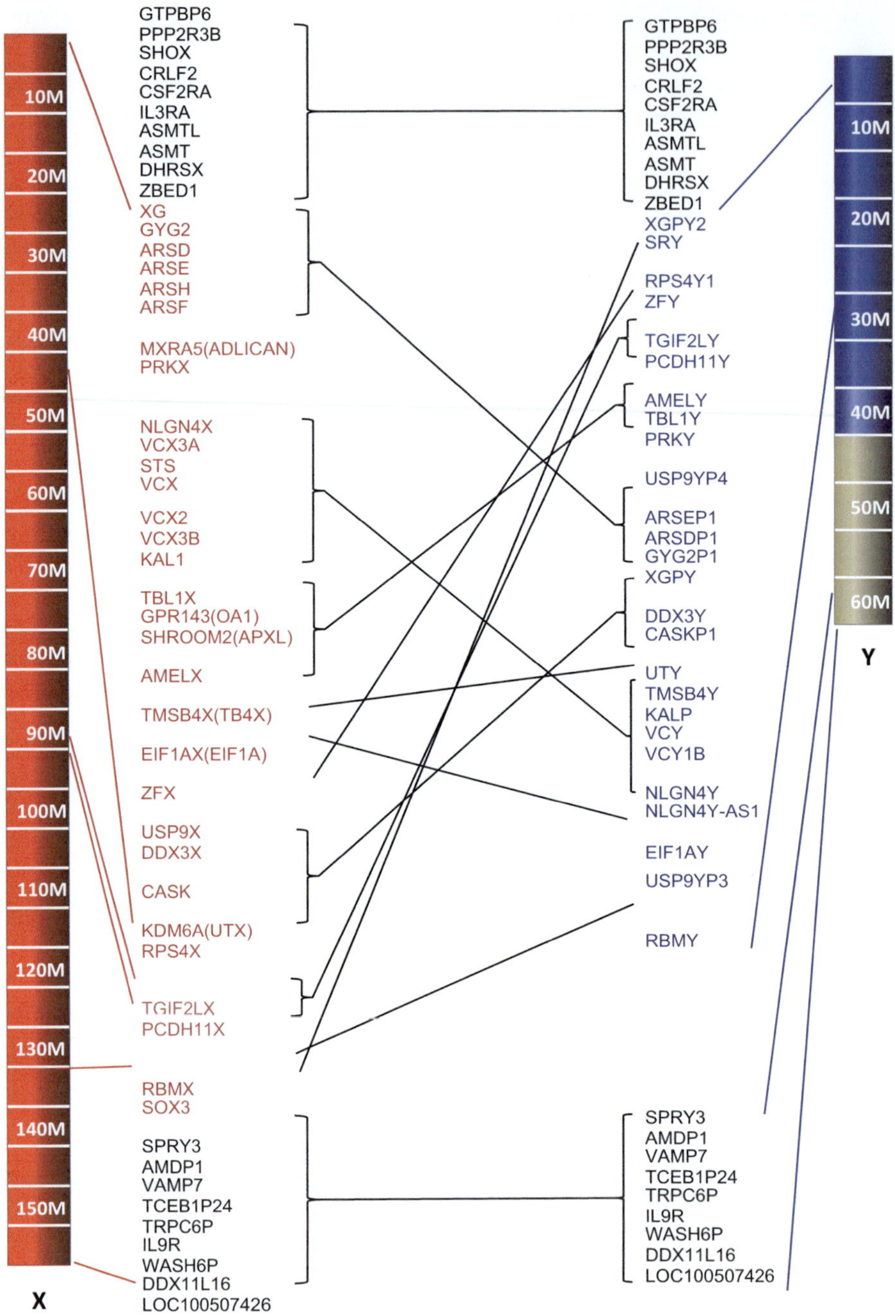

Fig. 8.1 Homologous gene maps on the human sex chromosomes. Red and blue bars represent the X and Y chromosome, respectively. Black part on the Y chromosome indicates a heterochromatin region, although it contains *PAR2* that is not in a heterochromatin. Gene names are given along each chromosome. The names in red and blue show sex-specific genes, whereas those in black stand for genes in *PAR1* or *PAR2*. Square brackets beside gene names indicate conserved blocks between the sex chromosomes. A line connecting X and Y blocks shows the gametologous relationship (This map is based on *Homo sapiens* Genome (Build 37.3))

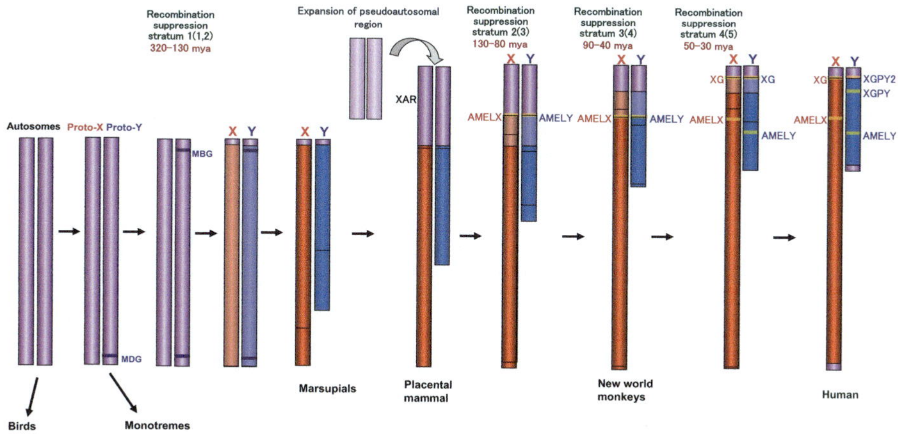

Fig. 8.2 Evolution of the human sex chromosomes. Chicken chromosome 4 and platypus chromosome 6 are an ancestoral type of the therian sex chromosome. Purple rectangle means a freely recombining (pseudoautosomal) region, whereas red and blue stand for non-recombining X and Y chromosomal regions, respectively. XAR is an X-added region. MDG means the male-determining gene (*SRY*). *AMELX* amelogenin X-linked, *AMELY* amelognin Y-linked, *XG* Xg blood group, *XGPY* XG pseudogene, Y-linked, *XGYP2* XG pseudogene, Y-linked 2

arm of the X chromosome. The third recombination arrest corresponded to the proximal half of the remaining part on the short arm and occurred in the common ancestral lineage of Eutheria. The most recent arrest occurred in a region that covers the distal half of the short arm in the primate common ancestor (Fig. 8.2). The third stratum is thus specific to Eutheria and the forth to primates. In addition to these recombination arrest events, it is believed that a pair of autosomes translocated after the differentiation of stratum 2 and attached to the then differentiating sex chromosomes. Strata 3 and 4 are thus of relatively recent origins and shared only among Eutheria.

However, recent works have made several revisions to this original evolutionary strata hypothesis. Researchers reexamined genomic sequences of these strata rather than CDS alone (Iwase et al. 2003; Skaletsky et al. 2003; Ross et al. 2007; Lemaitre et al. 2009, Wilson and Makova 2009; Katsura et al. 2012; Katsura and Satta 2012; Pandey et al. 2013). In the first such approach, Iwase et al. (2003) found that the boundary between strata 3 and 4 is located in the middle of *amelogenin X-linked* (*AMELX*) gene (Iwase et al. 2003, Fig. 8.3). *AMELX* encodes a protein that is involved in biomineralization during tooth enamel development. Furthermore, *AMELX* is a nested gene and is located in intron 1 of *Rho GTPase-activation protein 6* (*ARHGAP6*). This strongly suggested that the chromosomal inversion could not be responsible for the recombination arrest and a mechanism to have created stratum 3 (Iwase et al. 2003, 2007). This also holds true for the boundary between stratum 4 and *pseudo-autosomal region 1* (*PAR1*). *SRY* is located near to this boundary that is also in the middle of *Xg blood group* (*XG*) and *Xg pseudogene, Y-linked 2* (*XGPY*) gene on the X and Y chromosomes, respectively. These observations suggest that *SRY* translocated from the original *SOX3* gametologous

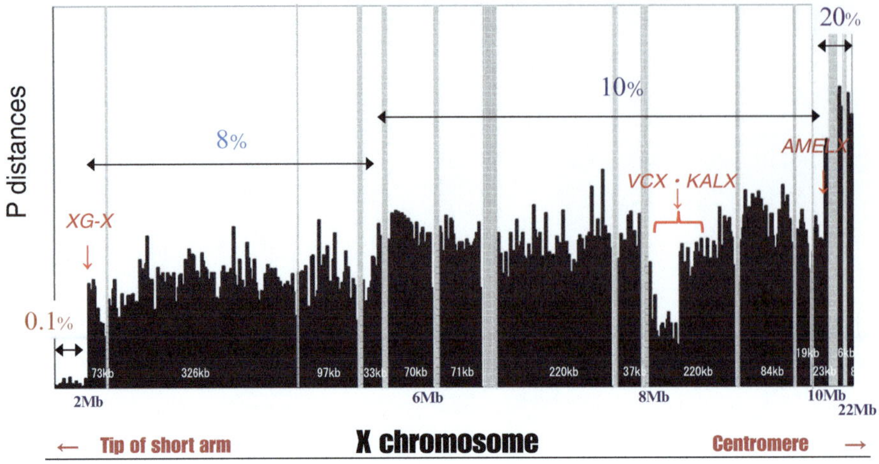

Fig. 8.3 The nucleotide sequence divergence between the short arms of the human X and Y chromosomes. The short arm of the X chromosome and its corresponding Y chromosomal region in the human genome are compared in terms of the nucleotide divergence or the P distance (y-axis) against the X chromosomal location (x-axis). The P distance increases in a stepwise manner from the terminal end (left most) to the centromere (right most). An interesting exception is the region containing *VCX* and *KALX*. Despite the fact that these belong to the region or stratum 4, their P distance of only 5% is much lower than the average over the region (10%). The total length of nucleotide sequences is 1.3 Mb, and both window and sliding size are assumed to be 1 kb

position and tightly linked other maleness related genes are a more likely cause for recombination suppression than chromosomal inversions as far as strata 3 and 4 are concerned.

Another analysis using genomic sequences from Metatheria and Eutheria revealed that stratum 2 might be a result of gene conversion and originally evolve as part of stratum 1. Katsura and Satta (2012) found the evidence for ancient gene conversion between two pairs of gametologs, *lysine (K)-specific demethylase 5D* (*SMCX/Y*) and *ubiquitin-activating enzyme E1* (*UBE1X/Y*) belonging to stratum 2. If the genomic regions apparently affected by gene conversion are excluded from the two genes, the remaining regions exhibit the K_S values typical to stratum 1, so that it is possible that recombination arrest occurred only once in the therian ancestor. Further analysis of genes on the X chromosome in Metatheria supports this conclusion; virtually no significant difference in the K_S values between marsupial strata 1 and 2. However, since other studies claim independent second suppression in Metatheria and Eutheria (Bellott et al. 2014; Cortez et al. 2014), further study is needed to make a unified view about the early phase in the evolutionary strata in mammals.

The current consensus is that although recombination suppression is the most likely mechanism for a stepwise nucleotide divergence between the mammalian sex chromosomes, the mechanism for recombination suppression requires further evolutionary analyses.

8.3 Origins of Genes on Human Sex Chromosomes

Genes on the human sex chromosomes are classified into three categories regarding to their origin. The first category is an ancient pair of gametologs, implying that their origins are alleles in the proto-sex chromosomes. Second and third category genes were translocated or retro-transposed from autosomes after the X and Y differentiation. In addition, some genes were amplified on the X or Y chromosome independently.

Typical genes for the first category are those used in the analysis for detecting strata as well as genes in *PAR1* (Fig. 8.1). Of the genes belonging to this category, *KALX/Y* and *VCX/Y* show interesting characteristics. Both genes are located in stratum 4, and the sequence divergences, including adjacent regions, are expected to be around 10%. However, the synonymous divergence between the X- and Y-genes is less than 5%. This low extent of sequence divergences is found not only in the coding regions of *KALX/Y* but also in intergenic regions that encompass the surrounding ~100 kb region and *VCX/Y* genes (Fig. 8.3). Iwase et al. (2010) examined this 100 kb region for both X and Y in humans, chimpanzees, and rhesus macaques. It turned out that there are two distinct regions (A and B) that exhibit different phylogenetic relationships. Region A contains exons 5–7 of *KALX/Y*, whereas region B contains exons 12–14 of the genes. The discontinuity of exons is due to a partial loss of sequences in primate *KALY* genes. In region A, the X and Y differentiation has initiated in the stem lineage of primates, consistent with the emergence of stratum 4. On the other hand, region B shows that X and Y differentiated twice and independently. Iwase et al. (2010) suggested roles of gene conversion for the explanation of this history of region B: the first conversion has occurred in the ancestor of only rhesus macaques and the other one in the common ancestor of humans and chimpanzees. Interestingly, the independent gene conversion involving *KALX/Y* was also found in gibbons (Iwase et al. 2010). The 100 kb region surrounding *KALX/Y* is thus prone to gene conversion between the sex chromosomes. It is tempting to speculate that the presence of a transposon (*LINE3A*) is responsible for this frequent conversion.

For the second category of translocation origins, only two genes are known, *TGFB-induced factor homeobox 2-like* (*TGIF2LX/Y*) and *protocadherin 11 X- or Y-linked* (*PCDH11X/Y*). These two genes are located side by side in stratum 1 and separated by only 1.86 Mb. The whole 1.86 Mb region shows high similarity between the X and Y chromosomes (99%), suggesting that the region on the X has duplicated onto the Y quite recently. Interestingly, this duplication is observed only in humans. *TGIF2LX/Y* have an autosomal homolog *TGIF2*. This autosomal gene is located on the chromosome 20 and composed of three exons with a large intron between exons 2 and 3 (>10 kb). On the other hand, sex chromosome-linked *TGIF2LX/Y* have two exons, and CDS is encoded by only exon 2. In addition, the intron and exon 1 are very short (<100 bp and 30 bp, respectively). Interestingly, *TGIF2LX* is conserved within eutherian X chromosomes, but the Y gametolog is found only in humans. The synonymous nucleotide divergence between *TGIF2LX* and *TGIF2* is 0.84, indicating that the transposition of *TGIF2* to the X chromosome was older than the emergence of Eutheria. Since the marsupial X chromosome does not have *TGIF2LX*, it is likely that the *TGIF2* locus was duplicated and a copy was translocated on the long arm of the X chromosome in the eutherian ancestor. Much later in the human lineage, *TGIF2LX* produced a copy locus on the Y chromosome. The synonymous divergence is as low as 0.7%, and it is an interesting question whether Neanderthals have experienced the duplicated transposition of *TGIF2LY*.

PCDH11X has the history similar to that of *TGIF2LX*. The ortholog of *PCDH11X* is found in almost all mammals of which the genome sequences are available, whereas *PCDH11Y* is found also only in humans. However, the closest relative to *PCDH11X* on the human autosomes is *PCDH9*, and orthologs of *PCDH11X* are found on chromosome 4p in chicken. Chicken chromosome 4 is orthologous to the present mammalian X chromosome (Rens et al. 2007; Shevchenko et al. 2013). This implies that *PCDH11Y* should have existed on the therian proto-Y chromosome. However, the absence of *PCDH11Y* on Metatheria and Eutheria except for humans means that *PCDH11Y* has been deleted from the proto-Y chromosome at the early stage of the therian X and Y differentiation. It therefore appears only in the extant human that *PCDH11X* was duplicated and the duplicate (*PCDH11Y*) was located on the Y chromosome.

The third category includes genes on palindromic regions on the X or Y chromosomes (Fig. 8.4). It is well known that the approximately 10.2 Mb (about 44% of 23 Mb of male-specific region on Y, Skaletsky et al. 2003) region is called "ampliconic" and consists of eight large palindromes. These palindromes are named P1–P8, and the evolution of this region has been well documented (Fig. 8.5; Kuroda-Kawaguchi et al. 2001; Skaletsky et al. 2003; Hughes et al. 2010; Kuroki et al. 2006; Katsura et al. 2012; Bhowmick et al. 2007). On the Y chromosome, several genes are shared between different palindromes (Bhowmick et al. 2007): e.g., *X-linked Kx blood group related, Y-linked* (*XKRY*) are shared among P1, P4, and P5. *Chromodomain protein, Y-linked* (*CDY*) are shared among P1, P3, and P5. All of genes on Y-palindromes have their X-gametologs originated in the proto-sex chromosomes, except for *CDY*. *CDY* was duplicated by retro-transposition from an autosomal gene shared by Eutheria, and this transposition occurred before the Eutherian radiation, supported by the synonymous divergence of 0.44 between the Y chromosomal and autosomal genes. *CDX* is independently duplicated from the same autosomal gene as *CDY* at different timing, before the divergence of Old World monkeys (OWM) and humans. Interestingly, the presence of *CDX* only in human suggests that this *CDX* is lost several times from the primate X chromosome except for humans.

Genes in palindromic structures on the X chromosome have been examined by Katsura and Satta (2011) and Kim et al. (2012) among others. On the human X

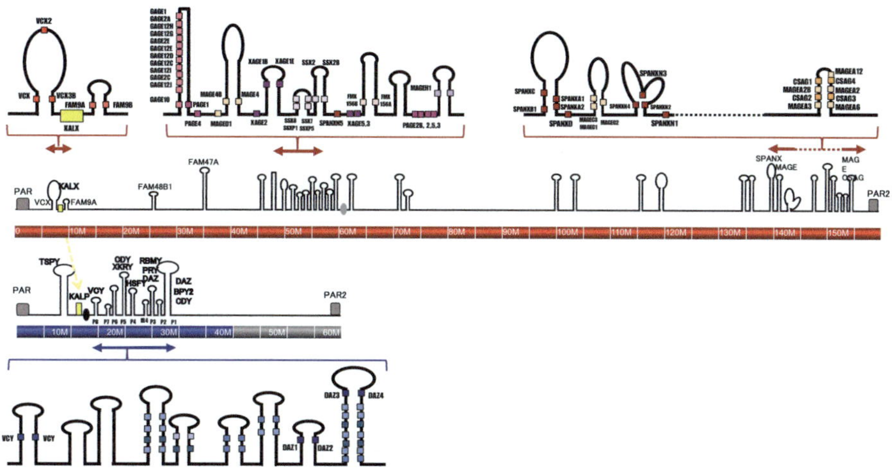

Fig. 8.4 Palindromes on the human sex chromosomes. Red and blue bars represent the human X and Y chromosome, respectively. A line on each chromosome represents a position of palindromes. Double-headed arrows show an enlargement of the region. Each colored rectangle in a palindrome indicates extensive expansion of members in a particular gene family, and the same color implies the same family

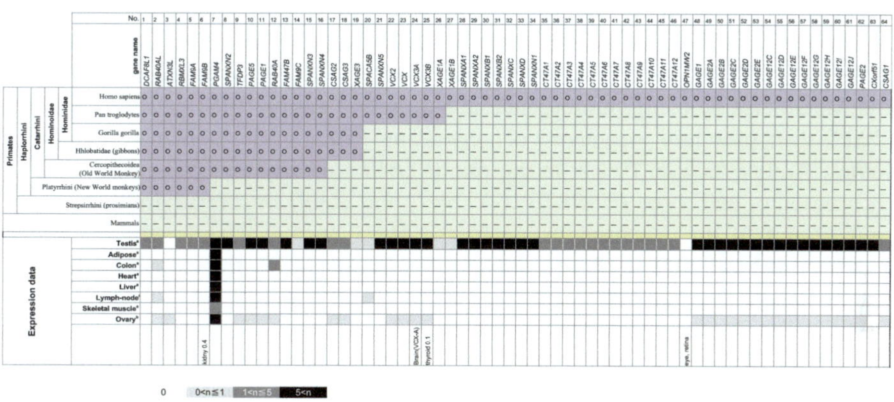

Fig. 8.5 Human-specific genes on the human X chromosome. The presence or absence of orthologs in nonhuman primates is indicated as a circle or a minus sign, respectively. Genes were identified as human specific in the comparison between human and mouse X chromosomes by Mueller et al. (2013)

chromosomes, the relatively large number (39) of palindromes is observed (Kim et al. 2012). Majority of genes on the palindromes are families of cancer and testis antigens (*CTAs*) such as melanoma antigen (*MAGE*), sarcoma antigen (*SAGE*), G antigen (*GAGE*), and X antigen (*XAGE*) families. They are highly expressed in several cancer cells and testis. The pattern of gene distribution and evolution of human X palindromic genes are different from those of Y palindromic genes.

Shared genes between different palindromes are rarely observed on the X: Rather a single palindrome contains a single *CTA* family. The origins of these families are not fully elucidated despite of their importance in medical and clinical fields.

8.4 Origins of Genes in *PAR* on Human X and Y Chromosome

At the terminal end of the short arm of sex chromosomes, recombination is ensured in *PAR1*. *PAR1* is observed in all eutherian genomes and plays an important role in X and Y chromosomal paring in the cell division. Gene contents and orders in *PAR1* may have been preserved since the time of the proto-sex chromosomes. Human *PAR1* is 2.6 Mb long and contains ten genes. However, the boundary between *PAR1* and stratum 4 differs among different families of mammals (Katsura et al. 2012). In contrast to *PAR1*, *PAR2* is located in the opposite terminal end of the sex chromosomes and is specific to humans. *PAR2* is 320 kb long but contains nine genes (*SPRY3*, *AMDP1*, *VAMP7* (*SYBL1*), *TCEB1P24*, *TRPC8P*, *IL9R*, *WASH6P* (*CXYorf1*), *DOX11L16*, and *LOC100507426*). Of these nine genes, four and five orthologs are found in the chimpanzee and macaque X chromosomes, respectively, but there are no orthologs on the Y chromosome in neither species.

Evolution of *PAR2* was investigated by Charchar et al. (2003) in which the presence or absence of orthologs to human *PAR2* genes was examined in Eutheria and Metatheria by using fluorescence in situ hybridization (FISH). Charchar et al. (2003) then proposed a scenario of generating *PAR2* in the human sex chromosomes. Among human *PAR2* genes, one (*SYBL1*) already existed on the proto-X chromosome, and three (*HSPRY3*, *CXYorf1*, and *IL9R*) on Metatheria autosomes were added to the X chromosome in the stem lineage of primates. Although Charchar et al. (2003) examined only four genes by FISH, the extensive in silico search for paralogs or orthologs of genes in *PAR2* supports this scenario. Some modifications were however necessary. One is that since *HSPRY3* has an ortholog only in the therian X chromosome, it too existed in the proto-X chromosome as in *SYBL1*. Homologs of *ILR9* are found in the primate X chromosome, but these are located on autosomes in non-primate mammals. It is also found that while *CXYorf1* exists in the mouse genome, there are no orthologs for *CXYorf1*, *TCEB1P24*, *TRPC8P*, *DOX11L16*, and *LOC100507426* on nonhuman primate X chromosomes.

8.5 Human-Specific Gene Gain and Loss

Recently, quite a few genes on human X chromosome have been identified as "specific" to humans in comparison with the mouse X chromosome (Mueller et al. 2013). Among these specific genes, we extensively reexamined their "specificity" with references to nonhuman primate genomes and revealed that approximately half of them have orthologs in other primates (Fig. 8.5). However, still we have 38 genes as human specific. Among these 38 genes, *SPANX*, *CT47A*, *XAGE*, *GAGE*, *PAGE*, and *CSAG* are members of multigene families. In other words, majority of these genes have specifically duplicated in the human genome. Many of these specifically duplicated genes are located on the X palindromes.

The origin of *SPNX*, *CSAG*, and *VCX* on X palindromes is unique. These genes have been partly derived from a single noncoding region on X chromosome. Examples of de novo generation of genes are little known (Mueller et al. 2013; Toll-Riera et al. 2008), and the detailed process of the de novo generation of the three genes will be described somewhere. It is noted that the homology between these genes and the original noncoding region is not high, so that the de novo generation of these genes should have occurred in a stem of Theria. Nevertheless, *SPNX* and *CSAG* are identified as human-specific genes (Fig. 8.5).

When the gene content on the human sex chromosomes is compared with that on the chimpanzee sex chromosomes, several unique features about gene gains and losses, especially to humans, have been revealed (Fig. 8.6). We examined the

Chromosome	human X		human Y	
Gene	pseudogenes	functional genes	pseudogenes	functional genes
Human loss = pseudogenes	158		46	
Common gene	127	762	5	44
Chimpanzee none	367	273	267	78
Uncharacterized LOC (ncRNA,Protein coding PREDICTED,Unknown etc.)		87		13
Total	652	1122	318	135

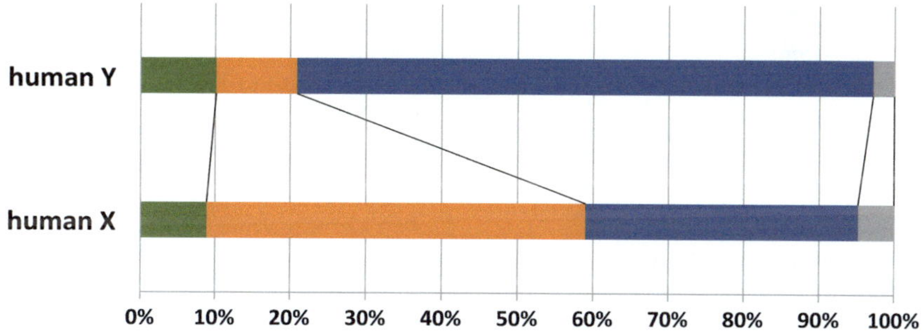

Fig. 8.6 Gene contents on human and chimpanzee sex chromosomes. The number of functional genes/pseudogenes on the human X and Y is 1122/652 and 135/318, respectively. "Common genes" indicate that orthologs to human genes are found in chimpanzees. "Human loss" means that the gene was pseudogene in humans but functional in chimpanzees, whereas "chimpanzee none" means that the orthologs of human genes are not detected in chimpanzees. The graph at the bottom shows the proportion of each category in the X and Y chromosomal genes

absence or presence using only the annotated genes. First, proportions of functional genes shared between two species on the total number of genes are approximately 11% for Y chromosome and 50% on X chromosome. This proportion for both sex chromosomes is quite low compared with that for autosomal genes. For pseudogenes, there are 652 and 318 on X and Y chromosomes, respectively. Among them, the proportion of shared pseudogenes is 19% and 2% for X and Y chromosome, respectively. On the other hand, human-specific ones are 24% on the X and 15% on the Y. The proportion of pseudogenes with no chimpanzee orthologs is quite high (56% of human X and 84% of human Y chromosome). The rate of gene deletion is also high for the sex chromosomes.

8.6 Conclusion

The mammalian sex chromosome differentiation initiated in the stem lineage of Theria (Eutheria and Metatheria). It appears that the differentiation was triggered when an allele of *SOX3* on the proto-sex chromosomes gained a new function for male determination and emerged as the primary male-determining gene *SRY*. The subsequent differentiation was driven by a stepwise suppression of recombination between these proto-sex chromosomes and eventually led the formation of four or more evolutionary strata. However, the mechanism of recombination suppression is still to be elucidated. We have discussed two possibilities, though not mutually exclusive; inversions on the Y chromosome or tight linkage among *SRY* and other sex determination genes.

Within each evolutionary stratum, the nucleotide divergence is almost similar and simply depends on when recombination of the stratum region was arrested. However, there are notable exceptions that are most likely caused by gene conversion-like events between the sex chromosomes or by transfer of a duplicate from the X to the Y chromosome. We have shown typical examples in strata 1 and 4. We have also suggested that LINE sequences may somehow be involved in such large-scale conversion and that some human-specific genes might have been generated by this mechanism.

Primate sex chromosomes have two different origins, although both were descended from two different pairs of autosomes. One pair of the proto mammalian sex chromosomes corresponds to extant chicken chromosome 1q/4p, and the other pair corresponds to chromosome 6 in the extant platypus genome. Most genes on the human X chromosome were originated from these autosomal genes, but some others were newly translocated from other autosomal genes. We have discussed the human *PAR2* region as a remarkable example, because to generate *PAR2*, partial transfer of an autosome to the X chromosome and duplication of this region onto the Y chromosome is required. However, this process of *PAR2* generation is specific for human but sufficiently complicated to be addressed in the future.

References

Bellott DW, Hughes JF, Skaletsky H, Brown LG, Pyntikova T, Cho T-J, Koutseva N, Zaghlul S, Graves T, Rock S, Kremitzki C, Fulton RS, Dugan S, Ding Y, Morton D, Khan Z, Lewis L, Buhay C, Wang Q, Watt J, Holder M, Lee S, Nazareth L, Rozen S, Muzny DM, Warren WC, Gibbs RA, Wilson RK, Page DC (2014) Mammalian Y chromosomes retain widely expressed dosage-sensitive regulators. Nature 508:494–499

Bhowmick BK, Satta Y, Takahata N (2007) The origin and evolution of human ampliconic gene families and ampliconic structure. Genome Res 17:441–450

Charchar FJ, Svartman M, El-Mogharbel N, Ventura M, Kirby P, Matarazzo MR, Ciccodicola A, Rocchi M, D'Esposito M, Graves JAM (2003) Complex events in the evolution of the human pseudoautosomal region 2 (PAR2). Genome Res 13:281–286

Charlesworth B, Charlesworth D (2000) The degeneration of Y chromosomes. Philos Trans R Soc Lond Ser B Biol Sci 355:1563–1572

Charlesworth D, Charlesworth B, Marais G (2005) Steps in the evolution of heteromorphic sex chromosomes. Heredity 95:118–128

Cortez D, Martin R, Toledo-Flores D, Froidevaux L, Waters PD, Grützver F, Kaessmann H (2014) Origins and functional evolution of Y chromosomes across mammals. Nature 508:488–493

Engelstädter J (2008) Muller's ratchet and the degeneration of Y chromosomes: a simulation study. Genetics 180:957–967

Graves J (2006) Sex chromosome specialization and degeneration in mammals. Cell 124:901–914

Grützner F, Rens W, Tsend-Ayush E, El-Mogharbel N, O'Brien PCM, Jones RC, Ferguson-Smith MA, Graves JAM (2004) In the platypus a meiotic chain of ten sex chromosomes shares genes with the bird Z and mammal X chromosomes. Nature 432:913–917

Hughes JF, Skaletsky H, Pyntikova T, Graves TA, van Daalen SKM, Minx PJ, Fulton RS, McGrath SD, Locke DP, Friedman C, Trask BJ, Mardis ER, Warren WC, Repping S, Rozen S, Wilson RK, Page DC (2010) Chimpanzee and human Y chromosomes are remarkably divergent in structure and gene content. Nature 463:536–539

Iwase M, Satta Y, Hirai Y, Hirai H, Imai H, Takahata N (2003) The amelogenin loci span an ancient pseudoautosomal boundary in diverse mammalian species. Proc Natl Acad Sci U S A 100:5258–5263

Iwase M, Kaneko S, Kim HL, Satta Y, Takahata N (2007) Evolutionary history of sex-linked mammalian amelogenin genes. Cells Tissues Organs 186:49–59

Iwase M, Satta Y, Hirai H, Hirai Y, Takahata N (2010) Frequent gene conversion events between the X and Y homologous chromosomal regions in primates. BMC Evol Biol 10:225. https://doi.org/10.1186/1471-2148-10-225

Katsura Y, Satta Y (2011) Evolutionary history of the cancer immunity antigen MAGE gene family. PLoS One 6.:Article ID:e20365

Katsura Y, Satta Y (2012) No evidence for a second evolutionary stratum during the early evolution of mammalian sex chromosomes. PLoS One 7:e45488

Katsura Y, Iwase M, Satta Y (2012) Evolution of genomic structures on mammalian sex chromosomes. Curr Genomics 13:115–123

Kim HL, Iwase M, Igawa T, Nishioka T, Kaneko S, Katsura Y, Takahata N, Satta Y (2012) Genomic structure and evolution of multigene families: "Flowers" on the human genome. Int J Evol Biol 2012:Article ID 917678

Kuroda-Kawaguchi T, Skaletsky H, Brown LG, Minx PJ, Cordum HS, Waterston RH, Wilson RK, Silber S, Oates R, Rozen R, Page DC (2001) The AZFc region of the Y chromosome features massive palindromes and uniform recurrent deletions in infertile men. Nat Genet 29:279–286

Kuroiwa A, Ishiguchi Y, Yamada F, Abe S, Matsuda Y (2010) The process of Y-loss event in XO/XO mammal, Ryukyu spiny rat. Chromosoma 119:519–526

Kuroki Y, Toyoda A, Noguchi H, Taylor TD, Itoh T, Kim DS, Kim DW, Choi SH, Kim IC, Choi HH, Kim YS, Satta Y, Saitou N, Yamada T, Morishita S, Hattori M, Sakaki Y, Park HS,

Fujiyama A (2006) Comparative analysis of chimpanzee and human Y chromosomes unveils complex evolutionary pathway. Nat Genet 38:158–167

Lahn B, Page D (1999) Four evolutionary strata on the human X chromosome. Science 286:964–967

Lemaitre C, Braga MDV, Gautier C, Sagot M-F, Tannier E, Marais GAB (2009) Footprints of inversion at present and past pseudoautosomal boundaries in human sex chromosomes. Genome Biol Evol 1:56–66

Matsuranaga S (2006) Sex chromosome-linked genes in plants. Genes Genet Syst 81:219–226

Mueller JL, Skaletsky H, Brown LG, Zaghlul S, Rock S, Graves T, Auger K, Warren WC, Wilson RK, Page DC (2013) Independent specialization of the human and mouse X chromosomes for the male germ line. Nat Genet 45:1083–1087

Muller H (1914) A factor for the fourth chromosome of drosophila. Science 39:906

Nei M (1969) Heterozygous effects and frequency changes of lethal genes in populations. Genetics 63:669–680

Ohno S (1967) Sex chromosome and sex linked genes. Springer, New York

Pandey RS, Sayres MAW, Azad RK (2013) Detecting evolutionary strata on the human X chromosome in the absence of gametologous Y-linked sequences. Genome Biol Evol (in press)

Rens W, CM O'Brien PCM, Grützner F, Clarke O, Graphodatskaya D, Tsend-Ayush E, Trifonov VA, Skelton H, Wallis MC, Johnston S, Veyrunes F, Graves JAM, Ferguson-Smith MA (2007) The multiple sex chromosomes of platypus and echidna are not completely identical and several share homology with the avian Z. Genome Biol 8:R243

Ross MT, Grafham DV, Coffey AJ, Scherer S, McLay K et al (2007) The DNA sequence of the human X chromosome. Nature 434:325–337

Ross J, Peichel CL (2008) Molecular cytogenetic evidence of rearrangements on the Y chromosome of the three spine stickleback fish. Genetics 179:2173–2182

Shevchenko AI, Zakharova IS, Zaikan SM (2013) The evolutionary pathway of X chromosome inactivation in mammals. ACTA Nat 5:40–53

Sinclair AH, Berta P, Palmer MS, Hawkins JR, Griffiths BL, Smith MJ, Foster JW, Frischauf A-M, Lovell-Badge R, Goodfellow PN (1990) A gene from the human sex-determining region encodes a protein with homology to a conserved DNA-binding motif. Nature 346:240–244

Skaletsky H, Kuroda-Kawaguchi T, Minx PJ, Cordum HS, Hillier L, Brown LG, Repping S, Pyntikova T, Ali J, Bieri T, Chinwalla A, Delehaunty A, Delehaunty K, Du H, Fewell G, Fulton L, Fulton R, Graves T, Hou SF, Latrielle P, Leonard S, Mardis E, Maupin R, McPherson J, Miner T, Nash W, Nguyen C, Ozersky P, Pepin K, Rock S, Rohlfing T, Scott K, Schultz B, Strong C, Tin-Wollam A, Yang SP, Waterston RH, Wilson RK, Rozen S, Page DC (2003) The malespecific region of the human Y chromosome is a mosaic of discrete sequence classes. Nature 423:825–837

Toll-Riera M, Bosch N, Bellora N, Castelo R, Armengol L, Estivill X, Alba MM (2008) Origin of primate orphan genes: a comparative genomics approach. Mol Biol Evol 26:603–612

Veyrunes F, Waters PD, Miethke P, Rens W, McMillan D, Alsop AE, Grützner F, Deakin JE, Whittington CM, Schatzkamer K, Kremitzki CL, Graves T, Ferguson-Smith MA, Warren W, Graves JAM (2008) Bird-like sex chromosomes of platypus imply recent origin of mammal sex chromosomes. Genome Res 18:965–973

Wilson MA, Makova KD (2009) Evolution and survival on eutherian sex chromosomes. PLoS Genet 5:e1000568

Chapter 9
Human Leukocyte Antigen (HLA) Region in Human Population Studies

Timothy A. Jinam

Abstract Human leukocyte antigen (HLA) genes function to present processed antigenic peptides to immune cells, hence their major importance in eliciting immune responses. HLA genes have been extensively studied for their disease associations and pre-/posttransplantation applications. Due to their highly polymorphic nature, HLA genes are also used for population genetic studies. This chapter summarizes the structure, functions, nomenclature, genotyping methods, and population genetic applications of the HLA system.

Keywords Human leukocyte antigen (HLA) · Major histocompatibility complex (MHC) · Immune system · Population genetics

9.1 Introduction

The major histocompatibility complex (MHC) is a cluster of genes that play important roles in innate and adaptive immune systems. The MHC is found in all jawed vertebrates (Kulski et al. 2002), and in humans it is referred to as the human leukocyte antigen (HLA). Located on the short arm of chromosome 6, the HLA region spans 3.6 Mb and contains 253 known genes (Shiina et al. 2009).

These genes are classified into classes I, II, and III. Their order within the genome is represented in Fig. 9.1. HLA class I and II genes are involved in presenting antigenic peptides to antigen-specific T-cell receptors, hence their importance in cell-mediated immune responses. The discrimination of self and foreign peptides also has important implications in transplantation and disease susceptibility. Although HLA class III genes play no active role in antigen presentation, they encode for other molecules with immune functions. Classical HLA genes are defined by their high level of allelic diversity, making them useful genetic

T.A. Jinam (✉)
Division of Population Genetics, National Institute of Genetics, Mishima, Japan
e-mail: tjinam@nig.ac.jp

N. Saitou (ed.), *Evolution of the Human Genome I*, Evolutionary Studies,
https://doi.org/10.1007/978-4-431-56603-8_9

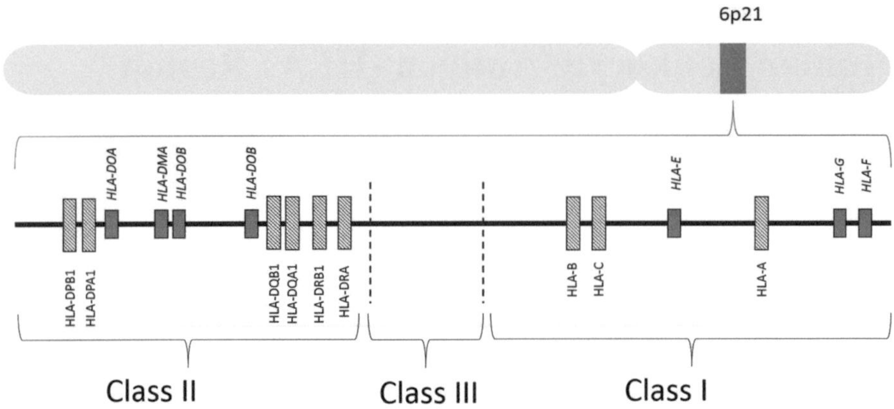

Fig. 9.1 Location of the human leukocyte antigen (HLA) region. Classical HLA genes are indicated by the shaded boxes, whereas nonclassical genes are indicated by gray boxes. Distances between genes are not to scale

markers for population studies. These HLA genes are found in class I and class II regions.

The class I region is on the telomeric side of chromosome 6, and within it are three classical HLA genes (HLA-A, HLA-B, HLA-C) and three nonclassical HLA genes (HLA-E, HLA-F, HLA-G). HLA class I genes encode transmembrane proteins found in all nucleated cells. Processed antigens presented by these HLA class I genes are recognized by CD4+ T cells (Williams 2001). Although not considered as classical genes, MICA and MICB genes are also highly polymorphic and have 30% protein homology with classical class I genes. These genes function as ligands for cells expressing a common activatory natural killer cell receptor (Stephens 2001).

The class III region lies between class I and class II regions and does not house any classical HLA genes. Instead, it contains genes that encode components for the formation of complements, cytokine genes, and others that do not appear to have direct immune-related function (Shiina et al. 2009). Some notable genes are C2 and C4, which encode complement proteins and TNF, which encodes a cytokine involved in various biological processes and has been linked to several diseases (Milner and Campbell 2001).

Class II region contains the classical HLA-DP, HLA-DQ, and HLA-DR genes. They encode transmembrane heterodimers that consist of alpha and beta chains. Genes that encode alpha and beta chains are suffixed with A and B, respectively, e.g., HLA-DPA1 and HLA-DPB1. Their expression is limited to certain types of cells, namely, B cells, macrophages, and dendritic cells. Antigenic peptides presented by these class II proteins are recognized by CD8+ T cells (Williams 2001). Nonclassical genes like HLA-DM play a role in promoting peptide loading of class II molecules in the lysosome (Alfonso and Karlsson 2000).

9.2 HLA Genotyping Methods

As of 2017, there are 16,251 known HLA alleles, with a majority belonging to class I genes (EMBL-EBI 2017). Because of this vast allelic diversity, HLA typing is very important especially in identifying matching alleles between transplantation recipient and donors. Early HLA typing methods involved detecting reactions between HLA antigens and their antisera in complement-mediated microlymphocytotoxicity assays, also known as serology (Terasaki and McClelland 1964).

Later, DNA-based HLA typing methods were developed. These methods essentially rely on polymerase chain reaction (PCR) to amplify the HLA region of interest. These PCR-amplified DNA fragments were then used to determine the specific allele type. In the sequence-specific primer (SSP) method, PCR primers were designed for specific alleles and the PCR products were subjected to gel electrophoresis to determine the HLA genotypes (Ando et al. 1996). On the other hand, sequence-specific oligonucleotide probe (SSOP) method uses labeled enzymes or colorimetric substrates for allelic identification (Levine and Yang 1994). The disadvantage of these methods is that the genotyped alleles depend on the predetermined set of PCR primers or oligonucleotide probes, meaning that novel HLA alleles that may have arisen due to mutations would not be detected.

Sequence-based typing (SBT) methods do away with this disadvantage by directly sequencing the relevant HLA genes and matching the output sequence with known alleles from a database. In this way, novel alleles can also be detected. Because only exons 2 and 3 of HLA genes are routinely sequenced in this method, alleles defined by mutations outside these two regions may not be properly identified. This issue may soon be irrelevant with the development of protocols using next-generation sequencers which have the ability to sequence the entire HLA gene (Bentley et al. 2009; Wang et al. 2012; Hosomichi et al. 2013).

9.3 Nomenclature of HLA Alleles

Unlike most other genes in the human genome, a naming system was developed specifically for the HLA genes to account for the vast number of alleles present. The allele name starts with the HLA gene followed by an asterisk and then a set of digits separated by colons, for example HLA-A*24:01:01:01. The first set of digits represent the antigen type identified using serology. The following set of digits are used to differentiate between non-synonymous changes that result in a different protein but with same serological properties. The third set of digits define synonymous changes whereas the last set of digits are used to show differences in noncoding regions of the gene. A final alphabet suffix is sometimes added to denote changes in the gene expression, e.g., N for null alleles.

The longer the allele name, the more information it conveys. Serological methods can only differentiate between the allele groups with the first set of digits, e.g., HLA-A*02, thus called a low-resolution typing method. DNA-based methods are also referred to as high-resolution typing methods because of its ability to discriminate alleles based on substitutions within the HLA gene. Examples of high-resolution HLA allele names are HLA-B*35:01 and HLA-DRB1*01:01:02.

9.4 Factors Affecting HLA Diversity

The classical HLA genes are the most polymorphic loci in the human genome, and this high level of allelic diversity is generated by mutation, gene conversion, or recombination. Allele frequencies in a population can fluctuate through the forces of natural selection or genetic drift. Since HLA molecules are involved in presenting foreign peptides to T cells to illicit an immune response, certain alleles can be advantageous in some human populations. For example, the HLA-B*53:01 allele was reported to be associated with protection against a severe type of malaria in African populations, and the allele frequency was considerably higher than in other non-African populations (Hill et al. 1991).

Another evolutionary factor affecting the diversity of HLA alleles is balancing selection, whereby multiple alleles are maintained in a population instead of a single allele. This can be brought about by heterozygote advantage in which heterozygotes have higher fitness than homozygotes. Individuals with homozygous HLA genotypes may be at a disadvantage compared to heterozygous individuals in combating infection due to a smaller repertoire of molecules to recognize and present antigenic peptides to immune cells. Evidence for balancing selection came from tests for neutrality which are based on the observed heterozygosity whereby certain HLA genes had higher than expected heterozygosity under neutrality (Meyer et al. 2006; Sanchez-Mazas 2007).

9.5 Application of HLA Genotyping

Demographic events such as migration and admixture may also shape HLA allelic diversity. Prior to the advancement of high-throughput single nucleotide polymorphism (SNP) genotyping methods, the HLA region was extensively used to study human population relationships. Phylogenetic analysis using just a single HLA gene was sometimes enough to infer population relationships, but in most cases, analysis using allele frequencies of multiple HLA genes would be more informative.

HLA genes are known to be in high linkage disequilibrium (LD) with each other. As such, certain allelic combinations tend to be inherited together as a haplotype. By treating this HLA haplotype as a single locus, they can also be utilized for phylogenetic analyses. Because HLA genotyping is done separately for each gene, HLA haplotypes have to be inferred using phasing algorithms such as those implemented in programs like fastPHASE (Scheet and Stephens 2006) or BEAGLE (Browning and Browning 2007).

HLA allele frequency data of various populations have been reported, and they vary in terms of the HLA gene genotyped and the allele resolution (low or high). This vast amount of information is readily available from databases such as the Allele Frequency Net Database (http://www.allelefrequencies.net). By gathering the allele frequency data from populations of interest, one can calculate the genetic distances between populations using measures such as Fst (Wright 1949), Nei's standard genetic distance (Nei 1972), or D_A distance (Nei et al. 1983). From the resulting genetic distance matrix, one may then visualize the relationship between populations using phylogenetic trees or Principal Component Analysis (PCA).

The relationships between various groups from different geographical locations are shown in Fig. 9.2. This phylogenetic tree was constructed from allele frequencies of HLA-A, HLA-B, and HLA-DRB1 genes downloaded from the Allele Frequency Net Database. The genetic distance between populations was calculated using Nei's standard genetic distance using the PHYLIP software package (Felsenstein 1981), and a neighbor-joining tree (Saitou and Nei 1987) was constructed. Even using just three HLA genes, a clear clustering pattern

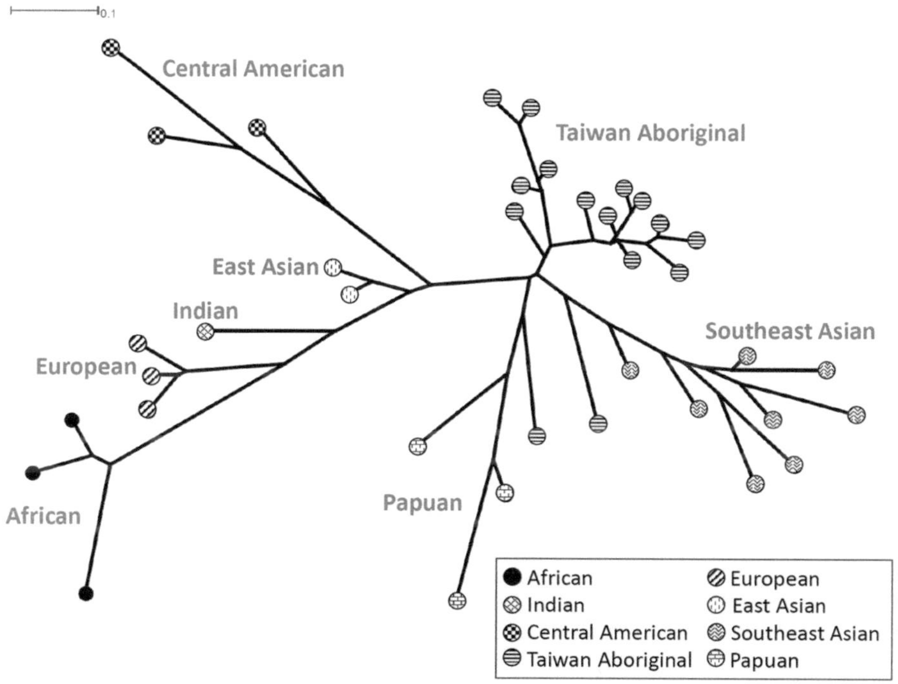

Fig. 9.2 Neighbor-joining tree constructed from Nei's standard genetic distance between populations. Allele frequencies of HLA-A, HLA-B, and HLA-DRB1 were used

corresponding to geographical locations was observed in the tree. In most cases, using the same number of other biallelic genetic loci could not resolve such clustering patterns. This is possible for the HLA locus because each HLA gene has many alleles; for Fig. 9.2, a total of 177 alleles across three genes were used. This demonstrates the utility of using the highly polymorphic HLA region for population genetic studies.

Some examples of a more targeted population study using the HLA loci include confirmation of the admixture model in the current Japanese population (Nakaoka et al. 2013). HLA haplotypes found in the Japanese were derived from the Korean Peninsula and from northern East Asians, which is in agreement with the admixture model for the origins of the Japanese (Hanihara 1991). In Southeast Asia and the Pacific islands, phylogenetic analysis of class II HLA genes was used to show that Papuans from highland areas are more closely related to Australian aboriginals than to Papuans from lowland areas (Mack et al. 2000).

While the high level of HLA polymorphism is beneficial for population genetic studies, it may prove a challenge when performing tissue or organ transplantation from a donor. If the donor's HLA genotype does not match that of the recipient's, there is a likelihood that the recipient's immune system would recognize the donor's HLA molecules as foreign, resulting in an immune response that would leave the transplanted tissue or organ destroyed. Chances of this happening can be reduced if the donor is a closely related family member or unrelated individuals with matching HLA profiles with the recipient. Thus HLA genotyping is routinely performed prior to transplantation to check the compatibility between the donor and the recipient.

References

Alfonso C, Karlsson L (2000) Nonclassical MHC class II molecules. Annu Rev Immunol 18:113–142

Ando H, Mizuki N, Ando R, Miyata Y, Miyata S, Wakisaka K, Inoko H (1996) HLA-C genotyping in the Japanese population using the PCR-SSP method. Tissue Antigens 48:55–58

Bentley G, Higuchi R, Hoglund B, Goodridge D, Sayer D, Trachtenberg EA, Erlich HA (2009) Highresolution, high-throughput HLA genotyping by next-generation sequencing. Tissue Antigens. 74:393–403

Browning SR, Browning BL (2007) Rapid and accurate haplotype phasing and missing-data inference for whole-genome association studies by use of localized haplotype clustering. Am J Hum Genet 81:1084–1097

EMBL-EBI (2017) IMGT/HLA statistics. Retrieved from http://www.ebi.ac.uk/ipd/imgt/hla/stats.html

Felsenstein J (1981) Maximum likelihood estimation of evolutionary trees from continuous characters. Am J Hum Genet 25:471–492

Hanihara K (1991) Dual structure model for the population history of the Japanese. Jpn Rev 2:1–33

Hill AV, Allsopp CE, Kwiatkowski D, Anstey NM, Twumasi P, Rowe PA, Bennett S, Brewster D, McMichael AJ, Greenwood BM (1991) Common west African HLA antigens are associated with protection from severe malaria. Nature 352:595–600

Hosomichi K, Jinam TA, Mitsunaga S, Nakaoka H, Inoue I (2013) Phase-defined complete sequencing of the HLA genes by next-generation sequencing. BMC Genomics 14:355
Kulski JK, Shiina T, Anzai T, Kohara S, Inoko H (2002) Comparative genomic analysis of the MHC: the evolution of class I duplication blocks, diversity and complexity from shark to man. Immunol Rev 190:95–122
Levine JE, Yang SY (1994) SSOP typing of the tenth international histocompatibility workshop reference cell lines for HLA-C alleles. Tissue Antigens 44:174–183
Mack SJ, Bugawan TL, Moonsamy PV, Erlich JA, Trachtenberg EA, Paik YK, Begovich AB, Saha N, Beck HP, Stoneking M, Erlich HA (2000) Evolution of Pacific/Asian populations inferred from HLA class II allele frequency distributions. Tissue Antigens 55:383–400
Meyer D, Single RM, Mack SJ, Erlich HA, Thomson G (2006) Signatures of demographic history and natural selection in the human major histocompatibility complex loci. Genetics 173:2121–2142
Milner CM, Campbell RD (2001) Genetic organization of the human MHC class III region. Front Biosci 6:D914–D926
Nakaoka H, Mitsunaga S, Hosomichi K, Shyh-yuh L, Sawamoto T, Fujiwara T, Tsutsui N, Suematsu K, Shinagawa A, Inoko H, Inoue I (2013) Detection of ancestry informative HLA alleles confirms the admixed origins of Japanese population. PLoS One 8:e60793
Nei M (1972) Genetic distance between populations. Am Nat 106:283–291
Nei M, Tajima F, Tateno Y (1983) Accuracy of estimated phylogenetic trees from molecular data. II. Gene frequency data. J Mol Evol 19:153–170
Saitou N, Nei M (1987) The neighbor-joining method: a new method for reconstructing phylogenetic trees. Mol Biol Evol 4:406–425
Sanchez-Mazas A (2007) An apportionment of human HLA diversity. Tissue Antigens. 1:198–202
Scheet P, Stephens M (2006) A fast and flexible statistical model for large-scale population genotype data: applications to inferring missing genotypes and haplotypic phase. Am J Hum Genet 78:629–644
Shiina T, Hosomichi K, Inoko H, Kulski JK (2009) The HLA genomic loci map: expression, interaction, diversity and disease. J Hum Genet 54:15–39
Stephens HA (2001) MICA and MICB genes: can the enigma of their polymorphism be resolved? Trends Immunol 22:378–385
Terasaki PI, McClelland JD (1964) Microdroplet assay of human serum cytotoxins. Nature 204:998–1000
Wang C, Krishnakumar S, Wilhelmy J, Babrzadeh F, Stepanyan L, Su LF, Levinson D, Fernandez-Vina MA, Davis RW, Davis MM, Mindrinos M (2012) High-throughput, high-fidelity HLA genotyping with deep sequencing. Proc Natl Acad Sci USA 109:8676–8681
Williams TM (2001) Human leukocyte antigen gene polymorphism and the histocompatibility laboratory. J Mol Diagn 3:98–104
Wright S (1949) The genetical structure of populations. Ann Eugenics 15:323–354

Chapter 10
Evolution of Genes for Color Vision and the Chemical Senses in Primates

Shoji Kawamura and Amanda D. Melin

Abstract Primates are generally regarded as visually oriented mammals, trading a sense of smell for good sight. However, recent studies have questioned this simplistic view, and it is not well understood the extent to which senses have evolved interactively or in concert with each other in primates including humans. For example, the number of olfactory receptor genes is not as clearly differentiated between species with different color vision as once asserted. Among senses, receptors of stimuli for vision, olfaction, and bitter/sweet/umami tastes all belong to the G-protein-coupled receptor (GPCR) family, for which the genetic mechanism of signal perception is well understood. Thus, it is now possible to explore the evolutionary correlation among different senses in primates by studying these receptor groups for interspecies divergence, intraspecies diversity, and functional differences among variants. In this chapter, we review recent findings on these receptors and senses in humans and other primates and discuss the future directions of studies on their sensory evolution.

Keywords Color vision · Chemical sense · Opsin · Olfactory receptors · TAS1Rs · TAS2Rs

S. Kawamura (✉)
Department of Integrated Biosciences, Graduate School of Frontier Sciences, The University of Tokyo, Kashiwa, Chiba, Japan
e-mail: kawamura@edu.k.u-tokyo.ac.jp

A.D. Melin (✉)
Department of Anthropology and Archaeology, Department of Medical Genetics, University of Calgary, Calgary, AB, Canada

Alberta Children's Hospital Research Institute, University of Calgary, Calgary, AB, Canada
e-mail: amanda.melin@ucalgary.ca

N. Saitou (ed.), *Evolution of the Human Genome I*, Evolutionary Studies,
https://doi.org/10.1007/978-4-431-56603-8_10

10.1 Introduction to Vision and Chemosensory Genes in Primates

Animals use their senses to find food, detect mates, recognize territories, avoid danger, and perceive social signals. The senses are deeply involved in survival and reproduction and have coevolved with various aspects of anatomy, physiology, and behavior of animals including humans. Different senses have also evolved interactively with each other, reflecting the selective demands operating on sensory systems through multimodal pathways. Thus, study of senses opens unique avenues leading to a better understanding of animal evolution.

Conventionally, senses are distinguished into several classes. Five senses are traditionally recognized for humans: sight (vision), hearing (audition), smell (olfaction), taste (gustation), and touch (mechanoreception). Among them, olfaction and gustation are senses for chemicals in the nasal cavity as odors and on the tongue as tastants, respectively, and are collectively known as chemical senses. The others are senses for physical stimuli: electromagnetic radiation (light) in vision, sound waves in audition, and pressure on the skin in mechanoreception. However, other senses beyond these five exist and include those for temperature (thermoception), pain (nociception), kinesthetic sense (proprioception), balance (equilibrioception), and various internal stimuli. Therefore, there is no firm agreement as to the exact number of different senses. Some senses are present in one species but are absent in another. For example, pheromones (chemicals released by a member of a species that elicits a stereotyped behavior and/or endocrinological response in another member of the same species) are sensed by many mammals through the vomeronasal organ but not by humans of which the organ is vestigial (Touhara and Vosshall 2009). Yet other species, including many fish, use electroreceptors and can detect electric fields (Bullock 1982).

In primates including humans, the genetic basis of stimulus reception is well characterized for vision, olfaction, and umami/sweet/bitter gustation among the traditional five senses. Vision receptors are known as opsins, olfaction as olfactory receptors (ORs), umami and sweet gustation as TAS1Rs (or T1Rs), and bitter gustation as TAS2Rs (or T2Rs), all of which are encoded by genes belonging to the G-protein-coupled receptor (GPCR) family having a seven transmembrane structure in common (Nei et al. 2008). This commonality allows us to apply interrelated experimental setups for forced expression of cloned genes and reconstitution of functional receptors in recipient cultured cells which are often foreign (heterologous) to expression of the gene of interest or production of the receptor encoded by the gene. In the case of opsins, the heterologous expression system enables us to directly determine the absorption spectra of the reconstituted photopigments, a primary feature of photoreceptors, to evaluate their chromatic sensitivity and consequently the color vision of the animal (Yokoyama 2000b). The wavelength of maximal absorption, known as λ_{max}, is widely used as a value representing the entire absorption spectrum. In the case of the chemosensors, the heterologous expression system enables us to search for the stimulus chemicals (ligands) and quantify the receptor's sensitivity to the ligands (Mombaerts 2004). Thus, by focusing these senses, it is now possible to conduct a solid study of receptor-based sensory evolution across different senses for primates.

Among these receptors, opsins have the oldest study history of biochemistry and biophysics (Wald 1968) and molecular genetics (Nathans and Hogness 1983). Gene identification of ORs (Buck and Axel 1991), TAS1Rs (Hoon et al. 1999; Li et al. 2002), and TAS2Rs (Adler et al. 2000) utilized the knowledge generated from study of opsins and was enabled more recently. While various analyses of opsin biochemistry and biophysics, such as conformational changes, photosensitivity, and G-protein activation, require specialized expertise and experimental settings (Shichida and Imai 1998), determination of λ_{max} is relatively feasible, requiring only a simple spectrometric measurement of absorbance for the reconstituted photopigments. On the other hand, sensitivity of chemical sensors to ligands can only be evaluated with an additional system, such as calcium imaging, which monitors whether and how much the cellular cascade of the G-protein-mediated signal transduction is activated upon binding of ligands to reconstituted receptors (Mombaerts 2004). Thus, in comparison to the λ_{max} measurement of opsins, specialized expertise and, often, extra considerations and creativity are required for successful monitoring in the heterologous expression system of chemosensors (Toda et al. 2011). Furthermore, heterologous production of functional ORs is generally less successful and often requires co-introduction of assisting genes such as those encoding receptor-transporting protein family (Zhuang and Matsunami 2008). Due to the longer history of study and experimental feasibility, our knowledge of the evolutionary diversity of primate senses had been primarily accumulated on opsins and color vision. In this review, we first expound the color vision evolution in primates and then introduce recent advancement of studies on evolutionary diversity of their OR, TAS1R, and TAS2R chemical sensors.

10.2 Primate Color Vision and Opsin Genes

10.2.1 A Basic Knowledge on Primate Color Vision

10.2.1.1 Vision Specialization of Primates

Primates are generally regarded as vision-oriented mammals. The visual system of primates, especially that of anthropoids (simians) [catarrhines (humans, apes, and Old World monkeys) and platyrrhines (New World monkeys)], is generally characterized by forward-facing eyes, a postorbital plate (a bony cup surrounding the eye), a fovea (a major central peak in density of cone photoreceptor cells in the retina), and increased representation of the visual centers in the brain cortex (Fleagle 2013). Forward-facing eyes are seen in all primates and enable stereoscopic vision. The postorbital plate is found in simians and prevents the chewing muscles from disrupting eye position, which could serve to improve visual acuity

(Heesy et al. 2007). The simian fovea also allows for very high visual acuity. These features are essential for agile movement and saltatory locomotion from branch to branch and appear to have evolved together with other primate pattern traits, such as grasping hands and feet with nails and an opposable thumb/ toe, retention of the collar bone allowing for a flexible forelimb movement, and the enlarged brain, for a predominantly arboreal and highly social life (Lambert 1987).

Among features of primate vision, of special interest to evolutionary biologists is the unique evolution of trichromatic color vision, which arose from a dichromatic ancestor. Color vision is based on the ability to discriminate light by differences in the wavelength (or hue). At least two different spectral classes of cone photoreceptors are necessary in the retina to perceive differences of wavelength compositions (i.e., colors). Generally speaking, the number of discriminable colors increases as the number of spectrally distinct photoreceptors increases and as the spectral overlap among them is reduced (Vorobyev 2004).

10.2.1.2 L/M and S Opsins and Trichromatic Color Vision in Primates

Vertebrate visual opsins are classified into five phylogenetic types, RH1 (rhodopsin or rod opsin for dim-light vision) and four cone opsins: RH2 (rhodopsin-like or green), SWS1 (short wavelength-sensitive type 1 or ultraviolet-blue), SWS2 (short wavelength-sensitive type 2 or blue), and M/LWS (middle- to long-wavelength-sensitive or red-green) (Yokoyama 2000a). After a brief controversy, these five types are now considered to have been present in the common ancestor of all vertebrates including jawless fish (Collin et al. 2003; Davies et al. 2012). Thus, early vertebrates could already have had four-dimensional color vision (tetrachromacy). Placental mammals maintain only two types of cone visual opsins, SWS1 and M/LWS, in addition to the RH1 rod opsin, and are hence dichromatic in color vision (Jacobs 1993). Conventionally, in the case of therian mammals, SWS1 opsin is called "blue" or "S" opsin, with λ_{max} at around 410–430 nm among primates, and M/LWS opsins are collectively called "red-green" or "L/M" opsins, with λ_{max} at around 530–560 nm among primates (Kawamura et al. 2012).

Primates are the only exception among placental mammals in attaining trichromatic vision. This is made possible by spectral diversification of the L/M opsin alleles of the single-locus X-linked gene in a few lemuriform primates (Tan and Li 1999; Veilleux and Bolnick 2009) and in a majority of New World monkeys (Matsumoto et al. 2014). In this system, all males are dichromatic, but females are either dichromatic or trichromatic ("allelic" or "polymorphic" trichromacy). In catarrhine primates and *Alouatta* (howler monkeys, a genus of New World monkeys) (Jacobs et al. 1996; Dulai et al. 1999; Araujo et al. 2008; Matsushita et al. 2014; Muniz et al. 2014; Silveira et al. 2014), trichromacy was achieved through juxtaposition (duplication) of the spectrally differentiated L/M opsin genes on the same X chromosome (Surridge et al. 2003; Jacobs and Nathans 2009). In this system, both males and females are trichromatic ("routine" trichromacy).

10.2.1.3 The "Three-Sites" Rule and Variation of Primate L/M Opsins

The majority of the spectral variation of primate L/M opsin subtypes is explained by amino acid composition at the residue 180, 277, and 285 ("three-sites" rule) (Hiramatsu et al. 2004; Yokoyama et al. 2008; Matsumoto et al. 2014). The λ_{max} of the L/M opsins with serine, tyrosine and threonine at residues 180, 277 and 285, respectively (denoted SYT), are expected to be approximately 560 nm (Yokoyama et al. 2008). The λ_{max} values of L/M opsins with other three-site combinations can be predicted by subtracting 5, 10, and 17 nm from 560 nm in the case of alanine, phenylalanine, and threonine at residues 180, 277, and 285, respectively (Yokoyama et al. 2008). In addition, interactions between these mutations are estimated to be −2 nm for S180A/T285A, +1 nm for Y277F/T285A, and +4 nm for S180A/Y277F/T285A (Yokoyama et al. 2008).

All eight possible combinations of the three sites are reported for primate L/M opsins (Fig. 10.1). Among them, major five subtypes which spread widely in New World monkeys (SYT, AYT, AFT, AYA, AFA) are estimated to have an antique origin in the common ancestor of New World monkeys (Boissinot et al. 1998). Among the five, AYT and AYA are shared with strepsirrhines and tarsiers, and SYT and AFA are shared with catarrhines (Fig. 10.1). Among the other three, SFT is specific to ateline New World monkeys (spider monkeys, woolly monkeys, and muriquis) (Hiramatsu et al. 2005; Talebi et al. 2006; Matsumoto et al. 2014); SYA is found only in howler monkeys as a relatively common recombinant variant (Matsushita et al. 2014) and in humans as a rare recombinant variant (Hayashi et al. 2006) between SYT and AFA; and SFA is found only in nonhuman catarrhines as a rare recombinant variant (Onishi et al. 1999) and in humans as a common recombinant variant (Deeb 2006) between SYT and AFA. Deviation of observed λ_{max} values from those expected from the "three-sites" rule has been at most 4 nm (Yokoyama et al. 2008; Matsushita et al. 2014). Exceptions to the "three-sites" rule are SYT, SFT, and AFT alleles of atelines (Fig. 10.1), which are short-wave shifted and devoid of the spectral effect of S180A due to unique mutations Y213D and N294 K (Matsumoto et al. 2014).

10.2.2 Visual Opsins and the Evolutionary Origin of Primate Color Vision

10.2.2.1 S Opsin Loss and Monochromacy

Variation of S and L/M opsin genes results in various modes of color vision (monochromacy, dichromacy, polymorphism with dichromacy and trichromacy, polymorphic trichromacy, and uniform trichromacy) at various taxonomic levels among primates. The monochromacy (color blindness) arises in all species of Lorisiformes (lorises and galagos/bushbabies) and *Aotus* (owl monkeys, a genus of New World monkeys) and in various species of cheirogaleid prosimians of Lemuriformes [in genera *Phaner* (fork-marked lemurs), *Cheirogaleus* (dwarf

				Prosimians			Simians (Anthropoidea)							
				Strepsirrhini		Haplorhini								
							Platyrrhini (NWM)						Catarrhini	
							Cebidae				Atelidae			
180	277	285	Expected λ_{max} (nm)	Lorisiformes	Lemuriformes	Tarsiiformes	Callitrichinae	Aotinae	Cebinae	Pitheciidae	Atelinae	Alouattinae	OWM & Apes	Humans
S	**Y**	**T**	**560**				✔		✔	✔	✔(553)	✔	✔	✔
S	**Y**	**T**	**555**		✔	✔	✔							✔
S	F	T	550								✔(538)			
A	**F**	**T**	**545**						✔	✔	✔(537)	✔		
S	F	A	543									✔		✔
A	**Y**	**A**	**536**	✔	✔	✔	✔	✔						
S	F	A	534										✔	✔
A	**F**	**A**	**532**						✔	✔	✔	✔	✔	✔

Fig. 10.1 L/M opsin subtypes in primates distinguished on the basis of the "three-sites" composition. Presence of a gene type is indicated with a tick (rare type is indicated with a smaller tick). At each amino acid site, the longer-wave residue is indicated with red, and the shorter-wave residue is indicated with green. The expected λ_{max} values are given to each subtype according to the "three-sites" rule. The major five subtypes found in Platyrrhini are boldfaced, and other subtypes are indicated with smaller font. *In Lemuriformes, only four species (*Varecia variegata*, *V. rubra*, *Propithecus coquereli*, and *Eulemur macaco flavifrons*) have been reported to retain two subtypes as alleles, while other species examined to date have either one of the two (Tan and Li 1999; Veilleux and Bolnick 2009). †In Tarsiiformes, extant species have either one of the two subtypes, but their common ancestral species is suspected to have had both subtypes as alleles (Tan and Li 1999; Melin et al. 2013b). ‡In Atelinae of New World monkeys (*NWM*), λ_{max} values of SYT, SFT, and AFT are significantly short-wave shifted (λ_{max} indicated in parentheses) from the expectation due to additional mutations and are highlighted with blue (Matsumoto et al. 2014). In Atelinae, two alleles are typically found in each species: SYT and SFT in *Ateles*, SYT and AFT in *Lagothrix lagotricha* (Matsumoto et al. 2014), and SYT and SFT in *Brachyteles* (AFA is only found in *Brachyteles hypoxanthus*) (Talebi et al. 2006). §In Alouattinae, AFT and SYA are recombinant variants recently reported (Matsushita et al. 2014). In Old World monkeys (*OWM*) and apes, SFA is reported as a rare recombinant variant in a macaque species *Macaca fascicularis* (Onishi et al. 1999) and in chimpanzee *Pan troglodytes* (Terao et al. 2005). **In humans, a variety of variants are reported (Deeb 2005; Hayashi et al. 2006). AYT is reported as a recombinant variant for *Saimiri boliviensis* in Cebinae (Cropp et al. 2002) but omitted here for simplicity

lemurs), and *Allocebus* (hairy-eared mouse lemur)] due to loss of S cone or loss of functional S opsin gene by deleterious mutations (Kawamura and Kubotera 2004; Tan et al. 2005; Levenson et al. 2007; Veilleux et al. 2013). These species all share a nocturnal activity pattern.

However, nocturnality does not appear to be a sufficient condition to lose color vision; many other nocturnal primates retain S opsin gene and maintain dichromacy (Jacobs 2013). Rather, the functionality of S opsin gene appears to be maintained by purifying selection in these species (Kawamura and Kubotera 2004; Tan et al. 2005; Perry et al. 2007; Veilleux et al. 2013). Retaining dichromacy in nocturnal strepsirrhines and tarsiers has been suggested as an evidence against the conventional view that ancestral primates were nocturnal, for a hypothesis that the ancestral primates were diurnal or cathemeral and that nocturnality has evolved several times, first in the lorisiforms but much later in other lineages, reflecting different time periods of functional relaxation among lineages (Tan et al. 2005).

10.2.2.2 Dichromacy and Nocturnality

The hypothesis strictly linking S opsin function to diurnality or cathemerality is losing a support due in part to a recent population genetic study which detected a signature of ongoing purifying selection maintaining the S opsin gene in lemuriform nocturnal aye-ayes (*Daubentonia madagascariensis*) (Perry et al. 2007). Aye-ayes have a short-waved shifted S opsin (λ_{max} at 406 nm) (Carvalho et al. 2012), and it is reported that twilight is enriched in short-wavelength (bluish) light with sufficient intensity for aye-ayes with the short-wave-shifted S opsin to perform cone-mediated color vision for their twilight activities (Melin et al. 2012). Another study regards the openness of forest canopy to the sky and the nocturnal activity under moonlight as the main factor influencing the retention of S opsin and color vision in nocturnal lemuriform prosimians (Veilleux et al. 2013). Nocturnal light intensity, particularly short-wave light, is much greater in open canopy forests than in the understory of closed canopy forests (Veilleux and Cummings 2012). Veilleux et al. (2013) found that lemuriform nocturnal species under open canopy habitats generally experience strong purifying selection to maintain the S opsin gene, while, in contrast, those under closed canopy habitats experience weaker purifying selection or a relaxation of selection on it. These studies suggest that dichromatic color vision can be compatible with dim-light activity patterns of ancestral primates.

10.2.2.3 Origin of Trichromacy Under Dim Light

Among strepsirrhines, occasional trichromacy due to the allelic polymorphism of the single-locus X-linked L/M opsin gene has been observed in two diurnal lemurid species [black-and-white ruffed lemurs (*Varecia variegata*) and red ruffed lemurs (*V. rubra*)], one diurnal indriid species [Coquerel's sifaka (*Propithecus coquereli*)], and one cathemeral lemurid species [blue-eyed black lemurs (*Eulemur macaco flavifrons*)] (Fig. 10.1) (Tan and Li 1999; Veilleux and Bolnick 2009). Occasional trichromacy is also suspected in the last common ancestor of crown tarsiers (Fig. 10.1) which is considered to be active in low light, due to the existence of hyper-enlarged eye orbits in the genus (Melin et al. 2013b). With the findings that full moonlight and twilight in tropical forest are sufficient for cone-mediated color vision (Melin et al. 2012), origin of primate trichromacy has recently been suggested in activities under dim (mesopic) light conditions (Melin et al. 2013b). Although more data on genetic variation of opsin genes and color vision are necessary for diurnal, cathemeral, and nocturnal primates, these recent studies challenge the traditional and simplistic view of the diurnal origin of primate trichromacy.

10.2.3 2.3. Visual Opsin Variation in New World Monkeys and Evolutionary Significance of Primate Color Vision

10.2.3.1 Overview and General Implications

New World monkeys are known with their extensive inter- and intraspecies variation of color vision (Fig. 10.1). As introduced above, howler monkeys (*Alouatta*) attained the sex-independent trichromacy through juxtaposition of the spectrally differentiated L/M opsin genes on the same X chromosome (Jacobs et al. 1996; Dulai et al. 1999). On the other extreme, owl monkeys (*Aotus*), the sole nocturnal anthropoid primates, are the cone monochromacy due to the loss of functional S opsin gene (Levenson et al. 2007). While *Cacajao* (uacaris) of Pitheciidae is the last genus whose color vision or L/M opsins remains to be reported, all the other 13 genera encompassing all the three platyrrhine families (Cebidae, Atelidae, and Pitheciidae) (Wildman et al. 2009) are reported to have allelic polymorphism of the single-locus X-lined L/M opsin gene (Fig. 10.1) and exhibit color vision variation, i.e., a mixed population of female and male dichromats and female trichromats (Jacobs 2007; Matsumoto et al. 2014).

A wide variation of allelic composition occurs among them (Fig. 10.1), ranging from diallelic, seen typically in *Ateles* (spider monkeys) and *Lagothrix* (woolly monkeys) (Jacobs and Deegan II 2001; Hiramatsu et al. 2005; Hiwatashi et al. 2010; Matsumoto et al. 2014), up to pentallelic suspected for *Callicebus moloch* (dusky titi monkeys) (Jacobs and Deegan II 2005) [however, three alleles were identified in *Callicebus brunneus* (brown titi monkeys) (Bunce et al. 2011b)]. Triallelic composition is the most widely spread form (SYT, AFT, AFA in Cebinae and Pithciidae, SYT, AYT, AYA in Calitrichinae) (Fig. 10.1) (Matsumoto et al. 2014; de Lima et al. 2015).

Because of the extensive intraspecific diversity of color vision, New World monkeys are the excellent model to study the utility and evolutionary significance of primate color vision. Different L/M opsin alleles confer different phenotypes on trichromacy and on dichromacy. Generally, larger spectral separation between two L/M opsins in trichromats results in higher red-green chromatic resolution (Melin et al. 2009, 2014). Likewise, larger spectral separation between S and L/M opsins in dichromats results in higher blue-yellow chromatic resolution (Osorio et al. 2004). Thus, New World monkeys are suited to evaluate the performance difference not only between trichromacy and dichromacy but also between different trichromat phenotypes and between different dichromat phenotypes.

Based on a simplistic view of selective advantage on higher dimension and resolution of color vision in primates, a number of predictions can be made on the variation of L/M opsin subtypes, foraging performance, reproductive success, and so on for different color vision phenotypes in New World monkeys. Recent studies have tested these predictions, with some supported and others not. Emerging is a more complex and condition-dependent nature of utility and evolution of primate trichromatic color vision.

10.2.3.2 Unexpected Hybrid L/M Opsins in Howler Monkeys

The juxtaposition of L and M opsin genes in howler monkeys was originally reported to consist of the longest (λ_{max} at ~560 nm) and the shortest (λ_{max} at ~530 nm) wave subtypes of L/M opsins in primates, respectively, enabling "normal" and routine trichromacy as seen in catarrhines (Jacobs et al. 1996). Thus, the finding was taken as a supporting evidence of the evolutionary advantage of trichromacy in primates, regarding polymorphic color vision in most New World monkeys as an intermediate stage of primate evolution from dichromacy to trichromacy with the spectrally most separated L/M opsin subtypes (Jacobs et al. 1996).

However, a recent study of natural populations of mantled howlers in Costa Rica and Nicaragua (*Alouatta palliata*) and Yucatan black howlers in Belize (*A. pigra*) found a hybrid L/M opsin gene in each species ("Apa_ML" and "Api_LM", corresponding to AFT and SYA in Fig. 10.1, respectively) with ~10% of frequencies (Matsushita et al. 2014). The λ_{max} of the values of the reconstituted hybrid photopigments are ~546 nm (Matsushita et al. 2014; Melin et al. submitted), which should result in mildly "anomalous" trichromats in humans' term (Deeb 2005) and comparable to those seen in Cebinae carrying an intermediate-λ_{max} allele, who are successful in discriminating stimuli using Ishihara pseudo-isochromatic plates (Saito et al. 2005a). Thus, on the contrary to the prediction, the attained "normal" trichromacy is not maintained in howler monkeys.

10.2.3.3 Unequal Allele Frequencies of L/M Opsins

If trichromacy is simply the best in fitness, allele frequencies of the L/M opsin subtypes are expected to be equal to maximize the number of trichromats. In the platyrrhine species with the single-locus L/M opsin alleles and color vision polymorphism, however, accumulated data have proven that this is not the case. Studies on the wild populations of white-faced capuchins (*Cebus capucinus*) and black-handed spider monkeys (*Ateles geoffroyi*) (Hiramatsu et al. 2005; Hiwatashi et al. 2010) and on other spider monkeys (Jacobs and Deegan II 2001) and muriquis (Talebi et al. 2006) have found that the longest-wave allele is most frequent and the shortest-wave allele is least frequent. Though with different patterns, unequal allele frequencies are also observed in squirrel monkeys (Cropp et al. 2002) or callitrichines (marmosets and tamarins) (Surridge et al. 2005). Deviation from equality could result from selective neutrality among alleles (Hartl and Clark 2007) and also from a selection not simply maximizing the number of trichromacy. A population genetic study of nucleotide sequence variation revealed that the spectrally differentiated L/M opsin alleles are indeed maintained by selection in the wild populations of capuchin and spider monkeys (Hiwatashi et al. 2010).

The skewed allele frequencies toward longer-wave alleles across the spider monkeys, muriquis, and capuchins could imply that the selection consists of complex opposing processes. For trichromats, red-green color discrimination would be greater in individuals having the longest- and the shortest-wave-sensitive L/M alleles than in individuals having an intermediate-wave-sensitive allele. On the other hand, for dichromats, the blue-yellow color resolution would be worst in individuals having the shortest-wave-sensitive L/M allele and be best having the longest-wave-sensitive allele (Osorio et al. 2004). Thus, the longest-wave-sensitive allele would be favored by both trichromats and dichromats, whereas the shortest one would be favored only by trichromats and disfavored by dichromats. Thus, the observed common skew toward longer-wave alleles could indicate that trichromat benefit does not always surpass opposing dichromat benefit and that different alleles could be maintained by different demands among vision types.

However, caution is required to draw a general conclusion because shorter-wave alleles can be favored by dichromats in a context to distinguish bluish fruits from background leaves in their long-distance vision (Melin et al. 2014). Bluish fruits tend to be small, and their importance could be larger for small-bodied primates such as squirrel monkeys and callitrichines. This may explain why the longer-wave skewed pattern is not obvious in these species. Shorter-wave alleles are also found to be favorable over short distances in computer simulation studies of primate foraging tasks (Rowe and Jacobs 2007; Melin et al. 2013a), although increased utility of other senses, such as luminance vision and olfaction, could lessen their advantage during short-range foraging (Hiramatsu et al. 2008, 2009; Melin et al. 2009).

A complexity is also manifested in the evolutionary history of ateline L/M opsin alleles, which appears to favor trichromacy on one side but not on the other side. In most atelines, the shortest-wave allele AFA has been lost or is exceptionally rare (Fig. 10.1) (Matsumoto et al. 2014), which could imply that dichromat benefit surpasses the opposing trichromat benefit. On the other hand, the spectral separation between the remaining two alleles (SYT and SFT in *Ateles* and *Brachyteles*; SYT and AFT in *Lagothrix*) is enlarged by mutations occurred in the ateline common ancestor, resulting in significant improvement of discriminating conspicuous dietary fruits from leaves in the natural habitat of spider monkeys under both bright and dim-light conditions (Matsumoto et al. 2014), which would benefit trichromats. An explanation satisfying both could be that trichromats may tolerate the loss of the shortest-wave allele if the spectral separation of the longest- and intermediate-wave alleles is still sufficient in discriminating stimuli (Saito et al. 2005a) and in foraging performance (Melin et al. 2009) as shown in capuchin monkeys. Conversely, in callitrichines, the spectral separation between the longest and the intermediate alleles is comparable to deuteranomalous human trichromats severely impaired in red-green chromatic discrimination (Deeb 2006) (~5 nm; Fig. 10.1). This may also explain why the shortage of shortest-wave allele is not obvious in callitrichines.

10.2.3.4 Behavioral Studies and Evaluation of Trichromacy Advantage

Limited Support or Contradictive Observations for Trichromacy Advantage

Superior color discrimination abilities of trichromacy are demonstrated in behavioral experiments for captive New World monkeys (Caine and Mundy 2000; Saito et al. 2005a). Finding fruits amid tropical foliage has long been proffered as an adaptive explanation for primate trichromacy. Nevertheless, field observations of free-ranging animals have provided only limited support for simple trichromacy advantage. In a mixed-species troop of saddleback (*Saguinus fuscicollis*) and mustached (*S. mystax*) tamarins, trichromats are further from their neighbors during vigilance than their dichromatic conspecifics. This is explained as resulting from the potentially better perception of predation risk in trichromats (Smith et al. 2005). In a population of white-faced capuchin monkeys (*C. capucinus*), dichromats sniff more figs and take longer foraging sequences than trichromats, especially for cryptic figs, and the trichromat phenotype with the most spectrally separated L/M opsin alleles shows the highest acceptance index for conspicuous figs (Melin et al. 2009). However, there are no differences in feeding rates among phenotypes (Melin et al. 2009).

Results of other behavioral observations of wild New World monkeys have produced equivocal results or results contradictory to the predictions from the trichromat advantage hypothesis. In the wild mixed-species troops of tamarins, the color vision types (dichromatic or trichromatic) do not show a consistent effect on the leadership of the troops to feeding trees (Smith et al. 2003). In other social groups of tamarins (*S. imperator imperator* and *S. fuscicollis weddelli*), no significant difference is detected between females (thought to consist of trichromats and dichromats) and males (all dichromats) in their ability to locate or discriminate between feeding sites (Dominy et al. 2003a). In a population of capuchin monkeys (*C. capucinus*), no significant difference is detected between trichromats and dichromats in feeding or energy intake rates (Vogel et al. 2007). In another population of the same capuchin monkey species, no difference is detected between dichromats and trichromats in time spent foraging on different food types (Melin et al. 2008). In a free-ranging social group of black-handed spider monkeys, dichromats are not inferior to trichromats in frequency, accuracy, and unit-time intake efficiency of detecting fruits (Hiramatsu et al. 2008). This is explained because the luminance contrast of fruits to background leaves is the main determinant of fruit detection in both dichromats and trichromats on the basis of colorimetric measurement of fruits and background leaves (Hiramatsu et al. 2008). In this social group of spider monkeys, irrespective of color vision phenotypes, the monkeys sniff and reject visually cryptic fruits more often than visually conspicuous fruits, implying that color vision is not the sole determinant for ingestion or rejection of fruits (Hiramatsu et al. 2009).

Dichromat Advantage

Dichromat advantage is reported in foraging for camouflaged insects in wild capuchins (Melin et al. 2007, 2010), in wild and captive tamarins (Smith et al. 2012), and in foraging under low light intensity in captive marmosets (Caine et al. 2010). Importantly, dichromats are reported to superior to trichromats at breaking camouflage caused by variegated backgrounds (Saito et al. 2005b). These findings of observational studies in natural environments suggest that the superior ability of trichromats to see the red-green color contrast may not translate into a net selective advantage.

Direct Evaluation of Fitness Effect of Trichromacy

Fedigan et al. (2014) tested whether color vision phenotype is a significant predictor of female fitness in a population of wild capuchins, using 26 years of long-term survival and fertility data. No advantage to trichromats over dichromats for three fitness measures (fertility rates, offspring survival, and maternal survival) was found. This finding suggests that a selective mechanism other than simple trichromat advantage (heterozygote advantage) is operating to maintain the color vision polymorphism. More attention should be directed to field testing the alternative mechanisms of balancing selection proposed to explain opsin polymorphism: niche divergence, frequency dependence, and mutual benefit of association (Fedigan et al. 2014).

Revising Conditions of Trichromacy Advantage

Recent findings imply that the adaptive value of primate trichromacy is conditional rather than universal, depending on the specific ecological demands on animals in their environments (Kawamura et al. 2012). The question does remain about what exactly these conditions are. Fruits, young leaves, predators, and social signals are the main influential visual targets suggested for trichromacy evolution in primates (Sumner and Mollon 2000, 2003; Dominy and Lucas 2001; Surridge et al. 2003; Vorobyev 2004; Fernandez and Morris 2007; Kamilar et al. 2013; Pessoa et al. 2014). Viewing distance of these objects in tropical foliage is also an important factor in primate color vision (Sumner and Mollon 2000). Although trichromatic color vision is useful for short-range tasks (Parraga et al. 2002), detecting fruits from a distance has long been suggested to confer a more important selective advantage to trichromatic primates (Sumner and Mollon 2000). However, an observer cannot definitively know when a monkey has detected an object from a distance, and most investigations of primate color vision have been directed to inspection and ingestion behaviors of foods already at close (<2 m) to moderate (<6 m) distances (Caine and Mundy 2000; Vogel et al. 2007; Hiramatsu et al. 2008; Melin et al. 2009). Thus, the likely advantage of trichromacy on long-range foraging has been devoted to little attention.

In an effort to address this gap, Melin et al. (2014b) reevaluated the trichromacy advantage on fruit foraging of wild capuchin monkeys from a distance by theoretically analyzing computer-simulated conspicuity of fruits from background leaves. In the simulation, trichromatic phenotypes correctly discriminate ca. 70–80% of the total dietary fruit spectra in a tropical forest. In contrast, less than one third of the fruits were discriminable to any of dichromatic phenotypes. This general pattern held for the most heavily consumed diet items, preferred foods, or seasonally critical species eaten during periods of overall food dearth. Furthermore, modeled-trichromatic phenotypes are able to discriminate the vast majority of small patch species, anticipated to provide a high finder's reward. These small resources are suggested to play a critical role in the adaptive value of trichromacy (Bunce et al. 2011a; Melin et al. 2014).

10.2.4 Uniform and Normal Trichromacy in Catarrhine Primates and Exceptional Variation in Human Color Vision

10.2.4.1 Nonhuman Catarrhines Contrasting to Platyrrhines

The L and M opsin genes of catarrhine primates are highly similar in nucleotide sequence (~96% identity) and are closely juxtaposed (Nathans et al. 1986). Thus, they are intrinsically susceptible to recombination and gene conversion between them which could cause hybrid L/M opsin genes, gene loss, and gene multiplication (Drummond-Borg et al. 1989; Ibbotson et al. 1992; Winderickx et al. 1992, 1993; Dulai et al. 1994; Verrelli and Tishkoff 2004). Furthermore, a recombination hot-spot chi element is conserved in the exon 3 among primates (Winderickx et al. 1993). Nevertheless, contrasting to platyrrhine primates, the incidence of color vision variation is remarkably low in nonhuman catarrhine primates (Onishi et al. 1999; Jacobs and Williams 2001; Terao et al. 2005). Among 744 male long-tailed macaques (*Macaca fascicularis*) examined, only three were found to have a single hybrid L/M opsin gene (SFA in Fig. 10.1) and to be dichromats (Onishi et al. 1999; Hanazawa et al. 2001). Among 58 male chimpanzees (*Pan troglodytes*), one was found to have a hybrid L/M opsin gene (SFA in Fig. 10.1) in addition to one normal M opsin gene on the X chromosome and to be an anomalous (protanomalous) trichromat (Saito et al. 2003; Terao et al. 2005). Thus, frequencies of color vision variants in male long-tailed macaques and male chimpanzees can be calculated to be ~0.4% and ~1.7%, respectively. These frequencies could be overestimated because no variants were found in 455 male monkeys from other macaque species (Onishi et al. 1999), and the chimpanzees examined were from limited numbers of breeding colonies (Terao et al. 2005). Other researchers have reported an absence of color vision defects in Old World monkeys and apes (Jacobs and Williams 2001; Verrelli et al. 2008; Hiwatashi et al. 2011).

Multiple copies of M opsin genes are likely to increase the frequency of unequal recombination events. In humans multiple M copies are found in 66% of males of European origin (Drummond-Borg et al. 1989) and 56% of Japanese males (Hayashi et al. 2001). Regarding nonhuman catarrhines, some studies report that multiple M copies are rare (Onishi et al. 1999; Terao et al. 2005; Verrelli et al. 2008) yet other studies report that they are common (Ibbotson et al. 1992; Dulai et al. 1994; Hiwatashi et al. 2011). Thus, among Old World monkeys and apes, there seems to be no clear trend on the copy number variation of M opsin gene.

A study of gibbon population samples showed that gene conversion has homogenized L and M opsin genes in introns (Hiwatashi et al. 2011). However, purifying selection against the homogenization has protected the nucleotide difference between L and M opsin genes in centrally located exons, exons 3 and 5 in particular, which include the spectrally crucial "three sites" (Hiwatashi et al. 2011). This confirms that gene conversions (and perhaps other forms of recombination) do occur between L and M opsin genes in nonhuman catarrhines, but the genes are eliminated from the population by natural selection if gene conversions affect the gene region relevant to spectral difference between L and M opsins. In nonhuman catarrhine primates, even mildly anomalous trichromats have not been found, suggesting a severe selective disadvantage on color vision variants.

The strict conservation of the normal trichromacy in nonhuman catarrhines is in sharp contrast to New World monkeys. The higher frequency of anomalous trichromacy in New World howler monkeys implies that the selective pressure to maintain "normal" trichromacy is lower in the neotropics (Matsushita et al. 2014). However, in New World monkey species with polymorphic color vision, genetic studies have shown that the spectrally different alleles of the L/M opsin gene are actively maintained by balancing selection (Boissinot et al. 1998; Hiwatashi et al. 2010). It is an open question whether the selection is for maintaining (1) simply heterozygotes of L/M opsin alleles, i.e., trichromacy per se, (2) dichromacy and trichromacy, or (3) subtypic variation in dichromacy and/or trichromacy. It is also an open question as to whether the difference between nonhuman catarrhines (uniform and normal trichromacy) and platyrrhines (polymorphic color vision) is attributable to a (1) biogeographic differences among continents, e.g., the severity of seasonality, or a prevalence of drably colored fruits and asynchronous species, e.g., figs and palm fruits (Dominy et al. 2003b); (2) dietary variability, e.g., degree of dependence on insects, leaves, or colorful fruits and different food patch sizes (Melin et al. 2014); (3) variation in social color signals (Fernandez and Morris 2007; Kamilar et al. 2013).

10.2.4.2 Uniqueness of Human Color Vision

Among catarrhine primates with routine and normal trichromacy, humans constitute a notable exception, in which deletion and multiplication of L/M opsin genes and creation of hybrid L/M opsin genes cause relatively high incidence of dichromacy and anomalous trichromacy, approximately 3–8% of males being color vision "defects" (Deeb 2006). Many humans have multiple copies of the M opsin gene in the L/M opsin gene array where the most upstream gene is typically L and the others are M. Only the upper two genes are expressed and when a hybrid gene occupies either position, it causes anomalous trichromacy (Hayashi et al. 1999). When there is only one L/M opsin gene on an X chromosome or when the two positions are occupied by the same genes, this causes dichromacy (red-green color blindness: more specifically, protanope, when L is lost, and deuteranope, when M is lost). These are typically found in men because women have two X chromosomes and thus are more likely to have a "normal" gene array in either one. There are rare cases of individuals, irrespective of sex, who lack functional blue cones (tritanopes, <1:10,000) due to mutations in S opsin gene on chromosome 7 (Sharpe et al. 1999).

The high incidence of color vision variation in humans can be interpreted most conservatively as a result of relaxation of the selective constraint to maintain the spectral difference between the L and M opsin genes. Alternatively, by one step further, adaptive explanations may be possible. Females with a hybrid L/M opsin likely have normal L and M opsins and are able to perceive finer chromatic distinction (Jameson et al. 2001) (a question remains, however, why this is not selected in nonhuman catarrhines). Studies of New World monkeys provide a profound inference on human color vision variation. The persistence of dichromats in the human population may reflect, as noted above, some advantage in achromatic visual tasks and in having different color vision morphs in a population. Humans are atypical primates, having largely left the foliated environments of forest some million years ago. Then, approximately 2 million years ago, hominins started to devise stone tools and included hunted meat as a considerable portion of their diet. Finally, the increased brain size eventually led to the development of agriculture some thousand years ago and the building of a modern industrial society only a few hundred years ago. The persistence of color vision variation in humans could be related to any of these major events: life outside forest reduces the need for color vision, hunting might have benefited by the presence of dichromatic group members, gathering might have benefited by the presence of hybrid L/M opsins, or large agricultural or industrial societies could isolate humans from selection against dichromacy and anomalous trichromacy. It would be crucial to infer when the color vision polymorphism is spread into population as today in human evolution by analyzing L/M opsin nucleotide diversity collected from global ethnic groups.

Finally, color vision is not the sole sense showing diversity among and within species of primates including humans. Interplay of sensory modalities gathers a recent attention in the study of sensory evolution in primates (Hiramatsu et al. 2009). In the following sections, we review recent findings on evolutionary diversity of olfactory, bitter and sweet/umami receptors in primates.

10.3 Primate Olfaction

Over the past decade, there has been an increasing awareness of the importance of olfaction to primates and the diversity of olfactory abilities within the order, as evidenced by variation in olfactory behavior, gross anatomy, and olfactory receptor genes. Scientists have attempted to assess the olfactory capabilities of primates by integrating data from different sources, including: (1) examining the number and diversity of genes underlying olfaction and the proportion of those genes that are functional versus pseudogenized; (2) examining the size of the various olfactory structures, both in proportion to overall brain size and in absolute terms; (3) conducting behavioral experiments examining the breadth of volatile chemicals that can be perceived by humans and nonhuman primates and the sensitivity of primates to these different chemicals; (4) observing the foraging behaviors of primates in captivity and in the wild to measure the context and extent of olfactory behaviors (e.g., sniffing), as well as the odorants that are found in items of import, such as foods and scent marks; and (5) in vitro expression of olfactory receptors in heterologous gene expression systems. Here we synthesize current progress in these areas and suggest areas for profitable future research.

Mammals possess two distinct olfactory systems, with some limited functional overlap: the main olfactory system (MOS) and the vomeronasal system (VNS) (Fortes-Marco et al. 2013). Although absent in many primates, the VNS is possessed by most mammals and primarily involved in sociosexual communication via detecting pheromones, which are relatively heavy, nonvolatile compounds (Touhara and Vosshall 2009). The MOS appears to be adapted to detecting volatile (airborne) chemicals and is considered to be widely useful in ecological contexts, for example, during terrestrial chemotaxis toward odorants plumes of fruit trees (Hudson 1999). However, the MOS may also play a role in pheromone detection, especially among those primates lacking a VNS (Martinez-Garcia et al. 2009). In this chapter we primarily focus on the ecological contexts of olfaction, and therefore focus on the MOS, and the olfactory receptor (OR) genes, which are expressed in the neurons of the main olfactory bulb.

10.3.1 Evolutionary Trends Among Olfactory Receptor Genes: Numbers, Diversity, and Preservation of Function

The sense of smell is capacitated by olfactory receptors in the nasal epithelium, which send signals to the MOB in response to the binding of odorants (Hasin-Brumshtein et al. 2009). Many attempts to understand the olfactory capability of animals have therefore focused on the genes encoding these receptors. The olfactory receptor (OR) gene family is one of the largest in the mammalian genome (Buck and Axel 1991; Mombaerts 2004) and is currently recognized to cluster into

13–18 subfamilies, which are distributed among all but 1 autosome (20 in human) and also absent from the Y chromosome (Hayden et al. 2010). In mammals, OR genes fall into two distinct classes. Class I ORs tend to bind hydrophilic odorants and are most common among aquatic mammals, while Class II ORs hydrophobic odorants and are more numerous in the genomes of terrestrial mammals (Niimura and Nei 2005; Saito et al. 2009; Hayden et al. 2010; Niimura et al. 2014). OR genes tend to be densely clustered into groups ranging from 5 to more than 100 genes arranged tandem, typically without other genes interspersed, and separated by intergenic distances of 1100 bases on average. The structure of OR genes is beautifully simple. Each is about 1 kb in length and intronless (Young et al. 2003). Like opsins, ORs are seven transmembrane G-coupled protein receptors, and a single OR gene is expressed per sensory neuron (Ressler et al. 1993; Vassar et al. 1993). Despite this, a "many-to-many" relationship between odorant molecules and ORs leads to the results that a single molecule may trigger many receptors and a single receptor will bind many odorant molecules (DeMaria and Ngai 2010). This compounding effect predicts the discrimination of trillions of different odors in humans and potentially many more in some other mammals (Bushdid et al. 2014). Given the size and complexity of the OR family, this group of genes holds great promise to reveal in fine detail the evolutionary history and adaptive nature of the sense of smell, as well as present considerable challenges associated with data collection and analysis.

Frequent attempts to use OR genes as a tool for gaining insight into olfactory biology focused on examining the proportion of functional: pseudogenized OR genes possessed by different species. Due to data limitations, early research in this field was constrained to focus on a small number of species, on one or few individual(s) per species, and also on a subset of the total OR genes. Conclusions from this work done on primates were initially presented in favor of the hypothesis that a sensory trade-off between smell and vision in primates. Convergent decreases were reported in the proportion of functional OR genes in catarrhine primates and howler monkeys, which was interpreted as a consequence of the acquisition of routine trichromatic vision (Gilad et al. 2004, 2007). Later work, drawing on larger datasets and analyses focused on the total number of functional OR genes and on shared OR orthologs, refuted this hypothesis and argued that degeneration of olfactory receptor gene repertories in primates did not have a clear link to the acquisition of full trichromacy. Rather than uniform loss among primates with a broader spectrum of color vision, OR gene evolution appears to be highly species-specific and influenced by ecotypes (Go and Niimura 2008; Hayden et al. 2010; Matsui et al. 2010). Together, these data suggest that OR genes are subject to strong positive selection, can evolve quickly following speciation events, and are not uniformly linked to color vision variation (Go and Niimura 2008; Zhou et al. 2014). Despite this, the hypothesis of a simplistic trichromacy-olfaction "trade-off" unfortunately remains well cited (Hayden et al. 2010; Jones et al. 2013; Zhou et al. 2014). Although a clear link between acquisition of trichromatic color vision and reduction of olfactory ability is not strongly supported, the concept of a sensory trade-off in primate evolution may still hold credence (Niimura and Nei 2007). A strong negative correlation is described between visual acuity and the size of the vomeronasal organ, suggesting that in some primates, vision may in part have replaced the role of pheromone-based olfaction in sociosexual communication (Garrett 2015).

Earlier studies revealed that primates have many fewer OR genes than do mice or dogs and also a greater proportion of pseudogenized OR genes (Gilad et al. 2004; Niimura and Nei 2007; Nei et al. 2008). This observation contributed to hypotheses that primate olfaction has greatly decreased over evolutionary time, from an ancestral "macrosmatic" state, to a current "microsmatic" state (Smith et al. 2004). However, recent work is causing a reevaluation of this line of thinking.

Ancestral state reconstructions suggest that primates (along with afrotherians and rodents) by and large appear to have maintained the ancestral functional OR genes, despite significant variation in pseudogenes in these taxa (Hayden et al. 2010; Niimura et al. 2014). Among all mammals, rates of OR births and deaths are much faster than in other gene families, and rates vary among lineages. Birth of new genes is faster among rodents and elephants, while gene death rates are accelerated among primates (Niimura et al. 2014). Thus, rather than a large decrease in OR gene repertoires among primates, there may have been extensive births of new OR genes among rodents (Nei and Rooney 2005). In agreement with the previous study (Niimura and Nei 2007), it is reported that hundreds of gains and losses of Class II OR genes occurred independently in different lineages (Niimura et al. 2014). With increasing data, we are also seeing little to no trend between total OR gene diversity and function and the proportion of pseudogenized OR genes. Although mice and dogs have relatively few OR pseudogenes, relative to a large number of functional ORs, some species have huge numbers of functional OR genes as well as large numbers of pseudogenes (e.g., elephants, guinea pigs (Niimura et al. 2014)). Therefore, using the ratio of functional to pseudogenized OR genes as a useful metric of olfactory breadth or ability is inappropriate. Furthermore, different species with similar numbers of OR genes may possess OR gene repertoires that differ considerably, as can be seen between humans and chimpanzees – suggesting evolutionary processes acting on ORs are very dynamic (Matsui et al. 2010). Finally, a recent examination of the golden snub-nosed monkey genome revealed that a large subset of OR genes were rapidly evolving and likely to be under positive selection (Zhou et al. 2014). This monkey is believed to be generally sensitive to the perception of fruity, lemony, and floral/woody odors, consistent with its plant-based odor perception. At the same time, the overall OR gene repertoire was lower than in other catarrhines. Together these results might suggest narrowing olfactory niche specialization. In sum, these observations reject the hypothesis that primates have an exceptionally poor sense of smell, and the extent would be high to which natural selection acts on primate olfaction.

Overall, surveying the total number of functional genes is likely to be a revealing metric of olfactory breadth (but not sensitivity, see section on sensitivity below) and a wiser strategy than targeting a portion of the OR genes and examining the proportion of those that are pseudogenized as an approximation of olfactory ability. Examining the total functional ORs, it is unquestionable that other mammals have a better sense of smell than primates, but this may reflect large expansion of olfactory systems in the former rather than large declines in the latter. The dismissal of primate olfaction as an important sensory system, as well as the logistical difficulties in assaying its variation and function, has caused us to leave unexamined the insights that olfactory ecology can provide into primate adaptation and evolutionary radiation. With continuing advances in the field, laboratory, and bioinformatics analyses, hopefully we will continue to break free from this crippling paradigm.

10.3.2 Genetic Diversity of OR Gene Repertoire in Modern Humans

Just as primates were viewed as having exceptionally poor sense of smell among mammals, humans have traditionally been viewed as being exceptionally poor among primates in part because of human-specific OR gene losses relative to chimpanzees (Hasin et al. 2008). Yet, this opinion too is changing (Table 10.1). Genetic variation of the OR gene repertoire in humans is now known to be much higher than previously thought. Although some authors argue that patterns of copy number variation in human OR genes is suggestive of neutral evolutionary processes (Nozawa et al. 2007), a growing consensus is that the majority of functional OR genes are maintained by natural selection. It is now known that, although different in composition, humans have functional OR gene complements that are similar in number, ~400 (and in proportion pseudogenized, ~51%) to other Old World primates, including chimpanzees (Go and Niimura 2008; Matsui et al. 2010), which contradicts earlier reports that humans have a fewer functional OR gene

Table 10.1 OR gene numbers and repertoires in primates

Common name	Functional[a]	Pseudogenes[b]	Total	Prop. Pseudo.	References
Human	392	465	857	0.46	HORDE database (Nov 17, 2014)
Human	387	415	802	0.48	Go and Niimura (2008)
Human	387	466	853	0.55	Hughes et al. (2014)
Neanderthal		473			Hughes et al. (2014)
Denisovan		469			Hughes et al. (2014)
Chimpanzee	380	433	813	0.47	Go and Niimura (2008)
Orangutan	296	525	821	0.64	Niimura et al. (2014)
Rhesus macaque	309	297	606	0.51	Go and Niimura (2008)
Marmoset	366	258	624	0.41	Matsui et al. (2010)

Variation in number reported for humans likely reflects how segregating pseudogenes are handled and discovery of new genes over time

[a]Some variation expected due to a large number of segregating pseudogenes

[b]Includes truncated

complement than apes do (Gilad et al. 2004). Rather than humans having fewer OR genes than chimpanzees, we now know that functional OR gene repertoires have diminished in parallel in humans and chimpanzees from the time of their last common ancestor (Matsui et al. 2010). However, if OR gene function and persistence is a reliable indicator of the importance of smell or olfactory function (see below for discussion of this topic), then humans are expected to have a better sense of smell than some primates. Somewhat surprisingly, humans have also retained a greater number of ancestral OR genes present in the last common ancestor of anthropoid primates and are more similar in repertoire to New World monkeys than to other extant catarrhine primates (Matsui et al. 2010).

Other lines of evidence speak to a human olfactory system that is much better than previously conceived. The central olfactory regions of the brain that process input from the olfactory receptors are more extensive in humans than is typically recognized (Shepherd 2004). These areas include regions used in simple smell detection and recognition: the olfactory cortex, the olfactory tubercle, the entorhinal cortex, and regions of the amygdala, hypothalamus, mediodorsal thalamus, orbitofrontal cortex, and the insula (Neville and Haberly 2004). Additionally, olfactory memory and discrimination of complex stimuli is serviced by other areas of higher-level processing, including the temporal and frontal lobes (Buchanan et al. 2003). This expanded repertoire of higher brain mechanisms could indicate olfactory processing in humans has been shifted to more cognitive areas and may permit increased processing of complex smells, such as those produced by human cuisines. Yet another line of evidence comes from one of the more exciting reports of olfactory function among humans using psychophysical testing revealed that, contrary to previous estimates that humans can discriminate ~10,000 odors, the actual numbers are likely in the trillions (Bushdid et al. 2014). Such impressive performance means the olfactory system greatly outperforms the other sensory systems in number of stimuli detected. Furthermore, each person likely has a highly personalized olfactory experience, given the extensive OR genetic diversity (Keller et al. 2007; Olender et al. 2012; McRae et al. 2013).

The extent of interindividual OR genetic variation among humans in gene sequences, copy number, and functionality is remarkably large. Recent studies estimate that allelic variants nearly double the count of loci in the human genome and more than half of the OR genes are subject to copy number variation (CNV) (Hasin et al. 2008; Olender et al. 2012). In total, approximately 66% of the ~400 OR loci are afflicted by CNV and nonfunctional SNPs. Given the frequency and expansive nature of CNVs (which can span 11 OR loci), it is likely that this form of genomic variation may be responsible for a majority of phenotypic differences (Hasin et al. 2008). However, segregating nonsense SNPs (present in some members of the population, but not all) accounts for at least some documented difference in odor perception (Keller et al. 2007; Menashe et al. 2007). Given the unusually high genetic diversity and the allelic exclusion during expression in the receptors, the human sense of smell is likely to be highly individualistic, explaining the occurrence of specific anosmia or inability to smell certain odors (Olender et al. 2012).

Advances in ancient DNA extraction, sequencing, and analysis have recently given us a window into potential olfactory adaptations of extinct relatives. Novel OR gene losses and gains have been described for both Neanderthal and Denisovan (Hughes et al. 2014). The presence of OR pseudogenes in Neanderthals and Denisovans that are segregating in humans (functional in some people and pseudogenized in others) was also detected and shared the same stop codon mutations. Overall, the authors conclude both Neanderthals and Denisovans had a larger number of OR pseudogenes than modern humans, and suggested that in Neanderthals, this might be reflected in small size of the olfactory bulb that had previously been noted (Bastir et al. 2011), although the authors acknowledge limitations of correlating OR genes, olfactory bulb size, and olfactory behavior and sensitivity (discussed in following sections below).

10.3.3 Olfactory Sensitivity of Primates

It is important to highlight the distinction between olfactory breadth and olfactory sensitivity. Here, we use the former to refer to the number of different odors an animal can perceive, while the latter refers to the concentration threshold at which a particular odor can be detected. While it is unclear which component(s) of olfactory ability the number of functional OR genes speaks to, it is likely that they reflect the breadth of an animal's olfactory repertoire more so than the sensitivity (Niimura et al. 2014), although it is conceivable that multiple ORs with similar structures could increase sensitivity to a given odorant.

What then determines olfactory sensitivity? The threshold of detection for a given odorant is attributed to the number of ORs in the nasal epithelium (NE) that can successfully bind that odorant. After being triggered, ORs send an action potential through their axons, which traverse the cribriform plate of the ethmoid bone, to the main olfactory bulb (MOB). A strong positive correlation between size of the MOB and the population size of ORs and surface area of the NE has long been established. As such, MOB and NE size and complexity (or their bony correlates in fossil animals) are often used as a proxy for olfactory sensitivity, although rich debate ensues over the extent to which size should be taken as an absolute metric versus scaled in proportion to body size or brain size and whether it should be compared relative to the size of other sensory systems (Smith et al. 2007). Consider the following illustration, it is true that the olfactory system of primates is smaller, relative to the visual system, than the olfactory system of many other mammals. This may suggest that olfactory sensitivity has decreased in importance over evolutionary time in primates. However, we can see that the primate visual system has drastically increased in size, so even if the size of the olfactory system remained unchanged from a common ancestor, we would still make this same observation that it was relatively small, even though olfaction might have remained equally important to primates over evolutionary history.

Another important point is effects of the choice of the comparative outgroup(s). If the organisms used to represent the state of last common ancestor have an olfactory system that is derived in their lineage – for example, has undergone expansion since the split with primates – then primates will appear to have decreased olfactory abilities in comparison. In a recent study, the olfactory bulb volume (and total brain volume) was estimated by Bayesian methods at ancestral nodes of major primate clades (Heritage 2014). It rejects a classic two-stepped, grade-shift model of decreasing importance of olfaction first in the ancestor of all primates and then again following the split of strepsirrhines and haplorrhines. Rather, it finds evidence of increases in MOB size (both absolute and relative to brain size) in strepsirrhine evolution relative to the ancestral crown primate, and decreases uniquely in the haplorrhines, with further convergent decreases in platyrrhines and in cercopithecoids.

The relative sizes of olfactory structures and associated neural regions can reveal a great deal about olfactory sensitivity. Among nocturnal strepsirrhines, species that consume more fruit have a larger olfactory bulb (Barton et al. 1995), suggesting they have more sensitive olfaction than diurnal species, a suggestion that is reinforced by behavioral studies of olfactory behavior in nocturnal primates (reviewed in the next section). However, what metrics of NE, MOB size, and complexity cannot tell us is whether their associated ORs code a few or many different OR genes and thus whether an animal is extremely sensitive to certain odors, less sensitive to many odors, or any combination thereof (Shepherd 2004). An animal with a large NE could have good breadth but fairly low sensitivity to any particular odor or vice versa. Alternatively, both breadth and sensitivity could be high or low; these variables need not co-vary in concert; indeed the total number of functional OR genes was not linked to size of MOB in a diverse group of mammals (Hayden et al. 2010), and behavioral studies have revealed primates can be more sensitive to certain compounds than other mammals, despite having fewer functional OR genes.

One of the larger, recent surprises in the field of primate sensory ecology came when it was revealed that the sensitivity of monkeys to certain odors exceeded that of typically "macrosmatic" animals, including dogs and rats (Laska et al. 2000). The odorants to which primates were very sensitive are common in fruity odors, and also included ethanol, which may have played an important role in the evolution of primate frugivory (Carrigan et al. 2015). These findings indicate that olfaction may play a crucial and underestimated role in primate behavioral ecology.

10.3.4 Sensory Ecology and Cross-Modal Usage of Olfaction in Primates

In a satisfying response to a growing recognition that olfaction has long been the neglected sense in primatology (e.g., Heymann (2006)), the last decade has produced several studies that are beginning to reveal how, when, and why primates use their sense of olfaction. In this section, we review the contributions of some of these studies and of their predecessors. Although these efforts have generated an increased awareness of the importance of olfaction to the sensory ecology of primates, they just scratch the surface of this complex and fruitful area of research. With advances in genomics (reviewed above), gene assay systems (section below), and increased multidisciplinary efforts to integrate behavioral studies in captivity and the wild with those measuring genetic diversity, gene function, and odorant composition and propagation in nature, the next decade is anticipated to catapult forward this exciting avenue of research.

10.3.4.1 Uses of Olfactory Behavior

In additional to our rapidly expanding understanding of how olfaction is used in social contexts (Klailova and Lee 2014), we are beginning to investigate how useful olfactory cues are during foraging. Primates and other animals are able to use their sense of smell under natural conditions to locate many of the food items important to their diets, including fruits, flowers, and invertebrate prey (Siemers et al. 2007). Olfaction is predicted to work over distances further than touch and vision, but to not be as far-reaching as audition (Dominy et al. 2001), making it rather intermediate in functional distance relative to the other senses. Its effectiveness should be highly susceptible to interference from local environmental conditions, including wind and humidity. It is likely that animals are most successful in following odor plumes under still, relatively humid conditions, but few studies tackle the effect of environmental variation on the use of olfaction or measure the propagation of odorants in different environments. Such areas would be a rich playing field for future work.

Unlike vision, olfaction is a sense that functions well during the night, and nocturnal primates – like other nocturnal mammals – are reported to use olfactory signals more often that diurnal species. For example, studies manipulating feeding platforms have found that owl monkeys used olfaction more effectively than sympatric diurnal primates (Bicca-Marques and Garber 2004). Plants seem to respond to the sensory systems of their seed dispersers by investing in olfactory cues over visual ones if the most profitable consumers are nocturnal. Bat-dispersed figs, for example, are large, green, and highly odiferous, whereas bird-dispersed figs are red ripe and do not smell as strongly (Melin et al. 2009). Recent studies have also found other evidence that visual – olfactory signaling trade-offs occurs among fruiting plants with varying dispersal syndromes; in particular, fruits with blue pigmentation (a costly color to produce, but one visible to a majority of vertebrates) tend not to produce volatile odorants (Valenta et al. 2015). Although some evidence of sensory trade seem to be evident at both the level of plant signaling and – to a yet uncertain extent – in the physiology of primates, with some species being more vision dependent and others more reliant on olfaction, it is also becoming increasingly clear that primates use multiple sensory modalities during foraging

(Hiramatsu et al. 2009; Melin et al. 2009), and that sensory integration in the brain is widespread (Kuang and Zhang 2014). Primates can often use one sense in lieu of another if the preferred mode becomes unavailable (Siemers et al. 2007; Rushmore et al. 2012). As such, we may not predict that the eventual loss of any sense is inevitable but that in many cases maintenance of multiple senses will be favored if together they confer higher fitness than their metabolic costs to maintain. Studies estimating the "running costs" versus foraging advantages attributed to different sensory machinery would shed some light in this area.

10.3.4.2 Sensory Integration: Olfaction, Vision, and the Other Senses (Taste, Touch, Sound)

The majority of research investigating the use of olfaction by primates has evaluated olfactory behavior relative to the use of other senses, primarily vision. Often visual and olfactory cues seem to provide synergistic information to foragers, and olfactory cues have been found to supplement, reinforce, or augment visual cues (Siemers et al. 2007; Melin et al. 2009; Valenta et al. 2013; Kuang and Zhang 2014). While gray mouse lemurs (*Microcebus murinus*) are able to locate invertebrate prey using acoustic, olfactory, or visual information alone, detection performance increases with the number of different senses used (Siemers et al. 2007).

The feeding ecology of the primates seems to play a major role in the relative use of different senses. Multiple studies have reported that primates that frequently consume fruits are more likely and able to use olfactory cues to identify preferred foods. Rushmore et al. (2012) found that frugivorous lemurs and generalist lemurs (frugivore/insectivores) could use olfaction along to complete a foraging task and that frugivores used olfactory cues with equal frequency to visual cues, while folivores required both vision and olfactory cues combined to reliably identify their preferred foods, and both folivores and generalists relied more heavily on vision. Similarly, Laska et al. (2007) report that highly frugivorous monkeys used their sense of smell more often than insectivorous squirrel monkeys, which relied more heavily on touch. Similar differences in the higher use of touch during fruit foraging by generalist capuchins relative to spider monkeys is known (Hiramatsu et al. 2008; Melin et al. 2009). The specialized hand morphology of spider monkeys that permits such efficient travel may also limit their ability to use touch and perhaps make them more reliant on their other senses. Overall the increased use of olfaction over vision by frugivores relative to folivores may indicate that (1) olfactory cues of fruits are more informative or reliable than olfactory cues of leaves or (2) visual cues of leaves are more informative or reliable than olfactory cues of fruits. The subtleties of these points should be investigated in detail. However, our preliminary evidence suggests that fruit chromaticity is not as informative as odorants are in signaling fruit nutritive quality, while leaf chromaticity appears to be a consistently reliable indicator of age, toughness, and nutrition (Melin and Kawamura, unpublished data).

Multimodal sensing appears to be especially important when assessing novel foods, and as familiarity increases, sensory streamlining tends to occur, which likely improves foraging efficiency (Laska et al. 2007). The use of multiple senses including increased use of olfaction, mechanosensation (biting and touching), and gustation (taste, i.e., licking) in addition to vision during food investigation sequences has also been reported to be more common in cases of deprivation of chromatic cues, dichromatic capuchins use olfaction, and other sensory investigations when selecting ripe figs more often than trichromatic capuchins do, and all monkeys were more likely to use multimodal sensing when fruit ripeness was not cued by color change (Melin et al. 2009). Similar reports are known for spider monkeys (Hiramatsu et al. 2009), suggesting some generalizability of this phenomenon.

A missing piece of the puzzle is identifying which odorants are useful in different contexts and for different food items. Recent work in the field has seen the use of field systems for capturing and transporting volatile odorants, which can then be measured via GCMS (Kawamura et al. unpublished (Valenta et al. 2013)). These studies are revealing associations between odor profiles and feeding behaviors. When coupled to specific abilities of ORs to bind different aromatics, we will be able to extend this to our study of olfactory receptor genes.

10.3.5 Heterologous Gene Expression Assay for Current and Ancestral ORs in Primates

The use of the heterologous gene expression assay is proving to be an invaluable tool in pursuing which odors the OR genes respond to, i.e., function and specificity of ORs. While determination of λ_{max} of visual opsins is relatively feasible and requires only a simple spectrometric measurement of absorbance for the reconstituted photopigments, sensitivity of chemical sensors to ligands can only be evaluated with an additional system, such as calcium imaging (Touhara et al. 1999). This monitors whether and how much the cellular cascade of the G-protein-mediated signal transduction is activated upon binding of ligands to reconstituted receptors (Mombaerts 2004). Thus, compared to the λ_{max} measurement of opsins, specialized expertise and often extra cares and inventions are required for successful monitoring in the heterologous expression system of chemosensors. Heterologous production of functional ORs has been greatly improved by elaborate co-introduction of assisting genes such as those encoding receptor-transporting protein family (Zhuang and Matsunami 2008).

Using the heterologous gene expression system, Keller et al. (2007) showed that a human odorant receptor, OR7D4, is selectively activated in vitro by androstenone and the related odorous steroid androstadienone and does not respond to many other odors examined. They demonstrated that genotypic variation in OR7D4 accounts for a significant proportion of the pleasantness or unpleasantness and intensity variance in perception of these steroidal odors (Keller et al. 2007).

Zhuang et al. (2009) reconstructed hypothetical ancestral orthologs of OR7D4 of primates. By functional analysis of these orthologs, they found an extremely diverse range of OR7D4 responses to the ligands in various primate species. They identified a number of nonsynonymous changes causing increases or decreases in sensitivity during evolution of Old World monkeys and hominoids. Interestingly, the majority of these changes were not predicted by the maximum likelihood analysis inferring positive Darwinian selection, stressing the importance of functional assay and verifying theoretical predictions (Zhuang et al. 2009).

Adipietro et al. (2012) extended the functional analysis to other OR genes for evolutionary changes. By examining 18 OR orthologs from chimpanzee and rhesus macaque and 17 mouse-rat orthologous pairs that are broadly representative of the OR repertoire, they found that 87% of human-primate orthologs and 94% of mouse-rat orthologs were different in receptor potency and efficacy to an individual ligand (Adipietro et al. 2012). These studies show that potency and/or efficacy of ORs dynamically change during evolution, even in closely related species, which likely reflects species-specific demands.

In most comprehensive of these heterologous expression studies, Saito et al. (2009) screened 93 odorants against 464 ORs and determined stimulants for 10 human and 52 mouse ORs. Their results demonstrate that some ORs are broadly tuned, while others are quite specialized (Saito et al. 2009). The specialization of ORs should be taken into account when assessing metrics of olfactory breadth among primates and other mammals.

10.3.6 Future Directions in Olfactory Research

Throughout this section, we have suggested some avenues that we feel would be worthwhile future endeavors in the pursuit of understanding the evolutionary pressure, current uses, and diversity of primate olfactory systems. We end by emphasizing one of the most important goals for the next decade – to make considerable advances in linking OR genotypes to sensory phenotypes (OR deorphanization) and understanding how the sensory phenotypes are adapted to local environments, including diet. The minute few OR genes with known phenotype include those that are associated with the detection of androstenone (a steroid related to testosterone) (Keller et al. 2007), aliphatic thiol, amyl butyrate (Gelis et al. 2012), beta-ionone, 2-heptanone, isobutyraldehyde, and beta-damascenone (McRae et al. 2013). With more than 90% of OR genes still orphaned, this is a daunting task, yet recent reviews suggest some promising ways forward (Peterlin et al. 2014). Gene expression assays are one direct approach. Genome-wide association studies are also guiding and refining our inquiries. Strong association is identified of sensitivity to some odorants with SNPs of some OR genes (McRae et al. 2013). Future studies that are successful in identifying which odorants are ecologically important, and which OR genes are stimulated by these odorants, will move this field of primate olfactory ecology forward with leaps and bounds.

10.4 Primate Bitter Taste Reception

Relative to ORs, the number of bitter taste receptors TAS2Rs is much fewer and less variable: ~30 function genes and ~10 pseudogenes in mammals including primates (Nei et al. 2008). Unlike ORs, most TAS2Rs are co-expressed in the same taste receptor cells of the tongue, suggesting that these cells are capable of responding to a broad array of bitter compounds (Adler et al. 2000). Human TAS2Rs are classified into four groups in terms of the breadth of agonist spectra (Meyerhof et al. 2010). The first group consists of TAS2R16 and TAS2R38 showing specificity to glucosinolates contained in cruciferous plants like cabbage, brussels sprouts, and cauliflower. The second group encompasses TAS2R3, TAS2R5, TAS2R8, TAS2R13, TAS2R49, and TAS2R50 which show limited agonist spectrum with common structural properties for receptor-agonist interactions. The third group comprises TAS2R1, TAS2R4, TAS2R7, TAS2R39, TAS2R40, TAS2R43, TAS2R44, and TAS2R47 which have quite broad agonist spectra, with the lack of clear common motifs that could be responsible for a specific recognition. The fourth group consists of TAS2R10, TAS2R14, and TAS2R46, which are very promiscuous, showing extremely wide molecular receptive ranges. These groups are each expected to be subjected to different selective regimes.

10.4.1 Neutral vs Non-neutral Genetic Variation of Bitter Taste Receptors in Humans and Chimpanzees

Recent population studies of TAS2R genes in human populations show signature of relaxed selection, speculated to be the result of human culture, including cooking and fire usage (Wang et al. 2004). However, a chimpanzee population study showed a seemingly similar pattern of selective relaxation (Sugawara et al. 2011). As found for the human genes, the functional constraint was relaxed for chimpanzee genes, but, paradoxically, weak balancing selection was also implied for the maintenance of the polymorphism in chimpanzee genes. Human TAS2R genes showing high Tajima's D value (a signature of balancing selection) are also high in chimpanzees, and those that are low in human are also low in chimpanzees, suggesting that human-specific culture, such as the use of fire, was not a cause of this phenomenon. Another chimpanzee population showed a different pattern of genetic diversity (Hayakawa et al. 2012), possibly reflecting subspecies-specific dietary repertoires. These studies reveal a previously unexpected complex pattern of functional and adaptive evolution in TAS2Rs.

10.4.2 TAS2R38 Nontaster Variations Independently Arising in Human, Chimpanzee, Macaques

Among TAS2Rs, the glucosinolates-sensitive TAS2R38 is well known as the agent of Mendelian variation in taste sensitivity to the bitter compound phenylthiocarbamide (PTC) in classic human genetics. Interestingly, the inactive alleles are also known to have occurred independently in chimpanzees (Wooding et al. 2006) and Japanese macaques (Suzuki et al. 2010) by different mutations. In humans, amino acid changes at three sites are responsible, in chimpanzees a mutation of the initiation codon resulting in the use of an alternative downstream start codon and production of a truncated receptor variant occurs, and in Japanese macaque another initiation codon mutation is found. It would be of great interest to investigate if the polymorphism is a local adaptation for specific dietary plants (Suzuki-Hashido et al. 2015).

Regarding TAS2R38 variation in humans, Campbell et al. (2012) reported a number of intriguing findings. They sequenced a 2975 bp region encompassing the gene in 611 Africans from 57 populations in West Central and East Africa with diverse subsistence patterns, as well as in a comparative sample of 132 non-Africans. They also examined the association between genetic variability at this locus and threshold levels of PTC bitterness in 463 Africans from the above populations to determine how variation influences bitter taste perception. They found a significant excess of novel rare nonsynonymous polymorphisms that recently arose only in Africa, high frequencies of haplotypes in Africa associated with intermediate bitter taste sensitivity, a remarkably similar frequency of common haplotypes across genetically and culturally distinct Africans, and an ancient coalescence time of common variation in global populations. Additionally, several of the rare nonsynonymous substitutions significantly modified levels of PTC bitter taste sensitivity in diverse Africans. While ancient balancing selection likely maintained common haplotype variation across global populations, they suggest that recent selection pressures may have also resulted in the unusually high level of rare nonsynonymous variants in Africa, implying a complex model of selection at the TAS2R38 locus in African populations. Furthermore, the distribution of common haplotypes in Africa is not correlated with diet, raising the possibility that common variation may be under selection due to their role in non-dietary biological processes. In addition, their data indicate that novel rare mutations contribute to the phenotypic variance of PTC sensitivity, illustrating the influence of rare variation on a common trait, as well as the relatively recent evolution of functionally diverse alleles at this locus (Campbell et al. 2012).

10.5 Umami/Sweet Taste Reception

In contrast to TAS2R bitter taste receptor genes, there are generally only three TAS1R genes among mammals: TAS1R1, TAS1R2, and TAS1R3 (Nei et al. 2008). The heterodimer comprising of TAS1R1 and TAS1R3 is the "umami" (amino acids and nucleic acids) receptor and that of TAS1R2 and TAS1R3 is the sweet (sugars) receptor (Hoon et al. 1999; Li et al. 2002). While the conserved number of TAS1Rs appears to reflect their fundamental function to detect essential nutrients (amino acids, nucleic acids, and sugars), recent studies have revealed great variation among mammals. TAS1R1 gene of the giant panda is pseudogenized coincided with its dietary switch from carnivorous to bamboo eating (Zhao et al. 2010). Loss of taste receptor function in mammals is widespread and directly related to feeding specializations: TAS1R2 in many carnivores including cats and TAS1R1 in some marine mammals which swallow foods not chew (Jiang et al. 2012).

There is also a large variation of ligand specificity: human TAS1R1/TAS1R3 specifically responds to L-glutamic acid, whereas mouse TAS1R1/TAS1R3 responds more strongly to other L-amino acids than to it. Site-directed mutagenesis and comparative studies of nonhuman primate TAS1R1 receptors have revealed variations relevant to their amino acid selectivity (Toda et al. 2013).

Kim et al. (2006) conducted a population study of human TAS1R genes and reported that TAS1R1 and TAS1R3 were less variable than the TAS1R2. The TAS1R3, the common subunit to both the sweet and umami receptors, was the most conserved. They proposed that human populations likely vary little with respect to umami perception, which is controlled by one major form of the receptor that is optimized for detecting glutamate, but may vary much more with respect to sweet perception (Kim et al. 2006). On the other hand, genetic variation and relevant sensitivity variation have been reported for human umami receptors (Shigemura et al. 2009). These studies have revealed previously unforeseen nature of genetic variation of this gene group for which conserved function is expected.

10.6 Ending Remarks

Our collective views on a simple trade-off between visual and olfaction, with a slow steady decrease in the importance of olfaction in primates, are changing. As we learn more about vision and the other chemical senses, we are seeing an increasingly complex and fascinating way in which primate sensory systems have evolved in response to selective agents. The next decade, with its continuing advances and decreasing costs in genomics, bioinformatics, and noninvasive, population-wide sampling, is expected to catapult forward our understanding of primate sensory ecology.

References

Adipietro KA, Mainland JD, Matsunami H (2012) Functional evolution of mammalian odorant receptors. PLoS Genet 8:e1002821

Adler E, Hoon MA, Mueller KL et al (2000) A novel family of mammalian taste receptors. Cell 100:693–702

Araujo AC, Didonet JJ, Araujo CS et al (2008) Color vision in the black howler monkey (*Alouatta caraya*). Vis Neurosci 25:243–248
Barton RA, Purvis A, Harvey PH (1995) Evolutionary radiation of visual and olfactory brain systems in primates, bats and insectivores. Phil Trans R Soc B 348:381–392
Bastir M, Rosas A, Gunz P et al (2011) Evolution of the base of the brain in highly encephalized human species. Nat Commun 2:588
Bicca-Marques JC, Garber PA (2004) Use of spatial, visual, and olfactory information during foraging in wild nocturnal and diurnal anthropoids: a field experiment comparing *Aotus*, *Callicebus*, and *Saguinus*. Am J Primatol 62:171–187
Boissinot S, Tan Y, Shyue SK et al (1998) Origins and antiquity of X-linked triallelic color vision systems in New World monkeys. Proc Natl Acad Sci USA 95:13749–13754
Buchanan TW, Tranel D, Adolphs R (2003) A specific role for the human amygdala in olfactory memory. Learn Mem 10:319–325
Buck L, Axel R (1991) A novel multigene family may encode odorant receptors: a molecular basis for odor recognition. Cell 65:175–187
Bullock TH (1982) Electroreception. Annu Rev Neurosci 5:121–170
Bunce JA, Isbell LA, Grote MN et al (2011a) Color vision variation and foraging behavior in wild neotropical titi monkeys (*Callicebus brunneus*): possible mediating roles for spatial memory and reproductive status. Int J Primatol 32:1058–1075
Bunce JA, Isbell LA, Neitz M et al (2011b) Characterization of opsin gene alleles affecting color vision in a wild population of titi monkeys (*Callicebus brunneus*). Am J Primatol 73:189–196
Bushdid C, Magnasco MO, Vosshall LB et al (2014) Humans can discriminate more than 1 trillion olfactory stimuli. Science 343:1370–1372
Caine NG, Mundy NI (2000) Demonstration of a foraging advantage for trichromatic marmosets (*Callithrix geoffroyi*) dependent on food colour. Proc R Soc Lond B 267:439–444
Caine NG, Osorio D, Mundy NI (2010) A foraging advantage for dichromatic marmosets (*Callithrix geoffroyi*) at low light intensity. Biol Lett 6:36–38
Campbell MC, Ranciaro A, Froment A et al (2012) Evolution of functionally diverse alleles associated with PTC bitter taste sensitivity in Africa. Mol Biol Evol 29:1141–1153
Carrigan MA, Uryasev O, Frye CB et al (2015) Hominids adapted to metabolize ethanol long before human-directed fermentation. Proc Natl Acad Sci USA 112:458–463
Carvalho LS, Davies WL, Robinson PR et al (2012) Spectral tuning and evolution of primate short-wavelength-sensitive visual pigments. Proc R Soc B 279:387–393
Collin SP, Knight MA, Davies WL et al (2003) Ancient colour vision: multiple opsin genes in the ancestral vertebrates. Curr Biol 13:R864–R865
Cropp S, Boinski S, Li W-H (2002) Allelic variation in the squirrel monkey X-linked color vision gene: biogeographical and behavioral correlates. J Mol Evol 54:734–745
Davies WI, Collin SP, Hunt DM (2012) Molecular ecology and adaptation of visual photopigments in craniates. Mol Ecol 21:3121–3158
de Lima EM, Pessoa DM, Sena L et al (2015) Polymorphic color vision in captive Uta Hick's cuxius, or bearded sakis (*Chiropotes utahickae*). Am J Primatol 77:66–75
Deeb SS (2005) The molecular basis of variation in human color vision. Clin Genet 67:369–377
Deeb SS (2006) Genetics of variation in human color vision and the retinal cone mosaic. Curr Opin Genet Dev 16:301–307
DeMaria S, Ngai J (2010) The cell biology of smell. J Cell Biol 191:443–452
Dominy NJ, Lucas PW (2001) Ecological importance of trichromatic vision to primates. Nature 410:363–366
Dominy NJ, Lucas PW, Osorio D et al (2001) The sensory ecology of primate food perception. Evol Anthropol 10:171–186
Dominy NJ, Garber PA, Bicca-Marques JC et al (2003a) Do female tamarins use visual cues to detect fruit rewards more successfully than do males? Anim Behav 66:829–837
Dominy NJ, Svenning JC, Li W-H (2003b) Historical contingency in the evolution of primate color vision. J Hum Evol 44:25–45

Drummond-Borg M, Deeb SS, Motulsky AG (1989) Molecular patterns of X chromosome-linked color vision genes among 134 men of European ancestry. Proc Natl Acad Sci USA 86:983–987

Dulai KS, Bowmaker JK, Mollon JD et al (1994) Sequence divergence, polymorphism and evolution of the middle-wave and long-wave visual pigment genes of great apes and old world monkeys. Vis Res 34:2483–2491

Dulai KS, von Dornum M, Mollon JD et al (1999) The evolution of trichromatic color vision by opsin gene duplication in New World and Old World primates. Genome Res 9:629–638

Fedigan LM, Melin AD, Addicott JF et al (2014) The heterozygote superiority hypothesis for polymorphic color vision is not supported by long-term fitness data from wild neotropical monkeys. PLoS One 9:e84872

Fernandez AA, Morris MR (2007) Sexual selection and trichromatic color vision in primates: statistical support for the preexisting-bias hypothesis. Am Nat 170:10–20

Fleagle JG (2013) Primate adaptation and evolution, 3rd edn. Academic Press, San Diego

Fortes-Marco L, Lanuza E, Martinez-Garcia F (2013) Of pheromones and kairomones: what receptors mediate innate emotional responses? Anat Rec 296:1346–1363

Garrett EC (2015) Was there a sensory trade-off in primate evolution? The vomeronasal groove as a means of understanding the vomeronasal system in the fossil record. The Graduate Center, City University of New York, New York

Gelis L, Wolf S, Hatt H et al (2012) Prediction of a ligand-binding niche within a human olfactory receptor by combining site-directed mutagenesis with dynamic homology modeling. Angew Chem Int Ed 51:1274–1278

Gilad Y, Przeworski M, Lancet D (2004) Loss of olfactory receptor genes coincides with the acquisition of full trichromatic vision in primates. PLoS Biol 2:e5

Gilad Y, Wiebe V, Przeworski M et al (2007) Correction: loss of olfactory receptor genes coincides with the acquisition of full trichromatic vision in primates (vol 2, pg 120, 2004). PLoS Biol 5:e148

Go Y, Niimura Y (2008) Similar numbers but different repertoires of olfactory receptor genes in humans and chimpanzees. Mol Biol Evol 25:1897–1907

Hanazawa A, Mikami A, Sulistyo Angelika P et al (2001) Electroretinogram analysis of relative spectral sensitivity in genetically identified dichromatic macaques. Proc Natl Acad Sci USA 98:8124–8127

Hartl DL, Clark AG (2007) Principles of population genetics, 4th edn. Sinauer Associates, Sunderland

Hasin Y, Olender T, Khen M et al (2008) High-resolution copy-number variation map reflects human olfactory receptor diversity and evolution. PLoS Genet 4:e1000249

Hasin-Brumshtein Y, Lancet D, Olender T (2009) Human olfaction: from genomic variation to phenotypic diversity. Trends Genet 25:178–184

Hayakawa T, Sugawara T, Go Y, Udono T, Hirai H, Imai H (2012) Eco-geographical diversification of bitter taste receptor genes (*TAS2Rs*) among subspecies of chimpanzees (*Pan troglodytes*). PLoS One 7:e43277

Hayashi T, Motulsky AG, Deeb SS (1999) Position of a 'green-red' hybrid gene in the visual pigment array determines colour-vision phenotype. Nat Genet 22:90–93

Hayashi S, Ueyama H, Tanabe S et al (2001) Number and variations of the red and green visual pigment genes in Japanese men with normal color vision. Jpn J Ophthalmol 45:60–67

Hayashi T, Kubo A, Takeuchi T et al (2006) Novel form of a single X-linked visual pigment gene in a unique dichromatic color-vision defect. Vis Neurosci 23:411–417

Hayden S, Bekaert M, Crider TA et al (2010) Ecological adaptation determines functional mammalian olfactory subgenomes. Genome Res 20:1–9

Heesy CP, Ross CF, Demes B (2007) Oculomotor stability and the functions of the postorbital bar and septum. In: Ravosa MJ, Dagosto M (eds) Primate origins: adaptations and evolution. Springer, New York, pp 257–283

Heritage S (2014) Modeling olfactory bulb evolution through primate phylogeny. PLoS One 9: e113904

Heymann EW (2006) The neglected sense-olfaction in primate behavior, ecology, and evolution. Am J Primatol 68:519–524
Hiramatsu C, Radlwimmer FB, Yokoyama S et al (2004) Mutagenesis and reconstitution of middle-to-long-wave-sensitive visual pigments of New World monkeys for testing the tuning effect of residues at sites 229 and 233. Vis Res 44:2225–2231
Hiramatsu C, Tsutsui T, Matsumoto Y et al (2005) Color-vision polymorphism in wild capuchins (*Cebus capucinus*) and spider monkeys (*Ateles geoffroyi*) in Costa Rica. Am J Primatol 67:447–461
Hiramatsu C, Melin AD, Aureli F et al (2008) Importance of achromatic contrast in short-range fruit foraging of primates. PLoS One 3:e3356
Hiramatsu C, Melin AD, Aureli F et al (2009) Interplay of olfaction and vision in fruit foraging of spider monkeys. Anim Behav 77:1421–1426
Hiwatashi T, Okabe Y, Tsutsui T et al (2010) An explicit signature of balancing selection for color-vision variation in New World monkeys. Mol Biol Evol 27:453–464
Hiwatashi T, Mikami A, Katsumura T et al (2011) Gene conversion and purifying selection shape nucleotide variation in gibbon L/M opsin genes. BMC Evol Biol 11:312
Hoon MA, Adler E, Lindemeier J et al (1999) Putative mammalian taste receptors: a class of taste-specific GPCRs with distinct topographic selectivity. Cell 96:541–551
Hudson R (1999) From molecule to mind: the role of experience in shaping olfactory function. J Comp Physiol A 185:297–304
Hughes GM, Teeling EC, Higgins DG (2014) Loss of olfactory receptor function in hominin evolution. PLoS One 9:e84714
Ibbotson RE, Hunt DM, Bowmaker JK et al (1992) Sequence divergence and copy number of the middle- and long-wave photopigment genes in Old World monkeys. Proc R Soc Lond B 247:145–154
Jacobs GH (1993) The distribution and nature of colour vision among the mammals. Biol Rev 68:413–471
Jacobs GH (2007) New World monkeys and color. Int J Primatol 28:729–759
Jacobs GH (2013) Losses of functional opsin genes, short-wavelength cone photopigments, and color vision-a significant trend in the evolution of mammalian vision. Vis Neurosci 30:39–53
Jacobs GH, Deegan JF II (2001) Photopigments and colour vision in New World monkeys from the family Atelidae. Proc R Soc Lond B 268:695–702
Jacobs GH, Deegan JF II (2005) Polymorphic New World monkeys with more than three M/L cone types. J Opt Soc Am A 22:2072–2080
Jacobs GH, Nathans J (2009) The evolution of primate color vision. Sci Am 300:56–63
Jacobs GH, Williams GA (2001) The prevalence of defective color vision in Old World monkeys and apes. Col Res Appl 26. (Suppl.):S123–S127
Jacobs GH, Neitz M, Deegan JF et al (1996) Trichromatic colour vision in New World monkeys. Nature 382:156–158
Jameson KA, Highnote SM, Wasserman LM (2001) Richer color experience in observers with multiple photopigment opsin genes. Psychon Bull Rev 8:244–261
Jiang P, Josue J, Li X et al (2012) Major taste loss in carnivorous mammals. Proc Natl Acad Sci USA 109:4956–4961
Jones G, Teeling EC, Rossiter SJ (2013) From the ultrasonic to the infrared: molecular evolution and the sensory biology of bats. Front Physiol 4:117
Kamilar JM, Heesy CP, Bradley BJ (2013) Did trichromatic color vision and red hair color coevolve in primates? Am J Primatol 75:740–751
Kawamura S, Kubotera N (2004) Ancestral loss of short wave-sensitive cone visual pigment in lorisiform prosimians, contrasting with its strict conservation in other prosimians. J Mol Evol 58:314–321
Kawamura S, Hiramatsu C, Melin AD et al (2012) Polymorphic color vision in primates: evolutionary considerations. In: Hirai H, Imai H, Go Y (eds) Post-genome biology of primates. Springer, Tokyo, pp 93–120

Keller A, Zhuang H, Chi Q et al (2007) Genetic variation in a human odorant receptor alters odour perception. Nature 449:468–472

Kim UK, Wooding S, Riaz N et al (2006) Variation in the human *TAS1R* taste receptor genes. Chem Senses 31:599–611

Klailova M, Lee PC (2014) Wild Western lowland gorillas signal selectively using odor. PLoS One 9:e99554

Kuang S, Zhang T (2014) Smelling directions: olfaction modulates ambiguous visual motion perception. Sci Rep 4:5796

Lambert D (1987) The Cambridge guide to prehistoric man. Cambridge University Press, Cambridge

Laska M, Seibt A, Weber A (2000) 'Microsmatic' primates revisited: olfactory sensitivity in the squirrel monkey. Chem Senses 25:47–53

Laska M, Freist P, Krause S (2007) Which senses play a role in nonhuman primate food selection? A comparison between squirrel monkeys and spider monkeys. Am J Primatol 69:282–294

Levenson DH, Fernandez-Duque E, Evans S et al (2007) Mutational changes in S-cone opsin genes common to both nocturnal and cathemeral *Aotus* monkeys. Am J Primatol 69:757–765

Li X, Staszewski L, Xu H et al (2002) Human receptors for sweet and umami taste. Proc Natl Acad Sci USA 99:4692–4696

Martinez-Garcia F, Martinez-Ricos J, Agustin-Pavon C et al (2009) Refining the dual olfactory hypothesis: pheromone reward and odour experience. Behav Brain Res 200:277–286

Matsui A, Go Y, Niimura Y (2010) Degeneration of olfactory receptor gene repertories in primates: no direct link to full trichromatic vision. Mol Biol Evol 27:1192–1200

Matsumoto Y, Hiramatsu C, Matsushita Y et al (2014) Evolutionary renovation of L/M opsin polymorphism confers a fruit discrimination advantage to ateline New World monkeys. Mol Ecol 23:1799–1812

Matsushita Y, Oota H, Welker BJ et al (2014) Color vision variation as evidenced by hybrid L/M opsin genes in wild populations of trichromatic *Alouatta* New World monkeys. Int J Primatol 35:71–87

McRae JF, Jaeger SR, Bava CM et al (2013) Identification of regions associated with variation in sensitivity to food-related odors in the human genome. Curr Biol 23:1596–1600

Melin AD, Fedigan LM, Hiramatsu C et al (2007) Effects of colour vision phenotype on insect capture by a free-ranging population of white-faced capuchins (*Cebus capucinus*). Anim Behav 73:205–214

Melin AD, Fedigan LM, Hiramatsu C et al (2008) Polymorphic color vision in white-faced capuchins (*Cebus capucinus*): is there foraging niche divergence among phenotypes? Behav Ecol Sociobiol 62:659–670

Melin AD, Fedigan LM, Hiramatsu C et al (2009) Fig foraging by dichromatic and trichromatic *Cebus capucinus* in a tropical dry forest. Int J Primatol 30:753–775

Melin AD, Fedigan LM, Young HC et al (2010) Can color vision variation explain sex differences in invertebrate foraging by capuchin monkeys? Curr Zool 56:300–312

Melin AD, Moritz GL, Fosbury RAE et al (2012) Why aye-ayes see blue. Am J Primatol 74:185–192

Melin AD, Kline DW, Hickey C et al (2013a) Food search through the eyes of a monkey: a functional substitution approach for assessing the ecology of primate color vision. Vis Res 87:87–96

Melin AD, Matsushita Y, Moritz GL et al (2013b) Inferred L/M cone opsin polymorphism of ancestral tarsiers sheds dim light on the origin of anthropoid primates. Proc R Soc B 280:20130189

Melin AD, Hiramatsu C, Parr NA et al (2014) The behavioral ecology of color vision: considering fruit conspicuity, detection distance and dietary importance. Int J Primatol 35:258–287

Menashe I, Abaffy T, Hasin Y et al (2007) Genetic elucidation of human hyperosmia to isovaleric acid. PLoS Biol 5:e284

Meyerhof W, Batram C, Kuhn C et al (2010) The molecular receptive ranges of human TAS2R bitter taste receptors. Chem Senses 35:157–170

Mombaerts P (2004) Genes and ligands for odorant, vomeronasal and taste receptors. Nat Rev Neurosci 5:263–278
Muniz JAPC, de Athaide LM, Gomes BD et al (2014) Ganglion cell and displaced amacrine cell density distribution in the retina of the howler monkey (*Alouatta caraya*). PLoS One 9: e115291
Nathans J, Hogness DS (1983) Isolation, sequence analysis, and intron-exon arrangement of the gene encoding bovine rhodopsin. Cell 34:807–814
Nathans J, Thomas D, Hogness DS (1986) Molecular genetics of human color vision: the genes encoding blue, green, and red pigments. Science 232:193–202
Nei M, Rooney AP (2005) Concerted and birth-and-death evolution of multigene families. Annu Rev Genet 39:121–152
Nei M, Niimura Y, Nozawa M (2008) The evolution of animal chemosensory receptor gene repertoires: roles of chance and necessity. Nat Rev Genet 9:951–963
Neville KR, Haberly LB (2004) Olfactory cortex. In: Shepherd GM (ed) The synaptic organization of the brain, 5th edn. Oxford University Press, New York, pp 415–454
Niimura Y, Nei M (2005) Evolutionary dynamics of olfactory receptor genes in fishes and tetrapods. Proc Natl Acad Sci USA 102:6039–6044
Niimura Y, Nei M (2007) Extensive gains and losses of olfactory receptor genes in mammalian evolution. PLoS One 2:e708
Niimura Y, Matsui A, Touhara K (2014) Extreme expansion of the olfactory receptor gene repertoire in African elephants and evolutionary dynamics of orthologous gene groups in 13 placental mammals. Genome Res 24:1485–1496
Nozawa M, Kawahara Y, Nei M (2007) Genomic drift and copy number variation of sensory receptor genes in humans. Proc Natl Acad Sci USA 104:20421–20426
Olender T, Waszak SM, Viavant M et al (2012) Personal receptor repertoires: olfaction as a model. BMC Genomics 13:414
Onishi A, Koike S, Ida M et al (1999) Dichromatism in macaque monkeys. Nature 402:139–140
Osorio D, Smith AC, Vorobyev M et al (2004) Detection of fruit and the selection of primate visual pigments for color vision. Am Nat 164:696–708
Parraga CA, Troscianko T, Tolhurst DJ (2002) Spatiochromatic properties of natural images and human vision. Curr Biol 12:483–487
Perry GH, Martin RD, Verrelli BC (2007) Signatures of functional constraint at aye-aye opsin genes: the potential of adaptive color vision in a nocturnal primate. Mol Biol Evol 24:1963–1970
Pessoa DM, Maia R, de Albuquerque Ajuz RC et al (2014) The adaptive value of primate color vision for predator detection. Am J Primatol 76:721–729
Peterlin Z, Firestein S, Rogers ME (2014) The state of the art of odorant receptor deorphanization: a report from the orphanage. J Gen Physiol 143:527–542
Ressler KJ, Sullivan SL, Buck LB (1993) A zonal organization of odorant receptor gene expression in the olfactory epithelium. Cell 73:597–609
Rowe MP, Jacobs GH (2007) Naturalistic color discriminations in polymorphic platyrrhine monkeys: effects of stimulus luminance and duration examined with functional substitution. Vis Neurosci 24:17–23
Rushmore J, Leonhardt SD, Drea CM (2012) Sight or scent: lemur sensory reliance in detecting food quality varies with feeding ecology. PLoS One 7:e41558
Saito A, Mikami A, Hasegawa T et al (2003) Behavioral evidence of color vision deficiency in a protanomalia chimpanzee (*Pan troglodytes*). Primates 44:171–176
Saito A, Kawamura S, Mikami A et al (2005a) Demonstration of a genotype-phenotype correlation in the polymorphic color vision of a non-callitrichine New World monkey, capuchin (*Cebus apella*). Am J Primatol 67:471–485
Saito A, Mikami A, Kawamura S et al (2005b) Advantage of dichromats over trichromats in discrimination of color-camouflaged stimuli in nonhuman primates. Am J Primatol 67:425–436
Saito H, Chi Q, Zhuang H et al (2009) Odor coding by a mammalian receptor repertoire. Sci Signal 2:ra9

Sharpe LT, Stockman A, Jagle H et al (1999) Opsin genes, cone photopigments, color vision, and color blindness. In: Gegenfurtner KR, Sharpe LT (eds) Color vision: from genes to perception. Cambridge University Press, Cambridge, pp 3–51

Shepherd GM (2004) The human sense of smell: are we better than we think? PLoS Biol 2:E146

Shichida Y, Imai H (1998) Visual pigment: G-protein-coupled receptor for light signals. Cell Mol Life Sci 54:1299–1315

Shigemura N, Shirosaki S, Sanematsu K et al (2009) Genetic and molecular basis of individual differences in human umami taste perception. PLoS One 4:e6717

Siemers BM, Goerlitz HR, Robsomanitrandrasana E et al (2007) Sensory basis of food detection in wild *Microcebus murinus*. Int J Primatol 28:291–304

Silveira LCL, Saito CA, da Silva FM et al (2014) *Alouatta* trichromatic color vision: cone spectra and physiological responses studied with microspectrophotometry and single unit retinal electrophysiology. PLoS One 9:e113321

Smith AC, Buchanan-Smith HM, Surridge AK et al (2003) Leaders of progressions in wild mixed-species troops of saddleback (*Saguinus fuscicollis*) and mustached tamarins (*S. mystax*), with emphasis on color vision and sex. Am J Primatol 61:145–157

Smith TD, Bhatnagar KP, Tuladhar P et al (2004) Distribution of olfactory epithelium in the primate nasal cavity: are microsmia and macrosmia valid morphological concepts? Anat Rec A 281A:1173–1181

Smith AC, Buchanan-Smith HM, Surridge AK et al (2005) Factors affecting group spread within wild mixed-species troops of saddleback and mustached tamarins. Int J Primatol 26:337–355

Smith TD, Rossie JB, Bhatnagar KP (2007) Evolution of the nose and nasal skeleton in primates. Evol Anthropol 16:132–146

Smith AC, Surridge AK, Prescott MJ et al (2012) Effect of colour vision status on insect prey capture efficiency of captive and wild tamarins (*Saguinus* spp.) Anim Behav 83:479–486

Sugawara T, Go Y, Udono T et al (2011) Diversification of bitter taste receptor gene family in western chimpanzees. Mol Biol Evol 28:921–931

Sumner P, Mollon JD (2000) Catarrhine photopigments are optimized for detecting targets against a foliage background. J Exp Biol 203:1963–1986

Sumner P, Mollon JD (2003) Colors of primate pelage and skin: objective assessment of conspicuousness. Am J Primatol 59:67–91

Surridge AK, Osorio D, Mundy NI (2003) Evolution and selection of trichromatic vision in primates. Trends Ecol Evol 18:198–205

Surridge AK, Suarez SS, Buchanan-Smith HM et al (2005) Color vision pigment frequencies in wild tamarins (*Saguinus* spp.) Am J Primatol 67:463–470

Suzuki N, Sugawara T, Matsui A et al (2010) Identification of non-taster Japanese macaques for a specific bitter taste. Primates 51:285–289

Suzuki-Hashido N, Hayakawa T, Matsui A et al (2015) Rapid expansion of phenylthiocarbamide non-tasters among Japanese macaques. PLoS One 10:e0132016

Talebi MG, Pope TR, Vogel ER et al (2006) Polymorphism of visual pigment genes in the muriqui (Primates, Atelidae). Mol Ecol 15:551–558

Tan Y, Li W-H (1999) Trichromatic vision in prosimians. Nature 402:36

Tan Y, Yoder AD, Yamashita N et al (2005) Evidence from opsin genes rejects nocturnality in ancestral primates. Proc Natl Acad Sci USA 102:14712–14716

Terao K, Mikami A, Saito A et al (2005) Identification of a protanomalous chimpanzee by molecular genetic and electroretinogram analyses. Vis Res 45:1225–1235

Toda Y, Okada S, Misaka T (2011) Establishment of a new cell-based assay to measure the activity of sweeteners in fluorescent food extracts. J Agric Food Chem 59:12131–12138

Toda Y, Nakagita T, Hayakawa T et al (2013) Two distinct determinants of ligand specificity in T1R1/T1R3 (the umami taste receptor). J Biol Chem 288:36863–36877

Touhara K, Vosshall LB (2009) Sensing odorants and pheromones with chemosensory receptors. Annu Rev Physiol 71:307–332

Touhara K, Sengoku S, Inaki K et al (1999) Functional identification and reconstitution of an odorant receptor in single olfactory neurons. Proc Natl Acad Sci USA 96:4040–4045

Valenta K, Burke RJ, Styler SA et al (2013) Colour and odour drive fruit selection and seed dispersal by mouse lemurs. Sci Rep 3:2424
Valenta K, Brown KA, Melin AD et al (2015) It's not easy being blue: are there olfactory and visual trade-offs in plant signalling? PLoS One 10:e0131725
Vassar R, Ngai J, Axel R (1993) Spatial segregation of odorant receptor expression in the mammalian olfactory epithelium. Cell 74:309–318
Veilleux CC, Bolnick DA (2009) Opsin gene polymorphism predicts trichromacy in a cathemeral lemur. Am J Primatol 71:86–90
Veilleux CC, Cummings ME (2012) Nocturnal light environments and species ecology: implications for nocturnal color vision in forests. J Exp Biol 215:4085–4096
Veilleux CC, Louis EE Jr, Bolnick DA (2013) Nocturnal light environments influence color vision and signatures of selection on the *OPN1SW* opsin gene in nocturnal lemurs. Mol Biol Evol 30:1420–1437
Verrelli BC, Tishkoff SA (2004) Signatures of selection and gene conversion associated with human color vision variation. Am J Hum Genet 75:363–375
Verrelli BC, Lewis CM Jr, Stone AC et al (2008) Different selective pressures shape the molecular evolution of color vision in chimpanzee and human populations. Mol Biol Evol 25:2735–2743
Vogel ER, Neitz M, Dominy NJ (2007) Effect of color vision phenotype on the foraging of wild white-faced capuchins, *Cebus capucinus*. Behav Ecol 18:292–297
Vorobyev M (2004) Ecology and evolution of primate colour vision. Clin Exp Optom 87:230–238
Wald G (1968) Molecular basis of visual excitation. Science 162:230–239
Wang X, Thomas SD, Zhang J (2004) Relaxation of selective constraint and loss of function in the evolution of human bitter taste receptor genes. Hum Mol Genet 13:2671–2678
Wildman DE, Jameson NM, Opazo JC et al (2009) A fully resolved genus level phylogeny of neotropical primates (Platyrrhini). Mol Phylogenet Evol 53:694–702
Winderickx J, Lindsey DT, Sanocki E et al (1992) Polymorphism in red photopigment underlies variation in colour matching. Nature 356:431–433
Winderickx J, Battisti L, Hibiya Y et al (1993) Haplotype diversity in the human red and green opsin genes: evidence for frequent sequence exchange in exon 3. Hum Mol Genet 2:1413–1421
Wooding S, Bufe B, Grassi C et al (2006) Independent evolution of bitter-taste sensitivity in humans and chimpanzees. Nature 440:930–934
Yokoyama S (2000a) Molecular evolution of vertebrate visual pigments. Prog Retin Eye Res 19:385–419
Yokoyama S (2000b) Phylogenetic analysis and experimental approaches to study color vision in vertebrates. Methods Enzymol 315:312–325
Yokoyama S, Yang H, Starmer WT (2008) Molecular basis of spectral tuning in the red- and green-sensitive (M/LWS) pigments in vertebrates. Genetics 179:2037–2043
Young JM, Shykind BM, Lane RP et al (2003) Odorant receptor expressed sequence tags demonstrate olfactory expression of over 400 genes, extensive alternate splicing and unequal expression levels. Genome Biol 4:R71
Zhao H, Yang JR, Xu H et al (2010) Pseudogenization of the umami taste receptor gene *Tas1r1* in the giant panda coincided with its dietary switch to bamboo. Mol Biol Evol 27:2669–2673
Zhou X, Wang B, Pan Q et al (2014) Whole-genome sequencing of the snub-nosed monkey provides insights into folivory and evolutionary history. Nat Genet 46:1303–1310
Zhuang H, Matsunami H (2008) Evaluating cell-surface expression and measuring activation of mammalian odorant receptors in heterologous cells. Nat Protoc 3:1402–1413
Zhuang HY, Chien MS, Matsunami H (2009) Dynamic functional evolution of an odorant receptor for sex-steroid-derived odors in primates. Proc Natl Acad Sci USA 106:21247–21251

Chapter 11
Global Landscapes of Human Phenotypic Variation in Inherited Traits

Ryosuke Kimura

Abstract Modern humans exhibit phenotypic variation among individuals, and phenotypes in some physical and physiological traits are highly differentiated between populations. This chapter focuses on genetic polymorphisms related to phenotypes that show interpopulation differentiation, which have traditionally attracted the attention of both anthropologists and human geneticists. Owing to the recent development of DNA technology, we have obtained powerful tools for use in identifying the genetic polymorphisms associated with phenotypes. In addition, the availability of genome diversity data associated with global populations has enabled us to identify the signatures of the local genetic adaptations that are engraved in our genomes. Using data associated with current phenotypic variation in humans, we can elucidate the history of human adaptation in response to the selective pressures of various environments.

Keywords Phenotypic variation · Genetic differentiation · Genetic adaptation · Selective sweep · Genome-wide association study · Physical and physiological traits

11.1 Introduction

The physical and physiological traits of humans vary among individuals. Some traits, such as skin color, hair morphology, and facial features, clearly differ between populations. While the study of human diversity is well established, exploration into the genetic basis of human phenotypic variation has only just

R. Kimura
Department of Human Biology and Anatomy, Graduate School of Medicine, University of the Ryukyus, Nishihara-cho, Okinawa, Japan
e-mail: rkimura@med.u-ryukyu.ac.jp

N. Saitou (ed.), *Evolution of the Human Genome I*, Evolutionary Studies,
https://doi.org/10.1007/978-4-431-56603-8_11

begun. This chapter, which introduces the genetic factors that affect certain traits, focuses on how and why human phenotypic variations are formed.

Over the past 100,000 years, anatomically modern humans have spread from Africa throughout the rest of the world and have occupied a variety of habitats with different environmental conditions. During the early stages of human expansion, adjustments to life in new environments were likely achieved by behavioral and physiological changes at the individual level, while genetic adaptations would have gradually occurred over successive generations within populations. Climatic and physical conditions, such as UV radiation, temperature, precipitation, humidity, and altitude, likely acted as selective forces shaping human phenotypic variation. Lifestyle, including diet and labor, could also have modified both genotypes and phenotypes over the course of generations. Epidemics could have resulted in strong positive selection for specific alleles related to disease tolerance. Potential selective pressures such as these indicate that the phenotypic variation in some inherited traits could have arisen as the result of recent adaptive evolution. The recent availability of genome-wide single nucleotide polymorphism (SNP) and sequence data for many human populations has enabled us to examine which genetic and phenotypic traits are the targets of natural selection and which are affected by neutral evolution.

11.2 What Drives Phenotypic Differentiation Between Human Populations?

Theoretical studies in population genetics have yielded a significant body of knowledge regarding patterns of genetic variation in human populations. Throughout the history of modern humans, divergence times between populations have been relatively short, which has resulted in ancestrally shared polymorphisms typically being observed among different populations (Fig. 11.1a, b). When a single genetic locus is examined, the gene tree associated with that particular locus typically differs from the average phylogenetic relationship and genetic closeness between populations. This discrepancy between gene trees and population trees is known as "incomplete lineage sorting." When examining unrelated individuals from a randomly mating population, the average tree developed using whole genome data exhibits a "star phylogeny," since each pair of unrelated individuals would be expected to have the same genetic distance (Fig. 11.1c). In humans, the average genetic distance between individuals from different populations is only slightly larger than the genetic distance between individuals from the same population.

Degrees of genetic differentiation, which are usually evaluated by a statistic F_{ST}, depend on the time since divergence and the migration rate between the populations. Because only a relatively short time has passed since human populations have diverged from one another, complete or nearly complete genetic differentiation has rarely been observed in human populations under neutral conditions. For example, Fig. 11.2 shows the distribution of F_{ST} values between three populations used for the HapMap project, and majority of SNP loci show very small

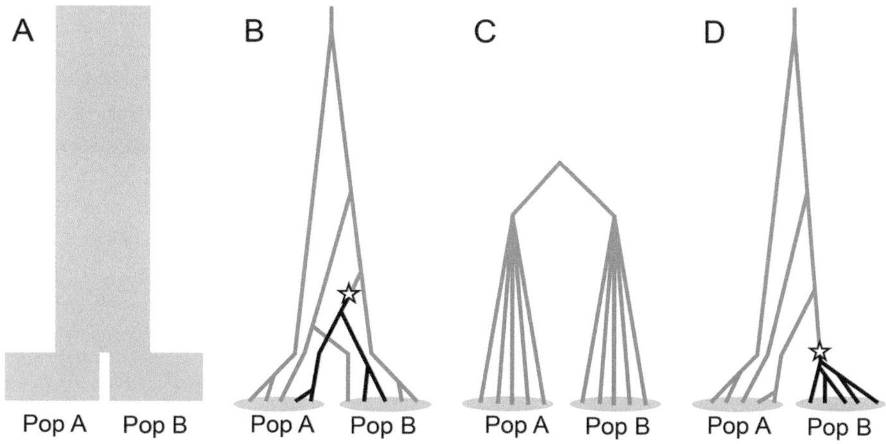

Fig. 11.1 Gene phylogeny versus population phylogeny. (**a**) Assumed population phylogeny of two populations. (**b**) Gene phylogeny of a single locus. Human populations generally do not have a monophyletic gene tree. A mutation (☆) causes a shared polymorphism between populations. (**c**) Dendrogram constructed from the expected average genetic distance. Analysis of the whole nuclear genome demonstrates that all unrelated individuals from a randomly mating population are expected to possess equal genetic closeness to one another. (**d**) Monophyletic gene tree in a population. Recent positive selection can result in a star phylogeny. A mutation (☆) creates a differentiated polymorphism between populations

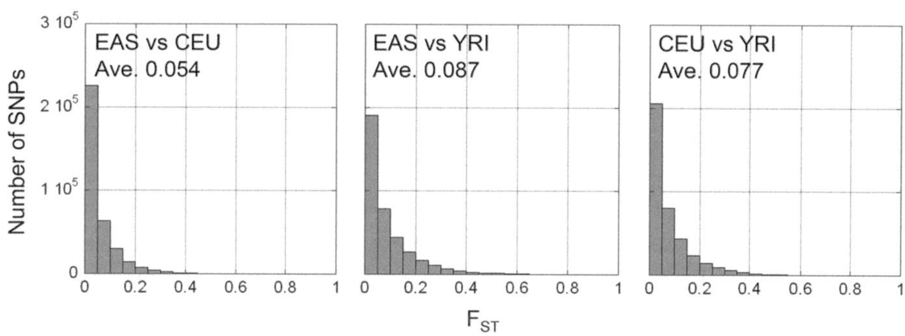

Fig. 11.2 F_{ST} between human populations determined using HapMap data (Chinese Han in Beijing, CHB; Japanese in Tokyo, JPT; Yorba in Ibadan, YRI; Utah residents with ancestry from northern and western Europe, CEU). EAS represents CHB+JPT

F_{ST} values. Therefore, in humans, loci exhibiting extremely large differentiation between populations can be identified as candidates for positively selected loci in one population (Fig. 11.1d). While phenotypic differentiation can be caused by neutral evolution, monogenic traits that are highly differentiated between human populations have likely been affected by some selective pressure. It is worth noting that positively selected loci can serve as population markers in humans although it is a typically held belief that neutral loci are more suitable for studying relationships between populations.

Polygenetic traits can result in large phenotypic differentiation between populations due to the summation of relatively small genetic differentiations at multiple loci (Fig. 11.3); however, even in the case of polygenetic traits, it is difficult for neutral evolution to result in large phenotypic differentiation, since changes in allele frequency occur in random directions under genetic drift. In contrast, selection acting on a phenotype can rapidly shift genotype distributions of multiple loci so that phenotype-related alleles increase within a given population. Therefore, large phenotypic differentiation in a polygenic trait can indicate the existence of some selective pressure; Q_{ST} is a metric of the degree of genetic differentiation indicated by quantitative traits, and Q_{ST}/F_{ST} comparisons provide a means to detect non-neutral differentiation in polygenic traits (Leinonen et al.

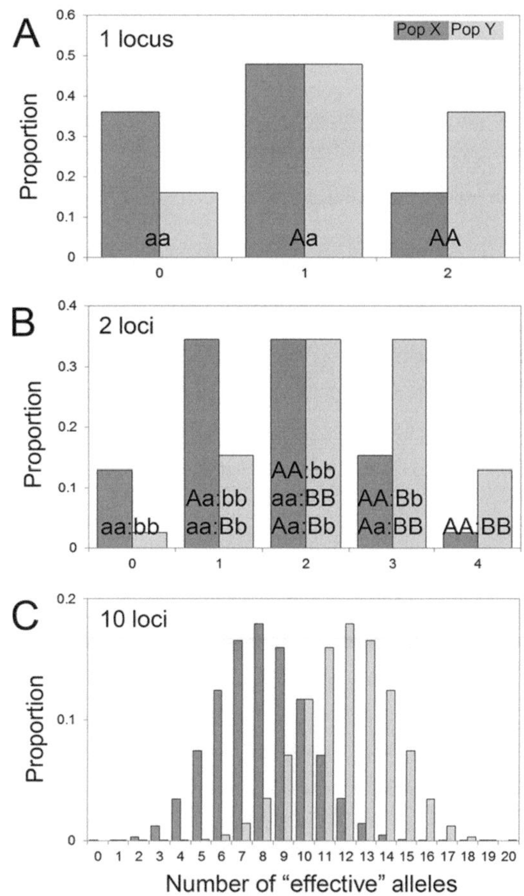

Fig. 11.3 Phenotypic distribution and differentiation in a polygenic quantitative trait. It is assumed that, at all loci associated with the trait, frequencies of the "effective" allele are 40% and 60% in populations A and B, respectively. (**a**) 1 locus. (**b**) 2 loci. (**c**) 10 loci

2013; Whitlock 2008). However, since phenotypes are also affected by environmental factors, it is difficult to isolate the effects of genetic factors in many cases.

11.3 Identification of Signatures of Adaptive Evolution in the Human Genome

Adaptive evolution occurs over a long time period and cannot be experimentally reproduced. Direct evidence of adaptive evolution could be obtained by observing differences in fitness between phenotypes, but proving that adaptive evolution occurred is usually impossible. Therefore, it is difficult to verify that a given phenotype is actually adaptive to an environment. In order to determine whether or not a phenotype is adaptive to a particular environment, one must accumulate indirect evidence by any means available, such as collecting information that can be gleaned from data regarding genetic diversity.

As described above, the extent of the differences in allele frequency between populations can act as a signature of positive selection. Another signature of positive selection is a reduction of long-ranged haplotype diversity, called "selective sweep." When a mutation occurs in a haplotype and that haplotype increases in a given population, alleles of surrounding sites that are tightly linked to the mutation also increase their frequencies. In this process, the links between the mutation and surrounding alleles are gradually broken over time through recombinations. Therefore, haplotype diversity can serve as an index for how rapidly a mutation proliferated in a given population, which would also serve as a measure of strength of positive selection (Fig. 11.4). Many studies have been undertaken regarding the methods and practical applications of using genome-wide data to identify signatures of selection (Grossman et al. 2010, 2013; Kimura et al. 2007; Pickrell et al. 2009; Sabeti et al. 2007; Tang et al. 2007; Voight et al. 2006; Wang et al. 2006; Williamson et al. 2007); however, previous approaches have not been adequate for the detection of selection on standing variation, known as "soft sweep" (Huang et al. 2009). When attempting to provide evidence of polygenic selection, gene set enrichment tests can prove useful, a point illustrated by Daub et al. (2013),

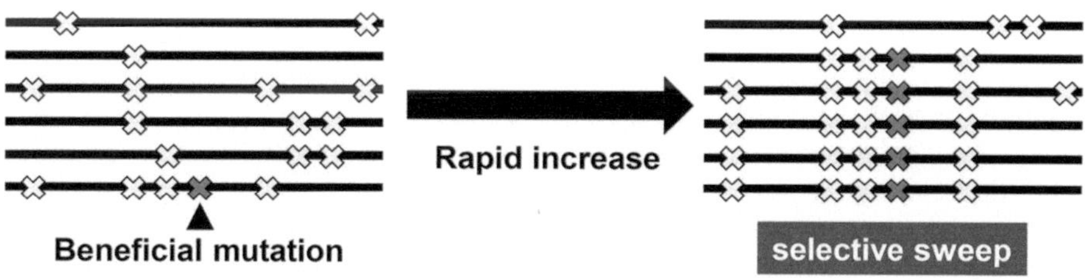

Fig. 11.4 Selective sweep. Rapid increase of a beneficial mutation results in a reduction in haplotype diversity in the region around the selected locus

who determined that the genes involved in immune-response pathways are enriched for signals of genetic adaptation.

Another strategy to obtain evidence for selection is to observe spatial patterns of allele frequencies. In heat and cold tolerance, genes involved in energy metabolism play a significant role, and Bergmann's and Allen's rules also assert the importance of both body mass and body proportion (Roberts 1953). Therefore, examining correlations between allele frequencies and climatic variables is one way to provide evidence for climatic adaptation in human populations. Studies examining these correlations have suggested that genetic variants associated with common metabolic disorders such as obesity, diabetes, and hypertension have been subjected to climate-mediated selective pressures (Hancock et al. 2011, 2008).

Genome-wide association studies (GWASs) have proven to be a powerful tool in the identification of genotype-phenotype correspondence and have been used to elucidate the genetic bases of medical and nonmedical traits in recent years (McCarthy et al. 2008; Plomin et al. 2009; Rosenberg et al. 2010). Some strong signatures of selection have been identified on phenotype-associated polymorphisms. The subsequent section provides a brief overview of the genetic bases of phenotypic variation that have attracted the interest of anthropologists.

11.4 Genes Associated with Common Phenotypic Variations I: Visible Traits

11.4.1 Height

Height is a highly heritable trait (heritability 70–90%). To date, GWASs have identified approximately 700 variants that have an association with height at a genome-wide significance level of $P < 5 \times 10^{-8}$. These include *ADAMTSL3*, *CDK6*, *DLEU7*, *GDF5*, *HHIP*, *HMGA2*, *LCORL*, and *ZBTB38* (Berndt et al. 2013; Lango Allen et al. 2010; Wood et al. 2014). Lango Allen et al. (2010) demonstrated that 180 loci explain approximately 10% of all phenotypic variation in height. More recently, Wood et al. (2014) estimated that 697 variants clustered in 423 loci explain one-fifth of the heritability, and all common variants in the genome together capture 60% of heritability. This suggests that the independent effects of common genetic variants cannot explain the majority of the contribution of genetic factors to variation in height. Studies targeting rare variants and epistasis effects are required in order to account for the "missing" heritability.

Average height varies among human populations, with African pygmies exhibiting particularly short statures, a feature believed to reflect past adaptation to a tropical environment. A genomic region on chromosome 3, which harbors a cluster of selection and association signals and includes genes such as *DOCK3* and *CISH*, could potentially explain, in part, the short stature observed in pygmy populations (Boyko 2011; Jarvis et al. 2012).

11.4.2 Obesity

Heritability of body mass index (BMI: weight kg/height m^2) is 40–70% (Maes et al. 1997; Stunkard et al. 1986). GWASs have detected a number of obesity-related polymorphisms, with strong association signals in *FTO*, *TMEM18*, *MC4R*, and *GNPDA2* (Berndt et al. 2013; Willer et al. 2009).

An ethnic difference exists in the prevalence of obesity, with Pima Indians and Polynesians being famous examples of exhibiting particularly high levels of obesity. The genetic backgrounds of these ethnic differences have not yet been determined, since the effects of currently identified polymorphisms are considerably smaller than the ethnic differences that have been observed. The "thrifty gene" hypothesis first proposed by Neel (1962) states that a genotype that efficiently stores energy would have been advantageous during times when food resources were scarce. Genome-wide scans for selection have nominated several candidates for "thrifty gene" in Polynesians (Kimura et al. 2008); however, further validation of phenotype-genotype association with respect to obesity in this population is required before any conclusions can be made. In studies on the Greenlandic Inuit and native Siberians, strong signatures of selective sweep were found on genes related with glucose uptake and fatty acid metabolism such as *TBC1D*, *CPT1A*, and *FADSs*, which is evidence for genetic adaptation to cold climates and specialized diets rich in protein and fatty acids (Clemente et al. 2014; Fumagalli et al. 2015; Moltke et al. 2014).

The risk of developing lifestyle diseases such as diabetes, hypertension, and heart disease increases along with increasing BMI. There is also strong evidence suggesting that at any given BMI, these risks are higher in some ethnic groups than they are in others (Shai et al. 2006; Wen et al. 2009). Hancock et al. (2008) identified significant correlations between climatic variables and the frequencies of genetic variants associated with metabolic disorders including obesity.

11.4.3 Pigmentation

It has been well established that skin, hair, and eye color vary among human populations with geographic gradation from low to high latitudes, a pattern that clearly reflects the strength of UV radiation. Genes involved in melanogenesis are well known from numerous studies involving either animal models or human pigmentation disorders (Rees 2003). Furthermore, GWASs have identified many genetic variants that explain variation in pigmentation-related traits. These variants are located in the cording and regulatory regions of melanogenesis-related genes—*ASIP*, *IRF4*, *KITLG*, *MC1R*, *OCA2*, *SLC24A4*, *SLC24A5*, *TYR*, *TYRP1*, etc. (Eriksson et al. 2010; Han et al. 2008; Nan et al. 2009; Stokowski et al. 2007; Sulem et al. 2007, 2008); however, since the majority of studies on pigmentation-related traits have focused on populations of European ancestry, genetic bases regarding global patterns of pigmentation-related traits are not fully understood.

To date, only a few studies examining East Asian populations with respect to pigmentation have been undertaken, with these studies having demonstrated that *MC1R* and *OCA*2 were associated with skin color variation in the Asian populations examined (Akey et al. 2001; Edwards et al. 2010; Yamaguchi et al. 2012). In a study on the Melanesian Solomon Islanders, Kenny et al. (2012) demonstrated that their blond hair is associated with a nonsynonymous variant in *TYRP1*, which is independent of the genetic basis of blond hair in Europeans. Genetic variations in melanogenesis-related genes often have pleiotropic effects, resulting in correlations between skin, hair, and eye color; however, some genes exhibit a tissue-specific effect.

Genome scans for positive selection have indicated that positive selection clearly acted on variants associated with low pigmentation in non-African populations (Lao et al. 2007; Norton et al. 2007; Sturm and Duffy 2012; Williamson et al. 2007). For example, *SLC24A5* and *SLC45A2* have signatures of hard selective sweeps in European populations, whereas *KITLG* and *OCA2* show evidence for selection in both European and East Asian populations. This means that the relaxation of the evolutionary constraint of damage due to the effects of solar UV radiation alone cannot explain the global pattern of human pigmentation. Possible selective pressures acting on skin pigmentation are the need for vitamin D synthesis and protection from photolysis of folate (Jablonski and Chaplin 2000). Alternatively, sexual selection is a potential explanation for positive selection on human pigmentation traits (Aoki 2002).

11.4.4 Morphology of Hair, Teeth, and Other Skin Appendages

Along with skin pigmentation, hair morphology is one of the traits exhibiting a high degree of differentiation between populations on different continents. African and Melanesian populations typically possess frizzled hair, while European populations typically exhibit wavy/curly hair or straight hair with a much lower frequency of frizzled hair. The majority of East Asian populations exhibit straight, thick hair (Franbourg et al. 2003).

One variant associated with hair morphology is located in *EDAR* and partly explains Asian-specific thick hair (Fujimoto et al. 2008). The nonsynonymous variant 370Val>Ala is found almost exclusively in Asian and Native American populations and has been shown to be a target of strong positive selection. GWASs examining individuals with European ancestry have identified variants associated with hair curliness in *TCHH* and *WNT10A* (Eriksson et al. 2010; Medland et al. 2009).

The *EDAR* variant has also been linked to tooth morphology. This variant is associated with the grade of shovel-shape in the incisors, a well-known Asian-specific phenotype (Fig. 11.5) (Kimura et al. 2009; Scott and Turner 1997). Along with the grade of shovel-shaped incisors, tooth size and the number of cusps in the second molars have been shown to increase due to the Asian-specific *EDAR* variant

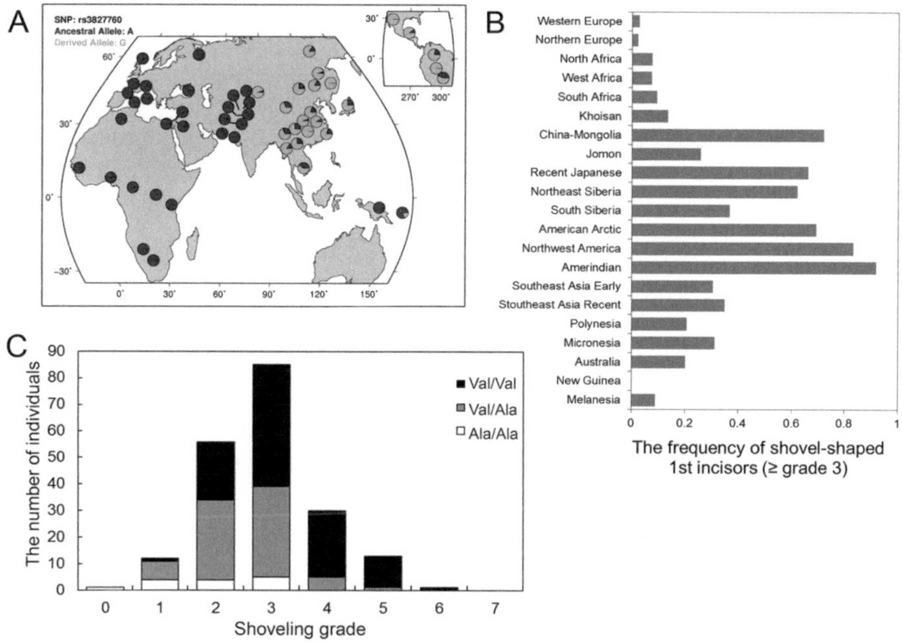

Fig. 11.5 *EDAR* 370Val>Ala and shovel-shaped incisors. (**a**) Global distribution of *EDAR* 370Val (ancestral) and 370Ala (derived) (rs3827760) (http://hgdp.uchicago.edu/cgi-bin/gbrowse/HGDP/). (**b**) The prevalence of strong, shovel-shaped incisors ($\geq$ grade 3) in world populations (Scott and Turner 1997). (**c**) The association of *EDAR* 370Val>Ala with shoveling of first incisors in the Japanese population (Kimura et al. 2009)

(Park et al. 2012). An animal experiment using knock-in mice expressing the human *EDAR* variant determined that the variant increases branch density of eccrine sweat glands and mammary glands, with the association of the variant and eccrine sweat gland density having also been confirmed in humans (Kamberov et al. 2013). Another study has shown that *EDAR* variants are associated with ear morphology (Adhikari et al. 2015).

The selective pressure that acted on the *EDAR* variant in Asia remains unclear, due in part to the pleiotropic effects of this variant. It is likely that only one of these phenotypes would have been under selective pressure, while the other phenotypic traits were seemingly selected for increased only as a by-product of this selection. Situations such as this indicate the importance of identifying genetic bases to aid in our understanding of phenotypic evolution.

11.4.5 Baldness

Androgenetic alopecia is the most common type of baldness and is characterized by progressive thinning and loss of hair on the scalp. Interpopulation differences in the prevalence of androgenetic alopecia exist, with populations of European ancestry exhibiting a greater prevalence of androgenetic alopecia than populations of both Asian and African ancestry (Hamilton 1951; Wang et al. 2010).

The *EDAR2-AR* region on X chromosome has been shown to be associated with androgenetic alopecia (Ellis et al. 2001), a finding confirmed by more recent GWAS investigations (Li et al. 2012). Recent GWASs have also identified genomic regions such as *PAX1* on chromosome 20, *HDAC4* on chromosome 2, and *HDAC9* on chromosome 7 as baldness-associated loci (Brockschmidt et al. 2011; Hillmer et al. 2008; Li et al. 2012; Richards et al. 2008).

In East Asian populations, a signature of selective sweep has been observed in the *EDAR2-AR* region, as most SNPs are fixed and genetic diversity is remarkably low (Hillmer et al. 2009). Since the region exhibiting low genetic diversity is very long and includes many genes, the identity of the selected gene and the selected phenotype remains unclear. Furthermore, this region alone cannot explain the observed differences between populations in prevalence of androgenetic alopecia.

11.4.6 Facial Morphology

Craniofacial shape is one of the most distinctive traits among humans. Facial morphology is highly heritable as we know that monozygotic twins have almost identical faces. Although numerous genes involved in craniofacial development have been identified through studies employing animal experiments or examining human genetic disorders, the search for the genetic factors responsible for common variation in human craniofacial morphology has only just begun. Recent advances in imaging technology as well as DNA technology have accelerated genetic studies regarding craniofacial morphology. A GWAS in Europeans identified SNPs associated with facial morphology on *PRDM16*, *PAX3*, *TP63*, *C5orf50*, and *COL17A1* (Liu et al. 2012). An association between *PAX3* variants and the shape of the nasal root was also detected in a separate GWAS study (Paternoster et al. 2012). In addition, new morphometric methods have enabled us to establish high-density semi-landmarks, instead of only using the limited number of anthropologically defined landmarks. Using these methods, a candidate gene study examining individuals of Chinese ancestry found that a genetic variant on *IRF6*, which is also a well-known risk factor associated with non-syndrome cleft lip, strongly affected mouth shape (Peng et al. 2013). Studies such as these could eventually allow for predictive modeling of faces based on sex, ancestry, and genes (Claes et al. 2014).

Some researchers have reported that either natural/sexual selection or assortative mating, rather than genetic drift, was responsible for facial differences between populations (Hubbe et al. 2009; Roseman and Weaver 2004; Roseman 2004; Guo et al. 2014). Identification of the genes associated with facial morphology will aid in our understanding of the selective pressures affecting facial shape in humans.

11.5 Genes Associated with Common Phenotypic Variations II: Physiological Traits

11.5.1 Lactase Persistence

Lactase persistence, the persistence of lactase activity from childhood into adulthood, is one of the most famous examples of genetic adaptation to a human dietary culture. The causal SNP for lactase persistence/nonpersistence in Europeans is located roughly 14 kb upstream from the *LCT* gene locus (Enattah et al. 2002). This genetic locus shows a strong signal for selective sweep in the European population (Bersaglieri et al. 2004). Pastoralism originated around 10,000 years ago, which resulted in lactase persistence becoming advantageous due to the availability of milk from domesticated animals as a food source for adults. A study based on approximate Bayesian computation has inferred that the lactase persistence allele first underwent selection among dairying farmers around 7500 years ago in a region between the central Balkans and Central Europe, possibly in association with the dissemination of the Neolithic Linearbandkeramik culture over Central Europe (Itan et al. 2009). Beja-Pereira et al. (2003) demonstrated geographic coincidence between distribution of the lactase persistence allele in contemporary Europeans, the increased genetic diversity in cattle milk genes, and locations of European Neolithic sites of cattle pastoralists. Although lactase persistence is also found in African populations, the causal SNPs differ from those of European populations (Ranciaro et al. 2014; Tishkoff et al. 2007). This indicates that multiple independent variants have allowed human populations to quickly modify *LCT* expression and that these variants have been strongly adaptive in adult milk-consuming populations.

11.5.2 Alcohol Intolerance

Harada et al. (1980) first demonstrated that a genetic variant in *ALDH2* responsible for aldehyde dehydrogenase deficiency is commonly found in Japanese populations. This variant, 504Glu>Lys, is strongly associated with the reduced alcohol tolerance commonly observed in Asian populations. In individuals with the deficiency variant, drinking a small amount of alcohol results in facial flushing, light-headedness, palpitations, and nausea. Homozygotes for the deficiency variant have an extremely reduced ability to metabolize alcohol and are likely to suffer from acute alcoholism. Instead, individuals with the deficiency variant have a lower risk of developing alcohol-related problems such as alcohol dependence and alcoholic liver disease due to their inability to consume alcoholic drinks (Goedde et al. 1983; Macgregor et al. 2009; Shibuya and Yoshida 1988). In East Asian populations, 30–40% of individuals carry at least one copy of *ALDH2*-deficient allele (Oota et al. 2004); however, the allele is extremely rare in people who are not of Asian descent.

The enzyme encoded by *ADH1B*, cytosolic alcohol dehydrogenase, also has functional variants. In East Asian populations, a nonsynonymous variant, 48His>Arg, can be observed at a frequency of approximately 75%, while this variant is almost never found in populations of African and European descent. In addition, a signature of selective sweep on the variant provides obvious evidence for positive selection acting on Asian populations. Another variant, 370Arg>Cys, is found almost exclusively in populations of African descent (Li et al. 2007; Osier et al. 2002; Peng et al. 2010). These two variants are known to metabolize ethanol at rates 100 times higher than the ancestral allele (Edenberg 2000; Hurley et al. 1994). It has also been reported that individuals possessing the 48His>Arg variant exhibit lower blood alcohol concentrations one day after a period of heavy drinking than individuals without the variant (Yokoyama et al. 2007). Recent GWASs have demonstrated strong evidence for the association of *ADH1B* polymorphisms with alcohol consumption, alcohol dependence, and other alcohol-mediated diseases (Frank et al. 2012; Kapoor et al. 2013; Park et al. 2013).

It is intriguing that the Asian-specific variants, *ALDH2* 504Glu>Lys and *ADH1B* 48His>Arg, are both associated with alcohol intolerance. One hypothetical explanation for positive selection on these genes could be that a high acetaldehyde concentration in blood may have antiprotozoal effects (Goldman and Enoch 1990).

11.5.3 Apocrine Gland Secretion

Wet/dry earwax types are a classical Mendelian trait (Matsunaga 1962); however, the genetic basis of these traits, a polymorphism on *ABCC11*, has been discovered relatively recently (Yoshiura et al. 2006). The ancestral allele, which is dominantly associated with wet earwax, is highly prevalent in African and European populations. On the other hand, in Asian populations, a variant, 180 Gly>Arg, which is recessively associated with dry earwax, is observed with a frequency of approximately 80%. In addition, it has been shown that the wet-type allele has a strong association with axillary osmidrosis (Matsunaga 1962; Nakano et al. 2009) and that dry-type earwax is associated with a lack of colostrum secretion from the mammary glands in women on the first postpartum day (Miura et al. 2007). Human *ABCC11* is thought to play a central role in the secretion of steroid metabolites from the apocrine glands.

Genetic diversity data show strong evidence for positive selection acting on the *ABCC11* variant in Asian populations. Since *ABCC11* is involved in the apocrine sweat function, the genetic adaptation may be toward Asian-specific climates. A simulation study has estimated that 180 Gly>Arg originated approximately 50,000 years ago (Ohashi et al. 2011).

11.5.4 Blood Types

Yamamoto et al. (1990) elucidated the molecular genetic basis of the ABO blood group system long after its discovery at the beginning of the twentieth century. The ABO gene encodes a glycosyltransferase that catalyzes the transfer of carbohydrates on the extracellular surface of red blood cell membranes. The proteins encoded by the A and B alleles transfer different carbohydrates, either N-acetylgalactosamine or galactose, into the H antigen to form either A or B antigens. The O allele is dysfunctional and produces neither the A nor the B antigen. ABH antigens are also highly expressed on a variety of cells and tissues other than red blood cells, such as platelets, the epithelium, the vascular endothelium, and sensory neurons (Eastlund 1998).

It has long been known that the ABO blood groups are associated with the plasma levels of the blood glycoprotein von Willebrand factor and factor VIII, which are involved in hemostasis (Moeller et al. 2001; O'Donnell and Laffan 2001; Preston and Barr 1964). A number of clinical and experimental studies have assessed the effects the ABO blood group on the risk factors for arterial or venous thrombotic events (Liumbruno and Franchini 2013). Recent GWASs have confirmed the association of non-O blood groups with a variety of vascular disorders including coronary heart disease, ischemic stroke, and venous thromboembolism (Dichgans et al. 2014; Heit et al. 2012; Reilly et al. 2011; Schunkert et al. 2011; Tang et al. 2012, 2013; Tregouet et al. 2009; Williams et al. 2013). It has also been shown that the *ABO* locus is associated with the levels of several biomarkers including cholesterols (Kim et al. 2011; Teslovich et al. 2010; Willer et al. 2013; Zhou et al. 2013), alkaline phosphatase (Chambers et al. 2011; Kamatani et al. 2010; Li et al. 2013; Yuan et al. 2008), and soluble adhesion molecules such as ICAM-1, E-selectin and P-selectin (Barbalic et al. 2010; Pare et al. 2008, 2011; Paterson et al. 2009; Qi et al. 2010).

The O allele has a reduced susceptibility to severe malaria caused by *Plasmodium falciparum* due to a reduced adhesion of red blood cells to the vascular endothelium (Band et al. 2013; Timmann et al. 2012). Decreased risks of gastric and pancreatic cancers related to the O allele due to its association with infection and activation of *Helicobacter pylori* have also been reported (Amundadottir et al. 2009; Edgren et al. 2010; Iodice et al. 2010; Risch et al. 2010; Wang et al. 2012; Wolpin et al. 2009). On the other hand, individuals possessing the O allele are more susceptible to severe infections caused by cholera (*Vibrio cholerae*) (Clemens et al. 1989; Glass et al. 1985; Harris et al. 2005) and *Escherichia coli* O157 (Blackwell et al. 2002; van Loon et al. 1991). It is hypothesized that the presence of A, B, and O alleles in human populations is maintained by balancing selection (Calafell et al. 2008), even though the distribution of these alleles varies among populations (Roychoudhury and Nei 1988). It is possible that the ABO blood group system is associated with susceptibility to many other infectious diseases, since bacterial and viral antigens have epitopes similar to ABH antigens and since the ABH antigens can be a receptor for binding by pathogens.

FUT2, which is classically known as the secretor factor locus (*Se*), encodes alpha-(1,2) fucosyltransferase (Rouquier et al. 1995). This enzyme regulates the secretion status of the ABH and Lewis (Le) antigens in tissues and body fluids other than blood cells (Oriol et al. 1981). Individuals with active *FUT2* express ABH and Le^b antigens, whereas individuals with inactive *FUT2* express only Le^a antigens. To date, several nonsynonymous variants that result in the nonsecretor phenotype have been identified (Kelly et al. 1995; Koda et al. 1996, 2000a; Kudo et al. 1996). Interestingly, different nonsecretor variants are found in Western and Eastern populations (Fig. 11.6). The frequency distribution of the *FUT2* variants is believed to be the result of balancing selection (Ferrer-Admetlla et al. 2009; Koda et al. 2000b, 2001). There are advantages and disadvantages associated with both secretor and nonsecretor alleles.

Lindesmith et al. (2003) reported that individuals possessing the nonsecretor allele are completely resistant to Norwalk virus infection; however, it has also been suggested that individuals possessing the nonsecretor allele are more susceptible to infections caused by *Haemophilus influenzae* (Blackwell et al. 1986a), *Neisseria meningitidis*, *Streptococcus pneumoniae* (Blackwell et al. 1986b), *V. cholerae* (Arifuzzaman et al. 2011), and *E. coli* (Sheinfeld et al. 1989). Recent GWASs revealed that the nonsecretor allele is associated with an increased risk of Crohn's disease (Franke et al. 2010; Jostins et al. 2012; McGovern et al. 2010) as well as increased plasma levels of vitamin B12 (Hazra et al. 2008, 2009; Lin et al. 2012; Tanaka et al. 2009). A recent study hypothesized that the FUT2 secretor variant may decrease plasma levels of B12 by influencing secretion of gastric intrinsic factor, a fucosylated glycoprotein that is required for the ileal uptake of vitamin B12 (Chery et al. 2013).

11.6 Closing Remarks

Recent studies examining the human genome have provided information regarding local genetic adaptations that have resulted in phenotypic differentiation between populations; however, determining what constitutes a true selective pressure remains difficult. Lifestyles, as well as climatic and physical conditions, could serve as selective pressures, and as we have learned from the study of blood types, one of the strongest selective forces could be infectious diseases. There exist many classical examples of malaria-resistant variants in *HBB*, *G6PD*, *DARC*, and other genes (Currat et al. 2002; Hamblin and Di Rienzo 2000; Ohashi et al. 2004; Sabeti et al. 2002; Zimmerman et al. 1999). Furthermore, HLAs have been well documented as targets of both positive and balancing selection (Hedrick et al. 1991; Hughes and Nei 1988; Prugnolle et al. 2005; Takahata et al. 1992). Recently, Fumagalli et al. (2011) suggested that pathogenic environments play a more important role in local adaptation than do climatic factors. It has also been supposed that autoimmune diseases in humans, such as celiac disease, type 1 diabetes, and multiple sclerosis, may have emerged as a by-product of adaptations that

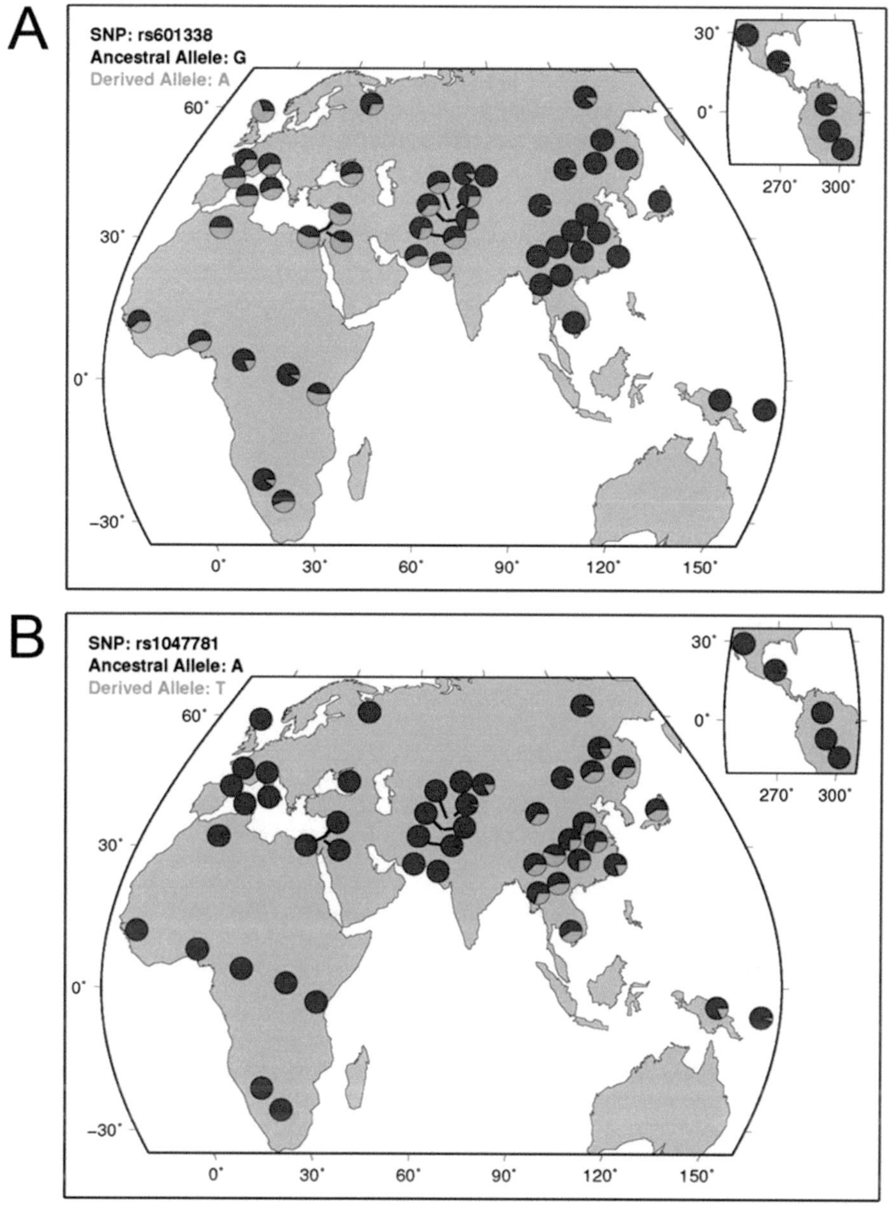

Fig. 11.6 Global distribution of nonsecretor variants in *FUT2*. (**a**) 154Trp>Ter (rs601338). (**b**) 140Ile>Phe (rs1047781) (http://hgdp.uchicago.edu/cgi-bin/gbrowse/HGDP/)

resulted from past pandemics of infectious diseases (Skoglund et al. 2011; Young et al. 2010). Recent genome scans for selective sweeps have detected a number of signatures on immunity-related genes. Together with the results of GWASs examining infectious diseases and immune functions, the findings outlined previously indicate that past endemics experienced by human populations and how those populations overcame them must be understood in greater detail. Further study of the phenotypes of modern humans will allow us to develop a greater understanding of the history of human conquests against the environments in which we have lived.

References

Adhikari K et al (2015) A genome-wide association study identifies multiple loci for variation in human ear morphology. Nat Commun 6:7500

Akey JM, Wang H, Xiong M, Wu H, Liu W, Shriver MD, Jin L (2001) Interaction between the melanocortin-1 receptor and P genes contributes to inter-individual variation in skin pigmentation phenotypes in a Tibetan population. Hum Genet 108(6):516–520

Amundadottir L et al (2009) Genome-wide association study identifies variants in the ABO locus associated with susceptibility to pancreatic cancer. Nat Genet 41(9):986–990

Aoki K (2002) Sexual selection as a cause of human skin colour variation: Darwin's hypothesis revisited. Ann Hum Biol 29(6):589–608

Arifuzzaman M et al (2011) Individuals with Le(a+b−) blood group have increased susceptibility to symptomatic vibrio cholerae O1 infection. PLoS Negl Trop Dis 5(12):e1413

Band G et al (2013) Imputation-based meta-analysis of severe malaria in three African populations. PLoS Genet 9(5):e1003509

Barbalic M et al (2010) Large-scale genomic studies reveal central role of ABO in sP-selectin and sICAM-1 levels. Hum Mol Genet 19(9):1863–1872

Beja-Pereira A et al (2003) Gene-culture coevolution between cattle milk protein genes and human lactase genes. Nat Genet 35(4):311–313

Berndt SI et al (2013) Genome-wide meta-analysis identifies 11 new loci for anthropometric traits and provides insights into genetic architecture. Nat Genet 45(5):501–512

Bersaglieri T, Sabeti PC, Patterson N, Vanderploeg T, Schaffner SF, Drake JA, Rhodes M, Reich DE, Hirschhorn JN (2004) Genetic signatures of strong recent positive selection at the lactase gene. Am J Hum Genet 74(6):1111–1120

Blackwell CC, Jonsdottir K, Hanson MF, Weir DM (1986a) Non-secretion of ABO blood group antigens predisposing to infection by Haemophilus influenzae. Lancet 2(8508):687

Blackwell CC, Jonsdottir K, Hanson M, Todd WT, Chaudhuri AK, Mathew B, Brettle RP, Weir DM (1986b) Non-secretion of ABO antigens predisposing to infection by Neisseria meningitidis and Streptococcus pneumoniae. Lancet 2(8501):284–285

Blackwell CC, Dundas S, James VS, Mackenzie DA, Braun JM, Alkout AM, Todd WT, Elton RA, Weir DM (2002) Blood group and susceptibility to disease caused by Escherichia coli O157. J Infect Dis 185(3):393–396

Boyko AR (2011) The domestic dog: man's best friend in the genomic era. Genome Biol 12(2):216

Brockschmidt FF et al (2011) Susceptibility variants on chromosome 7p21.1 suggest HDAC9 as a new candidate gene for male-pattern baldness. Br J Dermatol 165(6):1293–1302

Calafell F, Roubinet F, Ramirez-Soriano A, Saitou N, Bertranpetit J, Blancher A (2008) Evolutionary dynamics of the human ABO gene. Hum Genet 124(2):123–135

Chambers JC et al (2011) Genome-wide association study identifies loci influencing concentrations of liver enzymes in plasma. Nat Genet 43(11):1131–1138

Chery C et al (2013) Gastric intrinsic factor deficiency with combined GIF heterozygous mutations and FUT2 secretor variant. Biochimie 95(5):995–1001

Claes P et al (2014) Modeling 3D facial shape from DNA. PLoS Genet 10(3):e1004224

Clemens JD et al (1989) ABO blood groups and cholera: new observations on specificity of risk and modification of vaccine efficacy. J Infect Dis 159(4):770–773

Clemente FJ et al (2014) A selective sweep on a deleterious mutation in CPT1A in Arctic populations. Am J Hum Genet 95(5):584–589

Currat M, Trabuchet G, Rees D, Perrin P, Harding RM, Clegg JB, Langaney A, Excoffier L (2002) Molecular analysis of the beta-globin gene cluster in the Niokholo Mandenka population reveals a recent origin of the beta(S) Senegal mutation. Am J Hum Genet 70(1):207–223

Daub JT, Hofer T, Cutivet E, Dupanloup I, Quintana-Murci L, Robinson-Rechavi M, Excoffier L (2013) Evidence for polygenic adaptation to pathogens in the human genome. Mol Biol Evol 30(7):1544–1558

Dichgans M et al (2014) Shared genetic susceptibility to ischemic stroke and coronary artery disease: a genome-wide analysis of common variants. Stroke 45(1):24–36

Eastlund T (1998) The histo-blood group ABO system and tissue transplantation. Transfusion 38 (10):975–988

Edenberg HJ (2000) Regulation of the mammalian alcohol dehydrogenase genes. Prog Nucleic Acid Res Mol Biol 64:295–341

Edgren G, Hjalgrim H, Rostgaard K, Norda R, Wikman A, Melbye M, Nyren O (2010) Risk of gastric cancer and peptic ulcers in relation to ABO blood type: a cohort study. Am J Epidemiol 172(11):1280–1285

Edwards M, Bigham A, Tan J, Li S, Gozdzik A, Ross K, Jin L, Parra EJ (2010) Association of the OCA2 polymorphism His615Arg with melanin content in east Asian populations: further evidence of convergent evolution of skin pigmentation. PLoS Genet 6(3):e1000867

Ellis JA, Stebbing M, Harrap SB (2001) Polymorphism of the androgen receptor gene is associated with male pattern baldness. J Invest Dermatol 116(3):452–455

Enattah NS, Sahi T, Savilahti E, Terwilliger JD, Peltonen L, Jarvela I (2002) Identification of a variant associated with adult-type hypolactasia. Nat Genet 30(2):233–237

Eriksson N et al (2010) Web-based, participant-driven studies yield novel genetic associations for common traits. PLoS Genet 6(6):e1000993

Ferrer-Admetlla A, Sikora M, Laayouni H, Esteve A, Roubinet F, Blancher A, Calafell F, Bertranpetit J, Casals F (2009) A natural history of FUT2 polymorphism in humans. Mol Biol Evol 26(9):1993–2003

Franbourg A, Hallegot P, Baltenneck F, Toutain C, Leroy F (2003) Current research on ethnic hair. J Am Acad Dermatol 48(6 Suppl):S115–S119

Frank J et al (2012) Genome-wide significant association between alcohol dependence and a variant in the ADH gene cluster. Addict Biol 17(1):171–180

Franke A et al (2010) Genome-wide meta-analysis increases to 71 the number of confirmed Crohn's disease susceptibility loci. Nat Genet 42(12):1118–1125

Fujimoto A et al (2008) A scan for genetic determinants of human hair morphology: EDAR is associated with Asian hair thickness. Hum Mol Genet 17(6):835–843

Fumagalli M et al. (2011) Signatures of environmental genetic adaptation pinpoint pathogens as the main selective pressure through human evolution. PLOS Genet 7:e1002355

Fumagalli M et al (2015) Greenlandic Inuit show genetic signatures of diet and climate adaptation. Science 349(6254):1343–1347

Glass RI et al (1985) Predisposition for cholera of individuals with O blood group. Possible evolutionary significance. Am J Epidemiol 121(6):791–796

Goedde HW, Agarwal DP, Harada S, Meier-Tackmann D, Ruofu D, Bienzle U, Kroeger A, Hussein L (1983) Population genetic studies on aldehyde dehydrogenase isozyme deficiency and alcohol sensitivity. Am J Hum Genet 35(4):769–772

Goldman D, Enoch MA (1990) Genetic epidemiology of ethanol metabolic enzymes: a role for selection. World Rev Nutr Diet 63:143–160

Grossman SR et al (2010) A composite of multiple signals distinguishes causal variants in regions of positive selection. Science 327(5967):883–886
Grossman SR et al (2013) Identifying recent adaptations in large-scale genomic data. Cell 152 (4):703–713
Guo J et al (2014) Variation and signatures of selection on the human face. J Hum Evol 75:143–152
Hamblin MT, Di Rienzo A (2000) Detection of the signature of natural selection in humans: evidence from the Duffy blood group locus. Am J Hum Genet 66(5):1669–1679
Hamilton JB (1951) Patterned loss of hair in man; types and incidence. Ann N Y Acad Sci 53 (3):708–728
Han J et al (2008) A genome-wide association study identifies novel alleles associated with hair color and skin pigmentation. PLoS Genet 4(5):e1000074
Hancock AM, Witonsky DB, Gordon AS, Eshel G, Pritchard JK, Coop G, Di Rienzo A (2008) Adaptations to climate in candidate genes for common metabolic disorders. PLoS Genet 4(2): e32
Hancock AM et al (2011) Adaptations to climate-mediated selective pressures in humans. PLoS Genet 7(4):e1001375
Harada S, Misawa S, Agarwal DP, Goedde HW (1980) Liver alcohol dehydrogenase and aldehyde dehydrogenase in the Japanese: isozyme variation and its possible role in alcohol intoxication. Am J Hum Genet 32(1):8–15
Harris JB et al (2005) Blood group, immunity, and risk of infection with Vibrio cholerae in an area of endemicity. Infect Immun 73(11):7422–7427
Hazra A, Kraft P, Selhub J, Giovannucci EL, Thomas G, Hoover RN, Chanock SJ, Hunter DJ (2008) Common variants of FUT2 are associated with plasma vitamin B12 levels. Nat Genet 40(10):1160–1162
Hazra A, Kraft P, Lazarus R, Chen C, Chanock SJ, Jacques P, Selhub J, Hunter DJ (2009) Genome-wide significant predictors of metabolites in the one-carbon metabolism pathway. Hum Mol Genet 18(23):4677–4687
Hedrick PW, Whittam TS, Parham P (1991) Heterozygosity at individual amino acid sites: extremely high levels for HLA-A and -B genes. Proc Natl Acad Sci U S A 88(13):5897–5901
Heit JA, Armasu SM, Asmann YW, Cunningham JM, Matsumoto ME, Petterson TM, De Andrade M (2012) A genome-wide association study of venous thromboembolism identifies risk variants in chromosomes 1q24.2 and 9q. J Thromb Haemost 10(8):1521–1531
Hillmer AM et al (2008) Susceptibility variants for male-pattern baldness on chromosome 20p11. Nat Genet 40(11):1279–1281
Hillmer AM et al (2009) Recent positive selection of a human androgen receptor/ectodysplasin A2 receptor haplotype and its relationship to male pattern baldness. Hum Genet 126(2):255–264
Huang L, Li Y, Singleton AB, Hardy JA, Abecasis G, Rosenberg NA, Scheet P (2009) Genotype-imputation accuracy across worldwide human populations. Am J Hum Genet 84(2):235–250
Hubbe M, Hanihara T, Harvati K (2009) Climate signatures in the morphological differentiation of worldwide modern human populations. Anat Rec-Adv Integr Anat Evol Biol 292 (11):1720–1733
Hughes AL, Nei M (1988) Pattern of nucleotide substitution at major histocompatibility complex class I loci reveals overdominant selection. Nature 335(6186):167–170
Hurley TD, Bosron WF, Stone CL, Amzel LM (1994) Structures of three human beta alcohol dehydrogenase variants. Correlations with their functional differences. J Mol Biol 239 (3):415–429
Iodice S, Maisonneuve P, Botteri E, Sandri MT, Lowenfels AB (2010) ABO blood group and cancer. Eur J Cancer 46(18):3345–3350
Itan Y, Powell A, Beaumont MA, Burger J, Thomas MG (2009) The origins of lactase persistence in Europe. PLoS Comput Biol 5(8):e1000491
Jablonski NG, Chaplin G (2000) The evolution of human skin coloration. J Hum Evol 39 (1):57–106

Jarvis JP et al (2012) Patterns of ancestry, signatures of natural selection, and genetic association with stature in Western African pygmies. PLoS Genet 8(4):e1002641
Jostins L et al (2012) Host-microbe interactions have shaped the genetic architecture of inflammatory bowel disease. Nature 491(7422):119–124
Kamatani Y, Matsuda K, Okada Y, Kubo M, Hosono N, Daigo Y, Nakamura Y, Kamatani N (2010) Genome-wide association study of hematological and biochemical traits in a Japanese population. Nat Genet 42(3):210–215
Kamberov YG et al (2013) Modeling recent human evolution in mice by expression of a selected EDAR variant. Cell 152(4):691–702
Kapoor M et al (2013) A meta-analysis of two genome-wide association studies to identify novel loci for maximum number of alcoholic drinks. Hum Genet 132(10):1141–1151
Kelly RJ, Rouquier S, Giorgi D, Lennon GG, Lowe JB (1995) Sequence and expression of a candidate for the human secretor blood group alpha(1,2)fucosyltransferase gene (FUT2). Homozygosity for an enzyme-inactivating nonsense mutation commonly correlates with the non-secretor phenotype. J Biol Chem 270(9):4640–4649
Kenny EE et al (2012) Melanesian blond hair is caused by an amino acid change in TYRP1. Science 336(6081):554
Kim YJ et al (2011) Large-scale genome-wide association studies in East Asians identify new genetic loci influencing metabolic traits. Nat Genet 43(10):990–995
Kimura R, Fujimoto A, Tokunaga K, Ohashi J (2007) A practical genome scan for population-specific strong selective sweeps that have reached fixation. PLoS One 2(3):e286
Kimura R, Ohashi J, Matsumura Y, Nakazawa M, Inaoka T, Ohtsuka R, Osawa M, Tokunaga K (2008) Gene flow and natural selection in oceanic human populations inferred from genome-wide SNP typing. Mol Biol Evol 25(8):1750–1761
Kimura R et al (2009) A common variation in EDAR is a genetic determinant of shovel-shaped incisors. Am J Hum Genet 85(4):528–535
Koda Y, Soejima M, Liu Y, Kimura H (1996) Molecular basis for secretor type alpha(1,2)-fucosyltransferase gene deficiency in a Japanese population: a fusion gene generated by unequal crossover responsible for the enzyme deficiency. Am J Hum Genet 59(2):343–350
Koda Y, Soejima M, Johnson PH, Smart E, Kimura H (2000a) An Alu-mediated large deletion of the FUT2 gene in individuals with the ABO-Bombay phenotype. Hum Genet 106(1):80–85
Koda Y, Tachida H, Soejima M, Takenaka O, Kimura H (2000b) Ancient origin of the null allele se(428) of the human ABO-secretor locus (FUT2). J Mol Evol 50(3):243–248
Koda Y, Tachida H, Pang H, Liu Y, Soejima M, Ghaderi AA, Takenaka O, Kimura H (2001) Contrasting patterns of polymorphisms at the ABO-secretor gene (FUT2) and plasma alpha (1,3)fucosyltransferase gene (FUT6) in human populations. Genetics 158(2):747–756
Kudo T, Iwasaki H, Nishihara S, Shinya N, Ando T, Narimatsu I, Narimatsu H (1996) Molecular genetic analysis of the human Lewis histo-blood group system. II. Secretor gene inactivation by a novel single missense mutation A385T in Japanese nonsecretor individuals. J Biol Chem 271(16):9830–9837
Lango Allen H et al (2010) Hundreds of variants clustered in genomic loci and biological pathways affect human height. Nature 467(7317):832–838
Lao O, de Gruijter JM, van Duijn K, Navarro A, Kayser M (2007) Signatures of positive selection in genes associated with human skin pigmentation as revealed from analyses of single nucleotide polymorphisms. Ann Hum Genet 71(Pt 3):354–369
Leinonen T, McCairns RJ, O'Hara RB, Merila J (2013) Q(ST)-F(ST) comparisons: evolutionary and ecological insights from genomic heterogeneity. Nat Rev Genet 14(3):179–190
Li H et al (2007) Geographically separate increases in the frequency of the derived ADH1B*47His allele in eastern and western Asia. Am J Hum Genet 81(4):842–846
Li R et al (2012) Six novel susceptibility loci for early-onset androgenetic alopecia and their unexpected association with common diseases. PLoS Genet 8(5):e1002746
Li J et al (2013) Genome-wide association study on serum alkaline phosphatase levels in a Chinese population. BMC Genomics 14:684

Lin X et al (2012) Genome-wide association study identifies novel loci associated with serum level of vitamin B12 in Chinese men. Hum Mol Genet 21(11):2610–2617

Lindesmith L, Moe C, Marionneau S, Ruvoen N, Jiang X, Lindblad L, Stewart P, LePendu J, Baric R (2003) Human susceptibility and resistance to Norwalk virus infection. Nat Med 9 (5):548–553

Liu F et al (2012) A genome-wide association study identifies five loci influencing facial morphology in Europeans. PLoS Genet 8(9):e1002932

Liumbruno GM, Franchini M (2013) Beyond immunohaematology: the role of the ABO blood group in human diseases. Blood Transfus = Trasfusione del sangue 11(4):491–499

Macgregor S et al (2009) Associations of ADH and ALDH2 gene variation with self report alcohol reactions, consumption and dependence: an integrated analysis. Hum Mol Genet 18 (3):580–593

Maes HH, Neale MC, Eaves LJ (1997) Genetic and environmental factors in relative body weight and human adiposity. Behav Genet 27(4):325–351

Matsunaga E (1962) The dimorphism in human normal cerumen. Ann Hum Genet 25:273–286

McCarthy MI, Abecasis GR, Cardon LR, Goldstein DB, Little J, Ioannidis JP, Hirschhorn JN (2008) Genome-wide association studies for complex traits: consensus, uncertainty and challenges. Nat Rev Genet 9(5):356–369

McGovern DP et al (2010) Fucosyltransferase 2 (FUT2) non-secretor status is associated with Crohn's disease. Hum Mol Genet 19(17):3468–3476

Medland SE et al (2009) Common variants in the trichohyalin gene are associated with straight hair in Europeans. Am J Hum Genet 85(5):750–755

Miura K et al (2007) A strong association between human earwax-type and apocrine colostrum secretion from the mammary gland. Hum Genet 121(5):631–633

Moeller A, Weippert-Kretschmer M, Prinz H, Kretschmer V (2001) Influence of ABO blood groups on primary hemostasis. Transfusion 41(1):56–60

Moltke I et al (2014) A common Greenlandic TBC1D4 variant confers muscle insulin resistance and type 2 diabetes. Nature 512(7513):190–193

Nakano M, Miwa N, Hirano A, Yoshiura K, Niikawa N (2009) A strong association of axillary osmidrosis with the wet earwax type determined by genotyping of the ABCC11 gene. BMC Genet 10:42

Nan H, Kraft P, Hunter DJ, Han J (2009) Genetic variants in pigmentation genes, pigmentary phenotypes, and risk of skin cancer in Caucasians. Int J Cancer 125(4):909–917

Neel JV (1962) Diabetes mellitus: a "thrifty" genotype rendered detrimental by "progress"? Am J Hum Genet 14:353–362

Norton HL et al (2007) Genetic evidence for the convergent evolution of light skin in Europeans and East Asians. Mol Biol Evol 24(3):710–722

O'Donnell J, Laffan MA (2001) The relationship between ABO histo-blood group, factor VIII and von Willebrand factor. Transfus Med 11(4):343–351

Ohashi J, Naka I, Patarapotikul J, Hananantachai H, Brittenham G, Looareesuwan S, Clark AG, Tokunaga K (2004) Extended linkage disequilibrium surrounding the hemoglobin E variant due to malarial selection. Am J Hum Genet 74(6):1198–1208

Ohashi J, Naka I, Tsuchiya N (2011) The impact of natural selection on an ABCC11 SNP determining earwax type. Mol Biol Evol 28(1):849–857

Oota H et al (2004) The evolution and population genetics of the ALDH2 locus: random genetic drift, selection, and low levels of recombination. Ann Hum Genet 68(Pt 2):93–109

Oriol R, Le Pendu J, Sparkes RS, Sparkes MC, Crist M, Gale RP, Terasaki PI, Bernoco M (1981) Insights into the expression of ABH and Lewis antigens through human bone marrow transplantation. Am J Hum Genet 33(4):551–560

Osier MV et al (2002) A global perspective on genetic variation at the ADH genes reveals unusual patterns of linkage disequilibrium and diversity. Am J Hum Genet 71(1):84–99

Pare G, Chasman DI, Kellogg M, Zee RY, Rifai N, Badola S, Miletich JP, Ridker PM (2008) Novel association of ABO histo-blood group antigen with soluble ICAM-1: results of a genome-wide association study of 6,578 women. PLoS Genet 4(7):e1000118

Pare G et al (2011) Genome-wide association analysis of soluble ICAM-1 concentration reveals novel associations at the NFKBIK, PNPLA3, RELA, and SH2B3 loci. PLoS Genet 7(4): e1001374
Park J et al (2012) Effects of an Asian-specific nonsynonymous EDAR variant on multiple dental traits. J Hum Genet in press
Park BL et al (2013) Extended genetic effects of ADH cluster genes on the risk of alcohol dependence: from GWAS to replication. Hum Genet 132(6):657–668
Paternoster L et al (2012) Genome-wide association study of three-dimensional facial morphology identifies a variant in PAX3 associated with nasion position. Am J Hum Genet 90(3):478–485
Paterson AD et al (2009) Genome-wide association identifies the ABO blood group as a major locus associated with serum levels of soluble E-selectin. Arterioscler Thromb Vasc Biol 29 (11):1958–1967
Peng Y, Shi H, Qi XB, Xiao CJ, Zhong H, Ma RL, Su B (2010) The ADH1B Arg47His polymorphism in east Asian populations and expansion of rice domestication in history. BMC Evol Biol 10:15
Peng S, Tan J, Hu S, Zhou H, Guo J, Jin L, Tang K (2013) Detecting genetic association of common human facial morphological variation using high density 3D image registration. PLoS Comput Biol 9(12):e1003375
Pickrell JK et al (2009) Signals of recent positive selection in a worldwide sample of human populations. Genome Res 19(5):826–837
Plomin R, Haworth CM, Davis OS (2009) Common disorders are quantitative traits. Nat Rev Genet 10(12):872–878
Preston AE, Barr A (1964) The plasma concentration of factor Viii in the normal population. II. The effects of age, sex and blood group. Br J Haematol 10:238–245
Prugnolle F, Manica A, Charpentier M, Guegan JF, Guernier V, Balloux F (2005) Pathogen-driven selection and worldwide HLA class I diversity. Curr Biol 15(11):1022–1027
Qi L et al (2010) Genetic variants in ABO blood group region, plasma soluble E-selectin levels and risk of type 2 diabetes. Hum Mol Genet 19(9):1856–1862
Ranciaro A et al (2014) Genetic origins of lactase persistence and the spread of pastoralism in Africa. Am J Hum Genet 94(4):496–510
Rees JL (2003) Genetics of hair and skin color. Annu Rev Genet 37:67–90
Reilly MP et al (2011) Identification of ADAMTS7 as a novel locus for coronary atherosclerosis and association of ABO with myocardial infarction in the presence of coronary atherosclerosis: two genome-wide association studies. Lancet 377(9763):383–392
Richards JB et al (2008) Male-pattern baldness susceptibility locus at 20p11. Nat Genet 40 (11):1282–1284
Risch HA, Yu H, Lu L, Kidd MS (2010) ABO blood group, Helicobacter pylori seropositivity, and risk of pancreatic cancer: a case-control study. J Natl Cancer Inst 102(7):502–505
Roberts DF (1953) Body weight, race and climate. Am J Phys Anthropol 11(4):533–558
Roseman CC (2004) Detecting interregionally diversifying natural selection on modern human cranial form by using matched molecular and morphometric data. Proc Natl Acad Sci U S A 101(35):12824–12829
Roseman CC, Weaver TD (2004) Multivariate apportionment of global human craniometric diversity. Am J Phys Anthropol 125(3):257–263
Rosenberg NA, Huang L, Jewett EM, Szpiech ZA, Jankovic I, Boehnke M (2010) Genome-wide association studies in diverse populations. Nat Rev Genet 11(5):356–366
Rouquier S, Lowe JB, Kelly RJ, Fertitta AL, Lennon GG, Giorgi D (1995) Molecular cloning of a human genomic region containing the H blood group alpha(1,2)fucosyltransferase gene and two H locus-related DNA restriction fragments. Isolation of a candidate for the human secretor blood group locus. J Biol Chem 270(9):4632–4639
Roychoudhury AK, Nei M (1988) Human polymorphic genes: world distribution. Oxford University Press, New York
Sabeti PC et al (2002) Detecting recent positive selection in the human genome from haplotype structure. Nature 419(6909):832–837

Sabeti PC et al (2007) Genome-wide detection and characterization of positive selection in human populations. Nature 449(7164):913–918

Schunkert H et al (2011) Large-scale association analysis identifies 13 new susceptibility loci for coronary artery disease. Nat Genet 43(4):333–338

Scott GR, Turner CGI (1997) The anthropology of modern human teeth: dental morphology and its variation in recent human populations. Cambridge University Press, Cambridge

Shai I, Jiang R, Manson JE, Stampfer MJ, Willett WC, Colditz GA, Hu FB (2006) Ethnicity, obesity, and risk of type 2 diabetes in women: a 20-year follow-up study. Diabetes Care 29 (7):1585–1590

Sheinfeld J, Schaeffer AJ, Cordon-Cardo C, Rogatko A, Fair WR (1989) Association of the Lewis blood-group phenotype with recurrent urinary tract infections in women. N Engl J Med 320 (12):773–777

Shibuya A, Yoshida A (1988) Genotypes of alcohol-metabolizing enzymes in Japanese with alcohol liver diseases: a strong association of the usual Caucasian-type aldehyde dehydrogenase gene (ALDH1(2)) with the disease. Am J Hum Genet 43(5):744–748

Skoglund P, Gotherstrom A, Jakobsson M (2011) Estimation of population divergence times from non-overlapping genomic sequences: examples from dogs and wolves. Mol Biol Evol 28 (4):1505–1517

Stokowski RP et al (2007) A genomewide association study of skin pigmentation in a South Asian population. Am J Hum Genet 81(6):1119–1132

Stunkard AJ, Foch TT, Hrubec Z (1986) A twin study of human obesity. JAMA 256(1):51–54

Sturm RA, Duffy DL (2012) Human pigmentation genes under environmental selection. Genome Biol 13(9):248

Sulem P et al (2007) Genetic determinants of hair, eye and skin pigmentation in Europeans. Nat Genet 39(12):1443–1452

Sulem P et al (2008) Two newly identified genetic determinants of pigmentation in Europeans. Nat Genet 40(7):835–837

Takahata N, Satta Y, Klein J (1992) Polymorphism and balancing selection at major histocompatibility complex loci. Genetics 130(4):925–938

Tanaka T et al (2009) Genome-wide association study of vitamin B6, vitamin B12, folate, and homocysteine blood concentrations. Am J Hum Genet 84(4):477–482

Tang K, Thornton KR, Stoneking M (2007) A new approach for using genome scans to detect recent positive selection in the human genome. PLoS Biol 5(7):–e171

Tang W et al (2012) Genetic associations for activated partial thromboplastin time and prothrombin time, their gene expression profiles, and risk of coronary artery disease. Am J Hum Genet 91(1):152–162

Tang W et al (2013) A genome-wide association study for venous thromboembolism: the extended cohorts for heart and aging research in genomic epidemiology (CHARGE) consortium. Genet Epidemiol 37(5):512–521

Teslovich TM et al (2010) Biological, clinical and population relevance of 95 loci for blood lipids. Nature 466(7307):707–713

Timmann C et al (2012) Genome-wide association study indicates two novel resistance loci for severe malaria. Nature 489(7416):443–446

Tishkoff SA et al (2007) Convergent adaptation of human lactase persistence in Africa and Europe. Nat Genet 39(1):31–40

Tregouet DA et al (2009) Common susceptibility alleles are unlikely to contribute as strongly as the FV and ABO loci to VTE risk: results from a GWAS approach. Blood 113(21):5298–5303

van Loon FP et al (1991) ABO blood groups and the risk of diarrhea due to enterotoxigenic Escherichia coli. J Infect Dis 163(6):1243–1246

Voight BF, Kudaravalli S, Wen X, Pritchard JK (2006) A map of recent positive selection in the human genome. PLoS Biol 4(3):e72

Wang ET, Kodama G, Baldi P, Moyzis RK (2006) Global landscape of recent inferred Darwinian selection for homo sapiens. Proc Natl Acad Sci U S A 103(1):135–140

Wang TL et al (2010) Prevalence of androgenetic alopecia in China: a community-based study in six cities. Br J Dermatol 162(4):843–847

Wang Z, Liu L, Ji J, Zhang J, Yan M, Liu B, Zhu Z, Yu Y (2012) ABO blood group system and gastric cancer: a case-control study and meta-analysis. Int J Mol Sci 13(10):13308–13321

Wen CP, David Cheng TY, Tsai SP, Chan HT, Hsu HL, Hsu CC, Eriksen MP (2009) Are Asians at greater mortality risks for being overweight than Caucasians? Redefining obesity for Asians. Public Health Nutr 12(4):497–506

Whitlock MC (2008) Evolutionary inference from QST. Mol Ecol 17(8):1885–1896

Willer CJ et al (2009) Six new loci associated with body mass index highlight a neuronal influence on body weight regulation. Nat Genet 41(1):25–34

Willer CJ et al (2013) Discovery and refinement of loci associated with lipid levels. Nat Genet 45 (11):1274–1283

Williams FM et al (2013) Ischemic stroke is associated with the ABO locus: the EuroCLOT study. Ann Neurol 73(1):16–31

Williamson SH, Hubisz MJ, Clark AG, Payseur BA, Bustamante CD, Nielsen R (2007) Localizing recent adaptive evolution in the human genome. PLoS Genet 3(6):e90

Wolpin BM, Chan AT, Hartge P, Chanock SJ, Kraft P, Hunter DJ, Giovannucci EL, Fuchs CS (2009) ABO blood group and the risk of pancreatic cancer. J Natl Cancer Inst 101(6):424–431

Wood AR et al (2014) Defining the role of common variation in the genomic and biological architecture of adult human height. Nat Genet 46(11):1173–1186

Yamaguchi K et al (2012) Association of melanocortin 1 receptor gene (MC1R) polymorphisms with skin reflectance and freckles in Japanese. J Hum Genet 57(11):700–708

Yamamoto F, Clausen H, White T, Marken J, Hakomori S (1990) Molecular genetic basis of the histo-blood group ABO system. Nature 345(6272):229–233

Yokoyama A, Tsutsumi E, Imazeki H, Suwa Y, Nakamura C, Yokoyama T (2007) Contribution of the alcohol dehydrogenase-1B genotype and oral microorganisms to high salivary acetaldehyde concentrations in Japanese alcoholic men. Int J Cancer 121(5):1047–1054

Yoshiura K et al (2006) A SNP in the ABCC11 gene is the determinant of human earwax type. Nat Genet 38(3):324–330

Young JM, Massa HF, Hsu L, Trask BJ (2010) Extreme variability among mammalian V1R gene families. Genome Res 20(1):10–18

Yuan X et al (2008) Population-based genome-wide association studies reveal six loci influencing plasma levels of liver enzymes. Am J Hum Genet 83(4):520–528

Zhou L et al (2013) A genome wide association study identifies common variants associated with lipid levels in the Chinese population. PLoS One 8(12):e82420

Zimmerman PA et al (1999) Emergence of FY*A(null) in a plasmodium vivax-endemic region of Papua New Guinea. Proc Natl Acad Sci U S A 96(24):13973–13977

Chapter 12
Transcription Factor Genes

Mahoko Ueda Takahashi and So Nakagawa

Abstract Transcription factor (TF) genes encode DNA-binding proteins. In all organisms, TFs play central roles in transcription by regulating gene expression. TFs are involved in a variety of biological processes, such as development and cell cycle control. TFs comprise one of the largest known groups of genes. In the human genome, approximately 8% of genes encode TFs. Many TFs involved in developmental processes are well conserved among a wide range of species. This conservation indicates that alterations in TF function are a likely source of phenotypic diversity among species. In addition, numerous human diseases are known to arise from mutations that dysregulate the regulatory system. For example, more than 30% of human developmental disorders are caused by mutations in TFs (Boyadjiev and Jabs, Clin Genet 57:253–266, 2000). Therefore, the structures of TFs and the mechanisms regulating gene expression are areas of great interest in the fields of evolutionary studies and biomedicine. In this chapter, we review the basic concepts of TF structures and the biological processes through which TFs control gene expression by binding target sequences. We will additionally introduce the evolutionary patterns of TF gene families. We will further discuss the results of recent studies that have used new techniques, such as chromatin immunoprecipitation sequencing (ChIP-seq) and network motifs.

Keywords DNA-binding domain · Zinc finger · Homeodomain · Helix-loop-helix · Forkhead box genes · T-box genes · Whole-genome duplication

M.U. Takahashi (✉)
Micro/Nano Technology Center, Tokai University, Hiratsuka, Japan
e-mail: mahoko@tokai.ac.jp

S. Nakagawa
Micro/Nano Technology Center, Tokai University, Hiratsuka, Japan

Department of Molecular Life Science, Tokai University School of Medicine, Isehara, Japan
e-mail: so@tokai.ac.jp

N. Saitou (ed.), *Evolution of the Human Genome I*, Evolutionary Studies,
https://doi.org/10.1007/978-4-431-56603-8_12

12.1 General Features of Transcription Factor Genes

12.1.1 Structure

Typically, TFs are composed of distinct and separable functional domains, including the DNA-binding domain (DBD), transcriptional activation domain (TA), and signal-sensing domain (SSD) or ligand-binding domain (Fig. 12.1a). The primary role of the DBD is to target specific sequences (e.g., promoters, enhancers, and repressors). TFs are classified into distinct families according to the type of DBD. Although TFs usually have only one type of DBD, some TFs have multiple types of DBDs. The TA is involved in interactions with transcriptional co-regulatory proteins (cofactors). Although SSD is not always present, this domain allows signaling molecules to bind to TFs. This modularity allows TFs to create various combinations of functionally distinct domains and gain diverse functions (Fig. 12.1b).

12.1.2 Transcription Regulation

TFs regulate gene expression by binding to specific short sequences [six to eight base pairs (bp)], called cis-regulatory elements, usually in the upstream regions. The binding of TFs to a target site is the fundamental basis of gene regulatory networks. TFs that bind to cis-regulatory elements determine whether a particular gene will be turned "on" or "off" in a cell. Most TFs do not work alone. Instead,

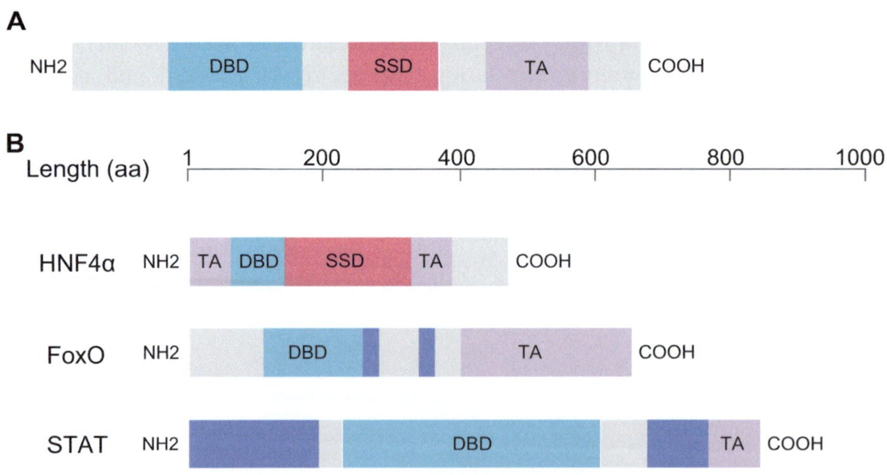

Fig. 12.1 Generalized structure of a transcription factor (TF). (**a**) A typical simplified TF structure. (**b**) Some examples of TF structure. *DBD* DNA-binding domain (green), *SSD* signal sensing domain (pink), *TA* transactivation domain (purple). Domains other than DBD, SSD, and TA are shown in dark blue

they cooperate with other proteins as activators and repressors within a complex and recruit RNA polymerase II to specific genes. In metazoan genomes, TFs (activators or repressors) can recognize and bind enhancer or silencer sequences to increase or reduce gene transcription, respectively. Enhancers are frequently used in specific differentiated cell types, are located up to 1 Mbp from the target gene (Sagai et al. 2005), and may occur either upstream or downstream of the gene or within an intron.

12.2 DNA-Binding Domain

As stated in Sect. 12.1.1, TFs are commonly classified according to the type of DBD. The structural DBD classes exhibit characteristic consensus amino acid sequences (motifs). As described in Sects. 12.1.1 and 12.1.2, the DBD can recognize and bind to specific DNA sequences (consensus binding sequences). Thus, the DBD classification indicates not only the structures but also the function and even the evolutionary history of a gene. In the human genome, three types of DBDs are predominant, accounting for 80% of the TF repertoire: the zinc finger (ZF), homeodomain, and helix-loop-helix (HLH) motifs (Fig. 12.2; Lander et al. 2001; Tupler et al. 2001; Vaquerizas et al. 2009). In particular, zinc-finger motifs comprise the largest family among metazoan TFs. In this section, we will describe the structures of these three examples of DBDs.

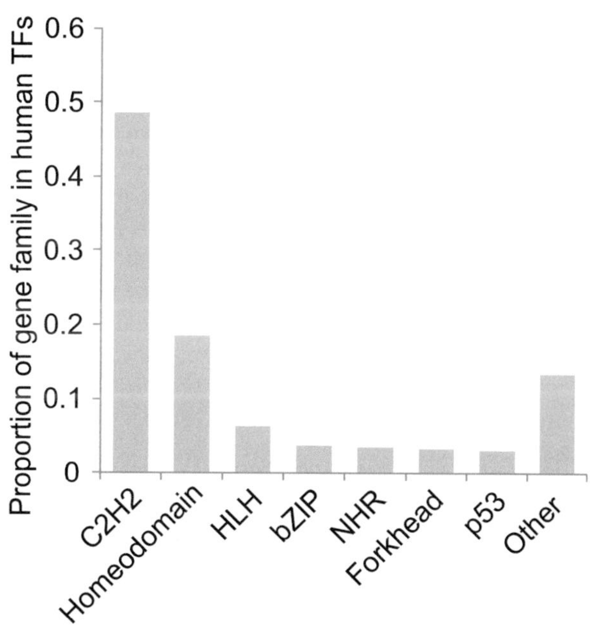

Fig. 12.2 Proportions of transcription factor (TF) gene families. A greater proportion indicates that a TF has undergone more frequent expansion (Data shown are from Vaquerizas et al. 2009). InterPro parent-child relationships between DNA-binding domains were used as the basis for TF family definition

12.2.1 Zinc Finger

This domain was first identified in the transcription factor IIIA from *Xenopus* oocytes (Miller et al. 1985). To date, many motifs have been identified in eukaryotes. A number of different ZF proteins feature multiple cysteine and/or histidine residues and zinc coordination. The ZF motif comprises multiple zinc-binding repeats of approximately 25 amino acids–40 amino acids. This domain requires the coordination of one or more zinc ions to stabilize its structure. Because of this characteristic, the term "zinc finger" is now widely used to define the domain. Most ZF domains bind specific DNA and RNA sequences, although some bind only RNA. Their functions are extraordinarily diverse and include gene transcription, protein translation, mRNA trafficking, cell adhesion, protein folding, and chromatin remodeling. ZF domains also feature structural diversity and thus can be further divided into several types based on the DBD domain (e.g., C_2H_2, C_3H, C_4, Gata-1, and GAL4). Among these domain types, C_2H_2 is the largest and the best-studied subfamily.

The C_2H_2 motif is characterized by two cysteines, two histidines, and several hydrophobic residues that stabilize the three-dimensional structure comprising a two-stranded antiparallel β-sheet and an α-helix surrounding a central zinc ion (Fig. 12.3a, b, Wolfe et al. 2000). The finger consensus sequence is ψ-X-Cys-X2-5-Cys-X3-ψ-X5-ψ-X2-His-X3-5-His, where X represents any amino acid and ψ represents a hydrophobic residue (Frankel et al. 1987). Zinc fingers are found in a wide range of species, from fungi to metazoans, including humans. However, the numbers of fingers and lengths of linkers (amino acid sequences) that separate fingers vary among lineages and kingdoms (Bohm et al. 1997; Clarke 1998). The DNA binding specificity depends on the number of fingers; generally, a greater number of fingers result in more fingers with specific affinities for different ligands (Elrod-Erickson and Pabo 1999; Iuchi 2000). The linker is an important structural element that controls the spacing of fingers along the DNA site and thus influences the DNA binding affinity (Wolfe 2000).

Zinc finger protein fingers interact with the major groove of B-DNA. For example, Zif268 (also known as Krox-24, NGFI-A, and Egrl) includes three zinc fingers, each of which contains a α-helix that fits directly into a guanine-rich consensus binding site (5′-GCGTGGGCG-3′). Amino acids at positions −1, 3, and 6 of the α-helix in each finger contact a 3-bp sequence in the primary DNA strand, and an amino acid at position 2 can interact with the complementary strand (Fig. 12.3a, Wolfe 2000). The highly conserved hydrophobic residues in the α-helix and β-sheets form critical hydrogen bonds with bases in the major groove (Fig. 12.3b, Pavletich and Pabo 1991). These bonds also help to stabilize the ZF motif. The Zif268-DNA complex is a prototypic structure for understanding DNA recognition by the ZF protein. However, ZFP-DNA complexes feature a variety of docking arrangements.

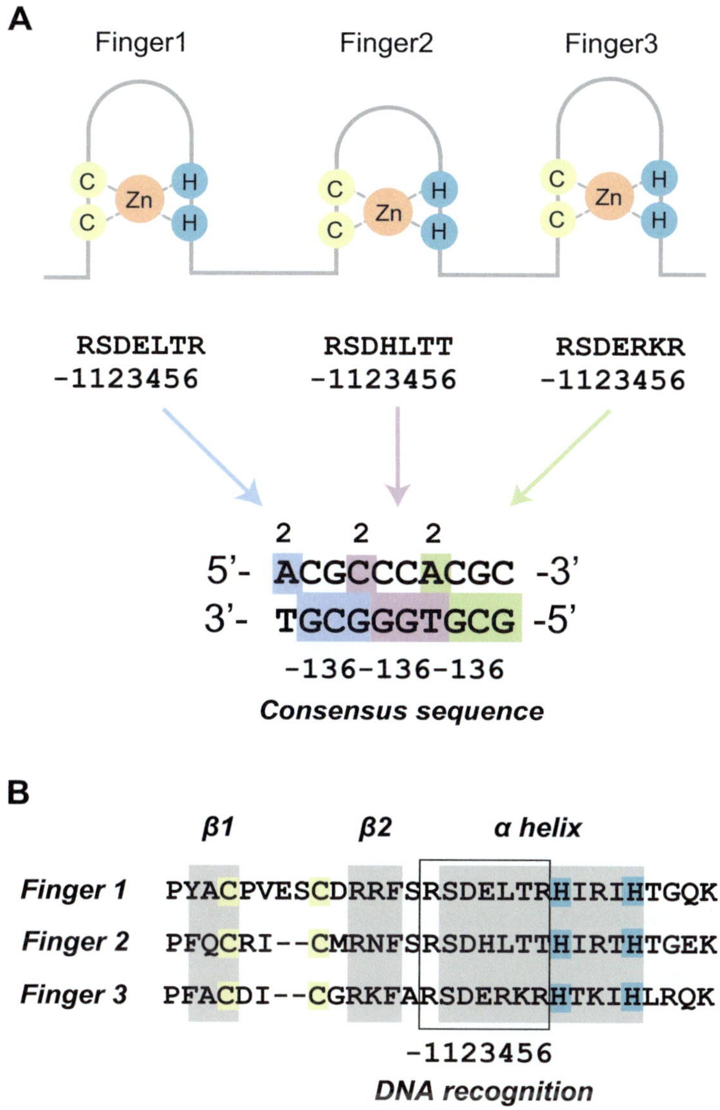

Fig. 12.3 Schematic diagram of a C2H2 zinc-finger motif. (**a**) Three fingers of murine Zif268 are shown. The paired cysteines (C), histidines (H), and zinc ion (Zn) are shown in yellow, blue, and orange, respectively. Amino acid codes under each finger indicate DNA recognition sites. Consensus binding sequences recognized by three fingers are shown at the bottom. Each arrow and shaded base color indicates that each finger forms specific contacts with bases in the consensus sequence. The positions of amino acids in the helix are also shown. (**b**) Amino acid alignments for fingers 1–3. Secondary structure elements (β-sheet and α-helix) are highlighted in gray

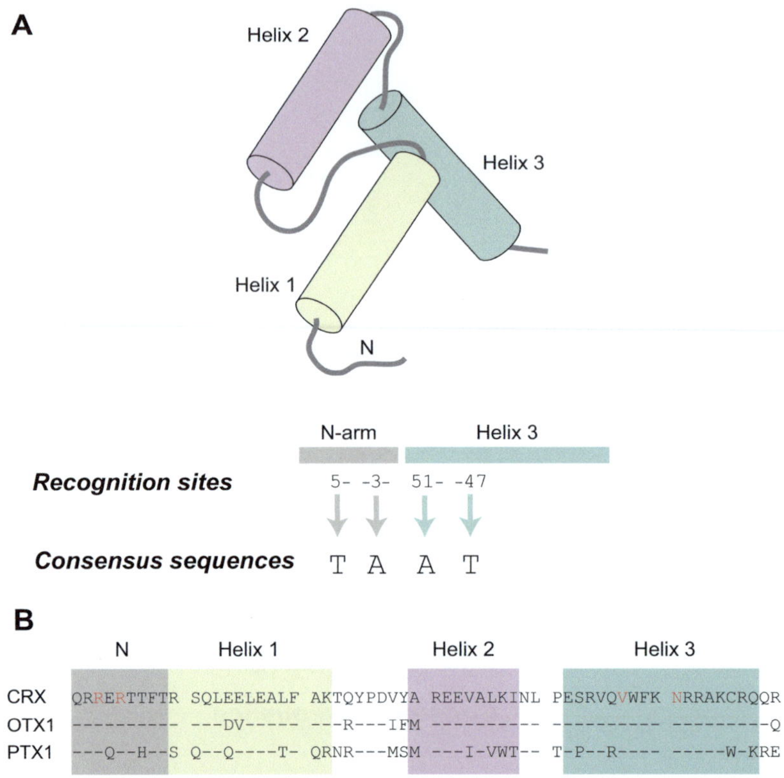

Fig. 12.4 Schematic diagram of a homeodomain motif. (**a**) A homeodomain and consensus recognition sequence are shown. The N-terminal arm is labeled "N." The amino acids at positions 3, 5, 51, and 47 of the homeodomain form contacts with DNA bases in the consensus sequence (TAAT). (**b**) Amino acid alignments of the homeodomains in human genes *CRX*, *OPT1*, and *PTX* are shown. Residues identical to those in *CRX* are indicated by dashes. Base-contacting residues are in red. Secondary structure elements (N-arm and helices 1–3) are highlighted in gray, yellow, pink, and green, respectively

12.2.2 Homeodomain

A homeobox (Hox) gene was first identified in *Drosophila* in 1978 (Lewis 1978), and subsequent cloning of the gene in several organisms revealed a common sequence motif, the homeodomain. This motif is a helix-turn-helix DBD and is known as the second-most abundant TF family in most metazoans, including mammals (Tupler et al. 2001; Vaquerizas et al. 2009). A typical homeodomain is 60 amino acids in length and contains a DNA recognition site for the consensus binding sequence (5′-TAAT-3′). This protein comprises three α-helices folded into a compact globular structure and an N-terminal arm located just adjacent to the first helix (Fig. 12.4a; Gehring et al. 1994). The first and second helices are connected by a loop, and the second and third helices form a helix-turn-helix (Banerjee-Basu and

Baxevanis 2001). Using the same numbering system for homeodomain amino acids as described in Qian et al. (1989), the base contacts made by basic residues in the N-terminal arm and the third helix are shown in Fig. 12.4a, b (Qian et al. 1989; Piper et al. 1999; LaRonde-LeBlanc and Wolberger 2003). In particular, the C-terminal of the third helix directly binds the DNA binding site in the major groove. Accordingly, the third helix is called the recognition helix and governs DNA binding specificity (Gehring et al. 1994). The first and second helices are oriented to helix 3 at an approximate 90° angle and lie above the recognition helix to stabilize the interactions of the third helix with the DNA bases (Vollmer and Clerc 1998).

Although the homeodomain structure is widely conserved across species, from flies to humans, the sequences exhibit significant variation. Many studies have attempted to subdivide homeodomains into small groups based on sequence conservation (Burglin 1994; Semenza 1998; Banerjee-Basu and Baxevanis 2001; Mukherjee and Burglin 2007). Banerjee-Basu and Baxevanis (2001), for example, reported six distinct homeodomain classes based on phylogenetic analysis and showed that this classification was consistent with the known functional and structural characteristics of these proteins. The DNA docking approaches of these domains are nearly identical despite low sequence identity even within subgroups (Kissinger et al. 1990; Wolberger et al. 1991).

12.2.3 Helix-Loop-Helix Domain

HLH gene products are found in organisms ranging from yeast to humans. These are involved in critical developmental processes, including nervous system and muscle development and cell proliferation. Helix-loop-helix gene products contain the basic HLH (bHLH) domain. This domain was first identified in the murine TFs E12 and E47 (Murre et al. 1989a). This domain is approximately 40–60 amino acids in length and contains a basic amino acid region (basic domain) followed by two amphipathic α-helices separated by a linker region of varying length (HLH motif) (Fig. 12.5a, Murre et al. 1994). The basic region is involved in binding to the consensus hexanucleotide sequences within the E-box (5′-CANNTG-3′) promoter region (Fig. 12.5b, Atchley et al. 1999).

Studies of mammalian bHLH proteins have shown that the conserved HLH structure is required for dimerization between two bHLH proteins (Ferré-D'Amaré et al. 1993; Ellenberger et al. 1994). The α-helices confer specificity for a particular partner protein. The helices of two bHLH proteins can interact to form homodimer or heterodimer complexes between different family members (Murre et al. 1989b; Kadesch 1993; Ellenberger et al. 1994). Some bHLH proteins contain a leucine zipper dimerization domain characterized by leucine repeats located at the C-terminal of the bHLH domain (Atchley and Fitch 1997).

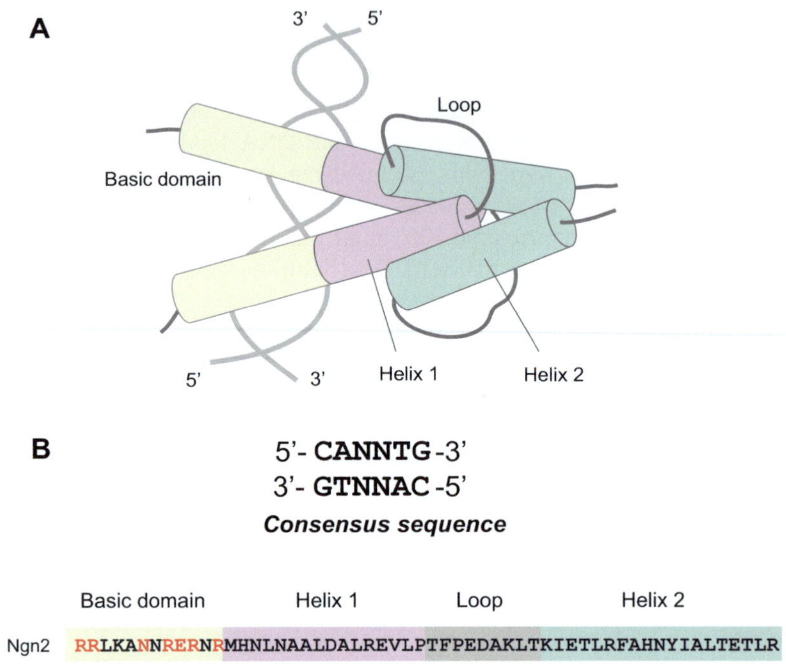

Fig. 12.5 Schematic diagram of a basic helix-loop-helix (bHLH) motif. (**a**) Three-dimensional view of the bHLH structure. (**b**) Consensus sequences and an amino acid alignment of the bHLH domain of the murine *Ngn2* gene. Each color in the sequence and structure of the bHLH dimer represents the same distinct region. The basic domain contacts the E-box in the major groove of the DNA. Residues that contact the E-box are highlighted in red

12.3 Evolution of TF Gene Families

A gene family is defined as a group of two or more genes in a genome with the same domain. Generally, all the genes in a family are homologous because they resulted from the duplication of an original gene. Whole-genome duplications contribute to gene number expansion in some lineages, such as vertebrates. Subsequent deletions and smaller-scale duplications such as segmental duplications (SDs) result in the elimination or creation of new gene copies and complicate the family history. Within a family, related TFs have remained together over long periods of evolution and are often located in contiguous clusters on a chromosome. These TFs often expand through tandem SDs (Demuth and Hahn 2009) and are under positive selection (Nowick and Stubbs 2010). Well-known examples of TF families include the NANOG, Hox (see Sects. 12.2.2 and 12.3.1), and KRAB-ZNF families (Nowick et al. 2009).

TF genes are present in all branches of the tree of life (bacteria, archaea, and eukaryotes). The emergence of complex multicellular organisms has been correlated with the complexity of gene regulatory mechanisms (Levine and Tjian 2003;

Table 12.1 Recently duplicated human transcription factors (TFs)

TF family	Number
KRAB-ZNF	54
Other ZnF-C_2H_2	7
Forkhead	6
bHLH	3
Beta-scaffold-STAT	2
Beta-scaffold-HMG	2

Data from Nowick and Stubbs (2010)

Moore 2005; de Mendoza et al. 2013). This can be observed in the dramatic expansions of TF families and combinatorial gene control by multiple TFs in higher organisms (Levine and Tjian 2003). TF genes often exhibit lineage-specific expansions, indicating that TF genes have evolved independently in different organisms. Some examples of TF genes in the human genome that have experienced a recent expansion are shown in Table 12.1. Studies of the evolutionary histories of TF genes across eukaryotic genomes have shown that these expansions occurred unevenly for TFs containing different types of DBDs (Vaquerizas et al. 2009; de Mendoza et al. 2013; see Fig. 12.2 for the uneven number of human TF types). These differences may have modified the TF expression patterns among lineages. In this section, we will provide some examples of TF gene family evolution from bacteria to humans.

12.3.1 Hox Cluster Genes

Hox genes have evolved through tandem duplications of one or more ancestral Hox genes. In other words, multiple copies of Hox genes formed clusters on a few chromosomes (called Hox clusters), thus providing the best example of evolutionarily conserved gene order within a family. The Hox genes in each cluster are associated with the embryonic development of basic body structures such as the head, trunk, and limbs. The order of Hox genes of each cluster corresponds to the order of the influenced body segments, a relationship known as colinearity.

Figure 12.6 shows the Hox genes and clusters found in metazoans. In tetrapods, four Hox clusters are usually observed on different chromosomes. In humans, 39 Hox genes are located in 4 Hox clusters (A–D) on chromosomes 7p15, 17q21.2, 12q13, and 2q31 (Fig. 12.6). On the other hand, the genomes of tunicates (e.g., sea squirts) and cephalochordates (e.g., amphioxus), both of which are closely related to vertebrates, only contain a single Hox cluster (Spagnuolo et al. 2003; Holland et al. 2008; Pascual-Anaya et al. 2013). These observations can be clearly explained by assuming that the four Hox clusters emerged through two rounds of genome duplications in vertebrate ancestors (Ohno 1970; McLysaght et al. 2002; Putnam et al. 2008; Holland et al. 2008; Pascual-Anaya et al. 2013). Indeed, all vertebrate Hox clusters contain parallel and overlapping sets of Hox genes with

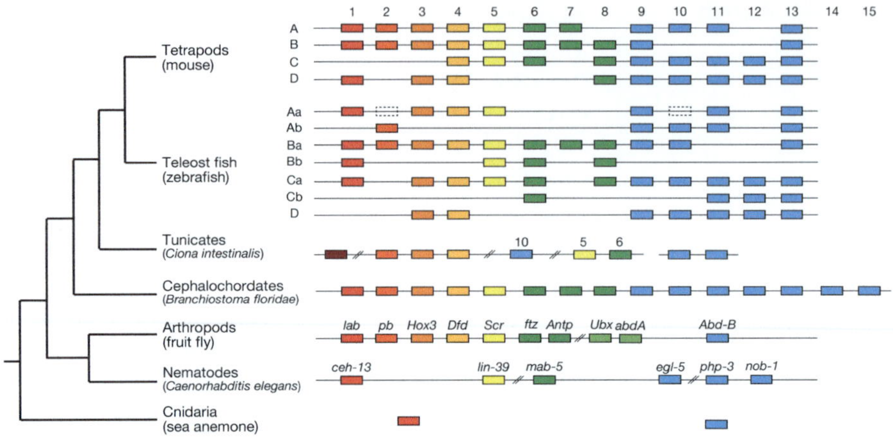

Fig. 12.6 Evolution of homeobox (Hox) genes. Phylogenetic tree of metazoan phyla and the distribution of Hox genes in metazoan phyla. Hox clusters from representative species in each taxonomic group are shown (This figure is based on the findings of Carroll et al. 2004 and Pascual-Anaya et al. 2013)

conserved orders (Fig. 12.6). Additionally, whole-genome duplication is known to have occurred in a teleost ancestor. Therefore, eight Hox clusters are usually observed in the genomes of teleosts such as zebrafish (Pascual-Anaya et al. 2013). In those species, massive losses of duplicated Hox genes, caused by redundancy, are also observed. Hox genes are developmentally important, and their structures are informative for an understanding of the evolution of genome organization.

12.3.2 Forkhead Box Genes

Forkhead (Fhk) box (Fox) proteins are a family of TFs defined by a monomeric DNA-binding domain of approximately 100 amino acid residues (Weigel et al. 1989; Weigel and Jäckle 1990; Kaestner et al. 2000; Benayoun et al. 2011). The canonical Fox domain usually consists of three or four α-helices and three β-strands, which form a helix-turn-helix motif (Clark et al. 1993; Kaestner et al. 2000; Benayoun et al. 2011). In humans, more than 100 Fox proteins, including alternative splicing variants, have been identified.

Fox proteins are involved in various biological processes, including organogenesis (Friedman and Kaestner 2006), energy metabolism (Coolican et al. 1997; Gross et al. 2009), and homeostasis (Zhang et al. 2006; Chen et al. 2009) and have also been linked to diseases such as cancer (Medema et al. 2000; Radhakrishnan et al. 2006; Nair et al. 2007), diabetes (Coolican et al. 1997; Nakae et al. 2002; Kitamura et al. 2002), and language disorders (Fisher et al. 1998; Lai et al. 2001; Enard et al. 2002).

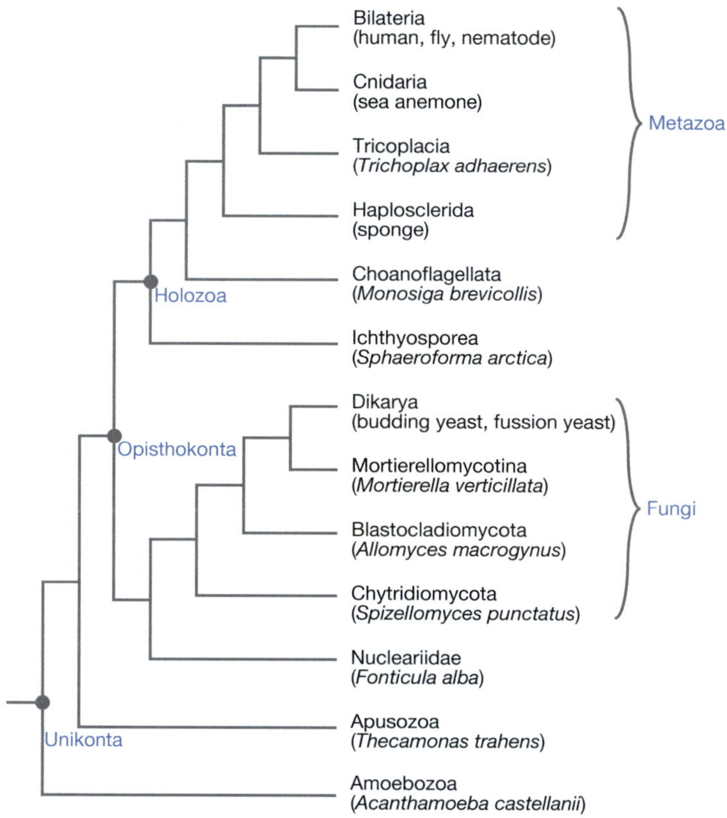

Fig. 12.7 Phylogeny of species containing forkhead-box (Fox) genes (This figure is based on the findings of Nakagawa et al. 2013)

To date, more than 2000 Fox proteins have been widely observed in a variety of species of opisthokonts, including metazoans, fungi, and several protists closely related to metazoan lineages such as Choanoflagellata and Ichthyosporea (Kaestner et al. 2000; Benayoun et al. 2011; Larroux et al. 2008; Benayoun et al. 2011). Figure 12.7 presents these phylogenetic relationships. The target sequences of Fox proteins are commonly known as the Fhk primary (FhkP) motif, RYAAAYA (R = A or G, Y = C or Y), and Fhk secondary (FhkS) motif, AHAACA (H = A, C, or T) (Pierrou et al. 1994; Overdier et al. 1994; Kaufmann et al. 1995; Badis et al. 2009; Zhu et al. 2012; Nakagawa et al. 2013). Recently, Fox proteins were also identified in Amoebozoa, a sister group to the opisthokonts (Sebé-Pedrós et al. 2011), and one of these proteins features a binding sequence pattern quite similar to the FhkP motif (Nakagawa et al. 2013). Several researchers have examined the evolutionary relationships among Fox proteins. Shimeld and colleagues conducted phylogenetic analyses of the Fox family in several metazoans and fungi, as well as the choanoflagellate *Monosiga brevicollis*, and reported that Fox families could be divided into two groups: Clades 1 and 2. Clade 1 includes 11 subfamilies, Fox A, B, C, D, E, F, G, H, I, L (L1 and L2), and Q (Q1 and Q2), whereas Clade 2 includes eight subfamilies: Fox J1, J2/3, K, M, N1/4, N2/3, O, and P (Larroux et al. 2008; Sebé-Pedrós et al. 2011). Note that these groups do not contain any fungal Fox genes. According to the sequence homologies and exon structures, Clade 1-type Fox domains emerged after the split in the metazoan lineage, as demonstrated by the absence of Clade 1 Fox domains in *M. brevicollis* (Larroux et al. 2008; Shimeld et al. 2010). The findings of Nakagawa et al. (2013) also supported the hypothesis suggesting a monophyletic group of Clade 1-type Fox genes in metazoan species. Wang et al. (2009) generated a phylogenetic tree of Fox domains obtained from metazoans and fungi and classified fungal Fox proteins into three subgroups: Fox 1, 2, and 3. However, details regarding the relationships among those Fox subfamilies are unclear for two major reasons (Nakagawa et al. 2013). First, the number of alignable amino acid residues within the Fox domain is too small to allow the phylogenic resolution of such a broadly and deeply divergent family, and regions outside the domain are not alignable even in the same subfamily members. Second, some Fox genes appear to have evolved through gene conversion and/or crossover events (Wang et al. 2009).

The typically known target DNA sequences of Fox proteins are the FhkP and FhkS motifs; however, another binding motif (FHL, GACGC) was also identified in several Fox proteins belonging to Fox M, N2/3, N1/4, R, and 3 (Zhu et al. 2009, 2012; Nakagawa et al. 2013). Note that among these subfamilies, Fox 3 was observed only in fungi, whereas the others were found in metazoans and closely related species. The relationships among these subfamilies were examined, and the FHL motif was found to have emerged independently at multiple points during Fhk protein evolution (Nakagawa et al. 2013). Canonical Fox base-contacting residues do not explain most alternate forms of specificity, and therefore, further studies are required to elucidate the Fhk binding specificity, which may shed light on the evolution of transcriptional networks regulated by Fox genes.

12.3.3 T-Box Genes

The T-box transcription family was identified on the basis of a conserved DNA-binding domain (called T-box) that was first identified in the mouse brachyury (T) gene product (Dobrovolskaïa-Zavadskaïa 1927; Herrmann et al. 1990). A mutation in the mouse T protein causes the brachyury ("short-tail") phenotype as a result of defective tail and axial development. Indeed, many T-box family genes are involved in early embryonic cell fate decisions, the regulation of extraembryonic structure development, embryonic patterning, and many aspects of organogenesis in all metazoans. Therefore, mutations in T-box genes affect various phenotypes, most of which are related to the diseases summarized in Table 12.2 (Naiche et al. 2005).

Table 12.2 Human T-box genes and related syndromes

Gene name	Syndrome
T subfamily	
T	Epithelial-mesenchymal transition in tumor (Fernando et al. 2010)
TBX19 (*TPIT*)	Recessive isolated ACTH deficiency (Pulichino et al. 2003)
Tbx1 subfamily	
TBX1	DiGeorge, craniofacial, glandular, vascular, and heart abnormalities (Baldini 2003)
BX10	Not known
TBX15	Cousin syndrome (Lausch et al. 2008)
TBX18	Not known
TBX20	Cardiac pathologies (Kirk et al. 2007)
TBX22	X-linked cleft palate with ankyloglossia (Braybrook et al. 2001)
Tbx2 subfamily	
TBX2	Not known
TBX3	Ulnar-mammary: hypoplastic mammary glands, abnormal external genitalia, limb abnormalities (Bamshad et al. 1997)
TBX4	Small patella (Bongers et al. 2004)
TBX5	Holt-Oram, heart and hand abnormalities (Basson et al. 1997)
Tbx6 subfamily	
TBX6	Spondylocostal dysostosis (Sparrow et al. 2013)
Tbr1 subfamily	
TBR1	Not known
EOMES	Multiple sclerosis (Patsopoulos et al. 2011; Parnell et al. 2014)
TBX21	Multiple sclerosis (Parnell et al. 2014)

Based on Naiche et al. (2005)

In humans, more than 20 T-box genes have been identified, all of which bind to the DNA consensus sequence TCACACCT (Casey et al. 1998). Until recently, T-box genes were thought to exist only in metazoans; however, Sebé-Pedrós et al. (2013) recently found that T-box genes are present in an outgroup of metazoans that includes ichthyosporeans, filastereans, and several fungi. When considering the phylogenetic tree of the T-box family, which includes members identified in non-metazoan species, brachyury is the most ancient member of the metazoan T-box family (Sebé-Pedrós et al. 2013). Interestingly, the T-box gene binding specificity is highly conserved between metazoans and non-metazoans. Additionally, T-box genes were found to have evolved in the last common ancestor of opisthokonts. These family members were secondarily lost in higher fungi and in choanoflagellates (Sebé-Pedrós et al. 2013). This evolutionary scenario differs considerably from previously described Fhk family scenarios (Larroux et al. 2008; Sebé-Pedrós et al. 2011; Nakagawa et al. 2013).

12.4 Recent Studies Related to TF Binding Sites

Changes in cis- or trans-regulatory regions are the major driving forces that underlie the evolution of gene expression. In the previous section, we learned about the structures and evolution of TF gene families, the main *trans* factors involved in transcriptional regulation. However, the identification of TF binding sites (TFBSs) is also essential for understanding these regulatory mechanisms. In this section, we will highlight recent studies of TFBSs that are cis-elements bound by TF proteins.

Traditionally, TFBSs and their corresponding TF targets have been screened using laborious biological experiments. However, novel approaches for the genome-wide mapping of transcriptional regulatory elements are now available. Such approaches include both computational and experimental methods. A common computational method for cis-element identification is phylogenetic footprinting, which identifies highly conserved DNA motifs present in a multiple sequence alignment. rVISTA (Loots et al. 2002) and Consite (Sandelin et al. 2004) are web-based tools that employ phylogenetic footprinting in TFBS detection. One of the most widely used methods involves motif discovery algorithms, such as MEME (Bailey et al. 2006). MEME and similar programs treat input sequences individually and search for repeated, ungapped sequence patterns that occur in input sequences. Some predictions generated using these methods have been tested and found to be functional in transgenic animals. However, not all computationally predicted TFBSs may be functional in vivo.

Additionally, direct in vitro methods, such as ChIP, are available. The ChIP-based approach was developed more than a decade ago. It was initially used in yeast and later applied to more complex organisms, such as humans. ChIP-based approaches identify target-binding regions using immunoprecipitated DNA to bind probes in a microarray (ChIP-chip) (Lieb 2003). However, for more complex organisms, particularly mammals, it has been difficult to make ChIP measurements with a high binding site resolution. However, ChIP-seq can provide high-resolution TFBS mapping (Johnson et al. 2007; Robertson et al. 2007; Mardis et al. 2007; Barski et al. 2007; Wold and Myers 2008). As a consequence, ChIP-chip is being rapidly displaced by ChIP-seq for the genome-wide discovery of TFBSs in gigabase-size genomes. Using these functional genomics approaches, studies of the broad differences in TF binding among species and the accumulation of these differences have become feasible. In this section, we will summarize the key findings from recent ChIP-based studies of transcription factor binding evolution.

Numerous recent studies using ChIP-based approaches have greatly increased our understanding of the mechanisms of TF binding evolution in metazoans. Studies of the embryos of related fruit flies have examined the binding of TFs involved in mesoderm development, such as Twist, Hunchback, and Bicoid (Fig. 12.8a, Bradley et al. 2010; He et al. 2011; Paris et al. 2013). In the closely related *Drosophila melanogaster* and *D. yakuba*, the binding positions of developmental regulators were highly conserved (Bradley et al. 2010; Villar et al. 2014). He et al. (2011) reported that 60% of the binding peaks for Twist were conserved

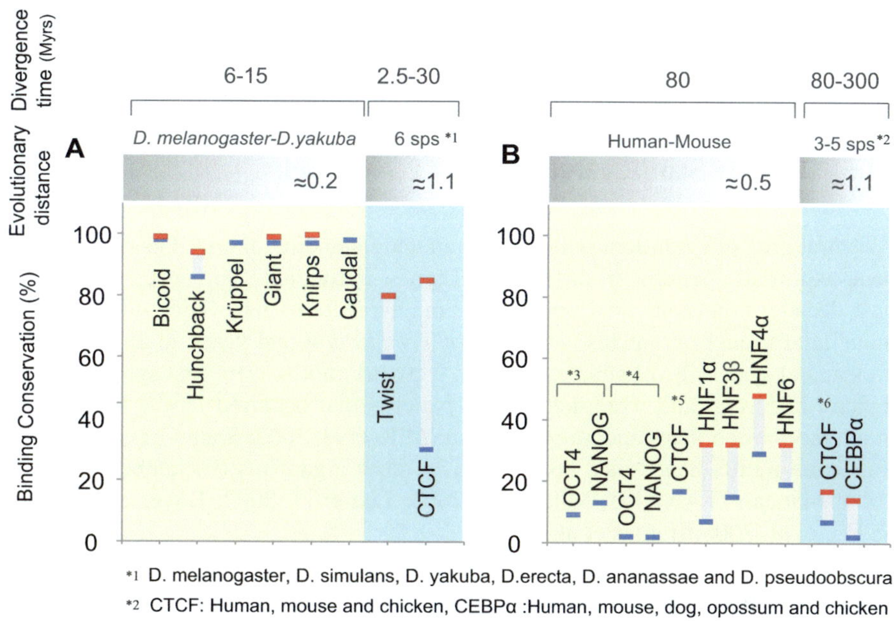

Fig. 12.8 Conservation of transcription factor (TF) binding in insects and mammals. The levels of TF binding position conservation as determined from recent cross-species chromatin immunoprecipitation studies of (**a**) *Drosophila* species and (**b**) mammals. Red and blue bars indicate the range of binding conservation. The conservation levels of yellow- and blue-colored genes are based on two-species and multiple species comparisons, respectively. At the top, the evolutionary distance (substitutions per fourfold degenerate site) indicates the pairwise distance between species with the longest branches. In *Drosophila*, information concerning the genes Bicoid to Caudal, Twist, and CTCF was reported by Bradley (2010), He et al. (2011), and Ni et al. (2012), respectively. In mammals, genes with (*3)(*4) and (*5), *HNF-1α* to *−6*, and *CEBPα* are from Loh et al. (2006)*3, Kunarso et al. (2010)*4, Martin et al. (2011)*5, Odom et al. (2007), and Schmidt et al. (2010), respectively

between *D. melanogaster* and *D. pseudoobscura*. This high conservation of TF binding across *Drosophila* species is interesting because their evolutionary distances have been estimated to be as divergent as those between humans and chickens (Stark et al. 2007).

In an experiment specifically designed to measure conserved tissue-specific TF binding in mammalian livers, the profiling of four TFs (HNF1α, HNF 3β, HNF 4α, and HNF 6) in humans and mice showed the large-scale turnover of in vivo binding (Odom et al. 2007). Despite the conserved functions of these factors, 41–89% of binding events appear to be species specific (Fig. 12.8b). Similarly, a comparison of ChIP-seq data for the liver-specific TFs CCAAT/enhancer-binding protein-α (CEBPα) and HNF4α in five vertebrates (human, macaque, mouse, opossum, and chicken) showed that only 4–14% of binding sites were conserved (Schmidt et al. 2010). These recent ChIP-based approaches have revealed that the evolution of TF binding in mammals contrasts remarkably with that in *Drosophila* species. Recent major findings regarding the evolution of TF binding are summarized in Fig. 12.8.

12.5 Recent Studies Related to TF Networks

A central goal of evolutionary biology is an understanding of how TFs can regulate gene expression patterns in different cell types, individuals, and species. Over the past decade, numerous analyses of TF-binding patterns have been conducted in unicellular model organisms, such as *Escherichia coli* and yeast, in an attempt to understand system-level gene regulation. Network motifs were first systematically defined in *E. coli* and were detected as patterns that occurred in the TF network much more often than randomly expected (Milo et al. 2002; Shen-Orr et al. 2002). The same motifs have since been found in other organisms (Eichenberger et al., 2004; Mangan et al. 2003; Milo et al. 2002; Lee et al. 2002; Boyer et al. 2005; Saddic et al. 2006; Iranfar et al. 2006).

Several studies reported that the simplest and most prevalent motif, the autoregulatory circuit, is enriched in bacterial TF networks (Fig. 12.9a, left; Shen-Orr et al. 2002; Rosenfeld et al. 2002; Isaacs et al. 2003). Another highly enriched transcriptional circuit is the feed-forward loop (FFL), which features a slightly more complex network architecture. This motif comprises three genes: a regulator, X, which regulates Y, and gene Z, which is regulated by both X and Y (Fig. 12.9a, right; Alon 2007). In both yeast and the more recently characterized human TF networks, the FFL is one of the most enriched motifs (Gerstein et al. 2012; Neph et al. 2012). For humans, system-level analyses of regulatory networks have been challenging because of the genome size and TF repertoire complexity. Recent large-scale datasets such as the Encyclopedia of DNA Elements (ENCODE) (The ENCODE Consortium 2012), which contain genome-wide binding sites of numerous TFs, enable these system-level network analyses. In this section, we will illustrate recent developments in the field of genomics that have facilitated the elucidation of human TF networks.

Gerstein et al. (2012) examined the global wiring patterns of 119 human TFs across all cell types. This analysis revealed distinct properties of the hierarchical TF network (Fig. 12.9b). TFs at the top of the hierarchy tend to have more general functions, whereas those at the bottom tend to have more specific functions. The mid-level TFs, which co-regulate targets to mitigate information-flow bottlenecks, exhibit the most regulatory collaborations between TFs. At this level, the FFL is the most enriched motif in the TF network. On the other hand, higher-level TFs have the greatest levels of connectivity with other networks.

Kim et al. (2007) examined the relationship between evolutionary selection and network topology in the human protein-protein interaction network and found that positively selected sites are preferentially located on the exposed surfaces of proteins. By contrast, network hub proteins are under purifying selection because their surfaces are involved in interactions. A recent study showed that selection acts not

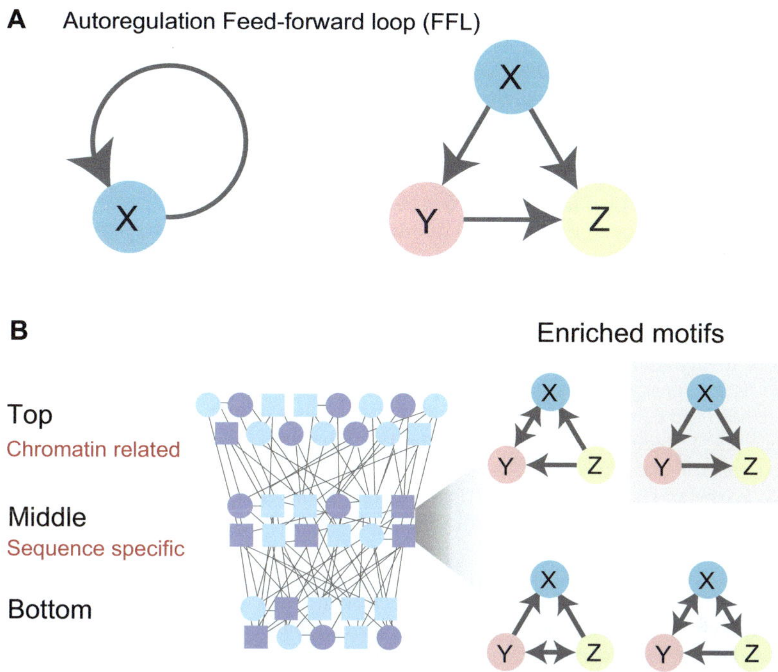

Fig. 12.9 Transcription factor (TF) hierarchy and motif enrichment. (**a**) Autoregulation (left) and feed-forward loop (right). (**b**) Left: TF network hierarchy is shown at left (red). Middle: Network nodes represent TFs. Sequence-specific TFs (TFSSs) and non-TFSSs are indicated by squares and circles, respectively. Right: Enriched network motifs at mid-level TFs are shown. Feed-forward loops, particularly the gray-boxed type, are enriched (Data were obtained from Gerstein et al. 2012)

only on proteins but also on their targets. Highly connected network elements (TFs and their targets) are both under strong selective constraint (Gerstein et al. 2012).

Further improvements in methodological and experimental approaches will provide opportunities to understand these increasingly complex TF regulatory networks. One great promise of these studies is a better understanding of how complex regulatory networks have evolved, how their evolution results in phenotypic change and speciation, and how organisms respond to their environment. Such knowledge will also allow us to identify disease-associated genes that contribute to clinical diagnosis and therapy.

References

Alon U (2007) Network motifs: theory and experimental approaches. Nat Rev Genet 8:450–461

Atchley WR, Fitch WM (1997) A natural classification of the basic helix-loop-helix class of transcription factors. Proc Natl Acad Sci U S A 94:5172–5176

Atchley WR, Terhalle W, Dress A (1999) Positional dependence, cliques, and predictive motifs in the bHLH protein domain. J Mol Evol 48:501–516

Badis G, Berger MF, Philippakis AA, Talukder S, Gehrke AR et al (2009) Diversity and complexity in DNA recognition by transcription factors. Science 324:1720–1723

Bailey TL, Williams N, Misleh C, Li WW (2006) MEME: discovering and analyzing DNA and protein sequence motifs. Nucleic Acids Res 34:W369–W373

Baldini (2003) DiGeorge's syndrome: a gene at last. Lancet 362:1342–1343

Bamshad M, Lin RC, Law DJ, Watkins WC, Krakowiak PA et al (1997) Mutations in human TBX3 alter limb, apocrine and genital development in ulnar-mammary syndrome. Nat Genet 16:311–315

Banerjee-Basu S, Baxevanis AD (2001) Molecular evolution of the homeodomain family of transcription factors. Nucleic Acids Res 29:3258–3269

Barski A, Cuddapah S, Cui K, Roh TY, Schones DE et al (2007) High-resolution profiling of histone methylations in the human genome. Cell 129:823–837

Basson CT, Bachinsky DR, Lin RC, Levi T, Elkins JA et al (1997) Mutations in human TBX5 cause limb and cardiac malformation in Holt-Oram syndrome. Nat Genet 15:30–35

Benayoun BA, Caburet S, Veitia RA (2011) Forkhead transcription factors: key players in health and disease. Trends Genet 27:224–232

Bohm S, Frishman D, Mewes HW (1997) Variations of the C2H2 zinc finger motif in the yeast genome and classification of yeast zinc finger proteins. Nucleic Acids Res 25:2464–2469

Bongers EM, Duijf PH, van Beersum SE, Schoots J, van Kampen A et al (2004) Mutations in the human TBX4 gene cause small patella syndrome. Am J Hum Genet 74:1239–1248

Boyadjiev SA, Jabs EW (2000) Online Mendelian Inheritance in Man (OMIM) as a knowledgebase for human developmental disorders. Clin Genet 57:253–266

Boyer LA, Lee TI, Cole MF, Johnstone SE, Levine SS et al (2005) Core transcriptional regulatory circuitry in human embryonic stem cells. Cell 122:947–956

Bradley RK, Li XY, Trapnell C, Davidson S, Pachter L et al (2010) Binding site turnover produces pervasive quantitative changes in transcription factor binding between closely related Drosophila species. PLoS Biol 8:e1000343

Braybrook C, Doudney K, Marçano AC, Arnason A, Bjornsson A et al (2001) The T-box transcription factor gene TBX22 is mutated in X-linked cleft palate and ankyloglossia. Nat Genet 29:179–183

Burglin T (1994) Comprehensive classification of homeobox genes. In: Duboule D (ed) Guidebook to the homeobox genes. Oxford University Press, Oxford

Carroll SB, Grenier JK, Weatherbee SD (2004) From DNA to diversity: molecular genetics and the evolution of animal design, 2nd edn. Wiley-Blackwell, NJ

Casey ES, O'Reilly MA, Conlon FL, Smith JC (1998) The T-box transcription factor Brachyury regulates expression of eFGF through binding to a non-palindromic response element. Development 125:3887–3894

Chen L, Xiao S, Manley NR (2009) Foxn1 is required to maintain the postnatal thymic microenvironment in a dosage-sensitive manner. Blood 113:567–574

Clark KL, Halay ED, Lai E, Burley SK (1993) Co-crystal structure of the HNF-3/fork head DNA-recognition motif resembles histone H5. Nature 364:412–420

Clarke B (1998) Zinc finger in Caenorhabditis elegans: finding families and probing pathways. Science 282:2018–2022

Coolican SA, Samuel DS, Ewton DZ, McWade FJ, Florini JR (1997) The mitogenic and myogenic actions of insulin-like growth factors utilize distinct signaling pathways. J Biol Chem 272:6653–6662

de Mendoza A, Sebé-Pedrós A, Sestak MS, Matejcic M, Torruella G et al (2013) Transcription factor evolution in eukaryotes and the assembly of the regulatory toolkit in multicellular lineages. Proc Natl Acad Sci U S A 110:E4858–E4866

Demuth JP, Hahn MW (2009) The life and death of gene families. BioEssays 31:29–39

Dobrovolskaïa-Zavadskaïa N (1927) Sur la mortification spontanée de la queue chez la souris noveau-née et sur l'existence d'un caractère (facteur) héréditaire "non-viable". C. R. Seanc. Soc Biol 97:114–116

Eichenberger P, Fujita M, Jensen ST, Conlon EM, Rudner DZ et al (2004) The program of gene transcription for a single differentiating cell type during sporulation in Bacillus subtilis. PLoS Biol 2:e328

Ellenberger T, Fass D, Arnaud M, Harrison SC (1994) Crystal structure of transcription factor E47: E-box recognition by a basic region helix-loop-helix dimer. Genes Dev 8:970–980

Elrod-Erickson M, Pabo CO (1999) Binding studies with mutants of Zif268. J Biol Chem 274:19381–19285

Enard W, Przeworski M, Fisher SE, Lai CS, Wiebe V et al (2002) Molecular evolution of FOXP2, a gene involved in speech and language. Nature 418:869–872

Fernando RI, Litzinger M, Trono P, Hamilton DH, Schlom J, Palena C (2010) The T-box transcription factor Brachyury promotes epithelial-mesenchymal transition in human tumor cells. J Clin Invest 120:533–544

Ferre-D'Amare AR, Prendergast GC, Ziff EB, Burley SK (1993) Recognition by Max of its cognate DNA through a dimeric b/HLH/Z domain. Nature 363:38–45

Fisher SE, Vargha-Khadem F, Watkins KE, Monaco AP, Pembrey ME (1998) Localisation of a gene implicated in a severe speech and language disorder. Nat Genet 18:168–170

Frankel AD, Berg JM, Pabo CO (1987) Metal-dependent folding of a single zinc finger from transcription factor IIIA. Proc Natl Acad Sci U S A 84:4841–4845

Friedman JR, Kaestner KH (2006) The Foxa family of transcription factors in development and metabolism. Cell Mol Life Sci 63:2317–2328

Gehring WJ, Qian YQ, Billeter M, Furukubo-Tokunaga K, Schier AF et al (1994) Homeodomain-DNA recognition. Cell 78:211–223

Gerstein MB, Kundaje A, Hariharan M, Landt SG, Yan KK et al (2012) Architecture of the human regulatory network derived from ENCODE data. Nature 489:91–100

Gross DN, Wan M, Birnbaum MJ (2009) The role of FOXO in the regulation of metabolism. Curr Diab Rep 9:208–214

He Q, Bardet AF, Patton B, Purvis J, Johnston J et al (2011) High conservation of transcription factor binding and evidence for combinatorial regulation across six Drosophila species. Nat Genet 43:414–420

Herrmann BG, Labeit S, Poustka A, King TR, Lehrach H (1990) Cloning of the T gene required in mesoderm formation in the mouse. Nature 343:617–622

Holland LZ, Albalat R, Azumi K, Benito-Gutierrez E, Blow M et al (2008) The amphioxus genome illuminates vertebrate origins and cephalochordate biology. Genome Res 18:1100–1111

Iranfar N, Fuller D, Loomis WF (2006) Transcriptional regulation of post-aggregation genes in Dictyostelium by a feed-forward loop involving GBF and LagC. Dev Biol 290:460–469

Isaacs FJ, Hasty J, Cantor CR, Collins JJ (2003) Prediction and measurement of an autoregulatory genetic module. Proc Natl Acad Sci U S A 100:7714–7719

Iuchi S (2000) Three classes of C2H2 zinc finger proteins. CMLS (Cell Mol Life Sci) 58:625–635

Johnson DS, Mortazavi A, Myers RM, Wold B (2007) Genome-wide mapping of in vivo protein-DNA interactions. Science 316:1497–1502

Kadesch T (1993) Consequences of heteromeric interactions among helix-loop-helix proteins. Cell Growth Differ 4:49–55

Kaestner KH, Knochel W, Martinez DE (2000) Unified nomenclature for the winged helix/forkhead transcription factors. Genes Dev 14:42–146

Kaufmann E, Müller D, Knöchel W (1995) DNA recognition site analysis of Xenopus winged helix proteins. J Mol Biol 248:239–254

Kim PM, Korbel JO, Gerstein MB (2007) Positive selection at the protein network periphery: evaluation in terms of structural constraints and cellular context. Proc Natl Acad Sci U S A 104:20274–20279

Kirk EP, Sunde M, Costa MW, Rankin SA, Wolstein O et al (2007) Mutations in cardiac T-Box factor gene TBX20 are associated with diverse cardiac pathologies, including defects of septation and valvulogenesis and cardiomyopathy. Am J Hum Genet 81:280–291

Kissinger CR, Liu B, Martin-Blanco E, Kornberg TB, Pabo CO (1990) Crystal structure of an engrailed homeodomain-DNA complex at 2.8 A resolution: a framework for understanding homeodomain-DNA interactions. Cell 63:579–590

Kitamura T, Nakae J, Kitamura Y, Kido Y, Biggs WH 3rd et al (2002) The forkhead transcription factor Foxo1 links insulin signaling to Pdx1 regulation of pancreatic β cell growth. J Clin Invest 110:1839–1847

Kunarso G, Chia NY, Jeyakani J, Hwang C, Lu X et al (2010) Transposable elements have rewired the core regulatory network of human embryonic stem cells. Nature Genet 42:631–634

Lai CS, Fisher SE, Hurst JA, Vargha-Khadem F, Monaco AP (2001) A forkhead-domain gene is mutated in a severe speech and language disorder. Nature 413:519–523

Lander ES, Linton M, Birren B (2001) Innitial sequencing and analysis of the human genome. Nature 409:809–921

LaRonde-LeBlanc NA, Wolberger C (2003) Structure of HoxA9 and Pbx1 bound to DNA: Hox hexapeptide and DNA recognition anterior to posterior. Genes Dev 17:2060–2072

Larroux C, Luke GN, Koopman P, Rokhsar DS, Shimeld SM et al (2008) Genesis and expansion of metazoan transcription factor gene classes. Mol Biol Evol 25:980–996

Lausch E, Hermanns P, Farin HF, Alanay Y, Unger S et al (2008) TBX15 mutations cause craniofacial dysmorphism, hypoplasia of scapula and pelvis, and short stature in cousin syndrome. Am J Hum Genet 83:649–655

Lee TI, Rinaldi NJ, Robert F, Odom DT, Bar-Joseph Z et al (2002) Transcriptional regulatory networks in Saccharomyces cerevisiae. Science 298:799–804

Levine M, Tjian R (2003) Transcription regulation and animal diversity. Nature 424:147–151

Lewis EB (1978) A gene complex controlling segmentation in Drosophila. Nature 276:565–570

Lieb JD (2003) Genome-wide mapping of protein-DNA interactions by chromatin immunoprecipitation and DNA microarray hybridization. Methods Mol Biol 224:99–109

Loh YH, Wu Q, Chew JL, Vega VB, Zhang W et al (2006) The Oct4 and Nanog transcription network regulates pluripotency in mouse embryonic stem cells. Nature Genet 38:431–440

Loots GG, Ovcharenko I, Pachter L, Dubchak I, Rubin EM (2002) rVista for comparative sequence-based discovery of functional transcription factor binding sites. Genome Res 12:832–839

Mangan S, Zaslaver A, Alon U (2003) The coherent feedforward loop serves as a sign-sensitive delay element in transcription networks. J Mol Biol 334:197–204

Mardis ER (2007) ChIP-seq: welcome to the new frontier. Nat Methods 4:613–614

Martin D, Pantoja C, Miñán AF, Valdes-Quezada C, Moltó E et al (2011) Genome-wide CTCF distribution in vertebrates defines equivalent sites that aid the identification of disease-associated genes. Nat Struct Mol Biol 18:708–714

McLysaght A, Hokamp K, Wolfe KH (2002) Extensive genomic duplication during early chordate evolution. Nat Genet 31:200–204

Medema RH, Kops GJ, Bos JL, Burgering BM (2000) AFX-like Forkhead transcription factors mediate cell-cycle regulation by Ras and PKB through p27kip1. Nature 404:782–787

Miller J, McLachlan AB, Klug A (1985) Repetitive zinc- binding domains in the protein transcription factor IIIA from Xenopus oocytes. EMBO J 4:1609–1614

Milo R, Shen-Orr S, Itzkovitz S, Kashtan N, Chklovskii D et al (2002) Network motifs: simple building blocks of complex networks. Science 298:824–827

Moore MJ (2005) From birth to death: the complex lives of eukaryotic mRNAs. Science 309:1514–1518

Mukherjee K, Burglin TR (2007) Comprehensive analysis of animal TALE Homeobox genes: new conserved motifs and cases of accelerated evolution. J Mol Evol 65:137–153

Murre C, Mc Caw PS, Baltimore D (1989a) A new DNA binding and dimerizing motif in immunoglobulin enhancer binding, daughterless, MyoD, and Myc proteins. Cell 56:777–783

Murre C, Mc Caw PS, Vaessin H, Caudy M, Jan LY et al (1989b) Interactions between heterologous helix-loop-helix proteins generate complexes that bind specifically to a common DNA sequence. Cell 58:537–544

Murre C, Bain G, van Dijk MA, Engel I, Furnari BA et al (1994) Structure and function of helix-loop-helix proteins. Biochim Biophys Acta 1218:129–135

Naiche LA, Harrelson Z, Kelly RG, Papaioannou VE (2005) T-box genes in vertebrate development. Annu Rev Genet 39:219–239

Nair S, Boczkowski D, Fassnacht M, Pisetsky D, Gilboa E (2007) Vaccination against the forkhead family transcription factor Foxp3 enhances tumor immunity. Cancer Res 67:371–380

Nakae J, Biggs WH 3rd, Kitamura T, Cavenee WK, Wright CV et al (2002) Regulation of insulin action and pancreatic beta-cell function by mutated alleles of the gene encoding forkhead transcription factor Foxo1. Nat Genet 32:245–253

Nakagawa S, Gisselbrecht SS, Rogers JM, Hartl DL, Bulyk ML (2013) DNA-binding specificity changes in the evolution of forkhead transcription factors. Proc Natl Acad Sci U S A 110:12349–12354

Neph S, Stergachis AB, Reynolds A, Sandstrom R, Borenstein E (2012) Circuitry and dynamics of human transcription factor regulatory networks. Cell 150:1274–1286

Ni X, Zhang YE, Nègre N, Chen S, Long M et al (2012) Adaptive evolution and the birth of CTCF binding sites in the Drosophila genome. PLoS Biol 10:e1001420

Nowick K, Huntley S, Stubbs L (2009) Rapid expansion and divergence suggest a central and distinct role for KRAB-ZNF genes in vertebrate evolution. In: Yoshida K (ed) Focus on zinc finger protein research. Research Signpost, Kerala, pp 13–29

Nowick K, Stubbs L (2010) Lineage-specific transcription factors and the evolution of gene regulatory networks. Brief Funct Genomics 9:65–78

Odom DT, Dowell RD, Jacobsen ES, Gordon W, Danford TW et al (2007) Tissue-specific transcriptional regulation has diverged significantly between human and mouse. Nat Genet 39:730–732

Ohno S (1970) Evolution by gene duplication. Springer-Verlag, New York

Overdier DG, Porcella A, Costa RH (1994) The DNA-binding specificity of the hepatocyte nuclear factor 3/forkhead domain is influenced by amino-acid residues adjacent to the recognition helix. Mol Cell Biol 14:2755–2766

Paris M, Kaplan T, Li XY, Villalta JE, Lott SE et al (2013) Extensive divergence of transcription factor binding in Drosophila embryos with highly conserved gene expression. PLoS Genet 9: e1003748

Parnell GP, Gatt PN, Krupa M, Nickles D, McKay FC et al (2014) The autoimmune disease-associated transcription factors EOMES and TBX21 are dysregulated in multiple sclerosis and define a molecular subtype of disease. Clin Immunol 151:16–24

Pascual-Anaya J, Aniello SD, Kuratani S, Garcia-Fernàndez J (2013) Evolution of Hox gene clusters in deuterostomes. BMC Dev Biol 13:26

Patsopoulos NA, The Bayer Pharma MS Genetics Working Group, the Steering Committees of Studies Evaluating IFNβ-1b and a CCR1-Antagonist, ANZgene Consortium, GeneMSA et al (2011) Genome-wide meta-analysis identifies novel multiple sclerosis susceptibility loci. Ann Neurol 70:897–912

Pavletich N, Pabo CO (1991) Zinc finger-DNA recognition: crystal structure of a Zif268-DNA complec at 2.1 Å. Science 252:809–817

Pierrou S, Hellqvist M, Samuelsson L, Enerbäck S, Carlsson P (1994) Cloning and characterization of seven human forkhead proteins: binding site specificity and DNA bending. EMBO J 13:5002–5012

Piper DE, Batchelor AH, Chang C-P, Cleary ML, Wolberger C (1999) Structure of a HoxB1-Pbx1 heterodimer bound to DNA: role of the hexapeptide and a fourth homeodomain helix in complex formation. Cell 96:587–597

Pulichino AM, Vallette-Kasic S, Couture C, Gauthier Y, Brue T et al (2003) Human and mouse TPIT gene mutations cause early onset pituitary ACTH deficiency. Genes Dev 17:711–716

Putnam NH, Butts T, Ferrier DEK, Furlong RF, Hellsten U et al (2008) The amphioxus genome and the evolution of the chordate karyotype. Nature 453:1064–1071
Qian YQ, Billeter M, Otting G, Muller M, Gehring WJ, Wuthrich K (1989) The structure of the Antennapedia homeodomain determined by NMR spectroscopy in solution: comparison with prokaryotic repressors. Cell 59:573–580
Radhakrishnan SK, Bhat UG, Hughes DE, Wang IC, Costa RH et al (2006) Identification of a chemical inhibitor of the oncogenic transcription factor forkhead box m1. Cancer Res 66:9731–9735
Robertson G et al (2007) Genome-wide profiles of STAT1 DNA association using chromatin immunoprecipitation and massively parallel sequencing. Nat Methods 4:651–657
Rosenfeld N, Elowitz MB, Alon U (2002) Negative autoregulation speeds the response times of transcription networks. J Mol Biol 323:785–793
Saddic LA, Huvermann B, Bezhani S, Su Y, Winter CM et al (2006) The LEAFY target LMI1 is a meristem identity regulator and acts together with LEAFY to regulate expression of CAULIFLOWER. Development 133:1673–1682
Sagai T, Hosoya M, Mizushina Y, Tamura M, Shiroishi T (2005) Elimination of a long-range cis-regulatory module causes complete loss of limb-specific Shh expression and truncation of the mouse limb. Development 132:797–803
Sandelin A, Wasserman WW, Lenhard B (2004) ConSite: web-based prediction of regulatory elements using cross-species comparison. Nucleic Acids Res 32:W249–W252
Schmidt D, Wilson MD, Ballester B, Schwalie PC, Brown GD et al (2010) Five-vertebrate ChIP-seq reveals the evolutionary dynamics of transcription factor binding. Science 328:1036–1040
Sebé-Pedrós A, de Mendoza A, Lang BF, Degnan BM, Ruiz-Trillo I (2011) Unexpected repertoire of metazoan transcription factors in the unicellular holozoan Capsaspora owczarzaki. Mol Biol Evol 28:1241–1254
Sebé-Pedrós A, Ariza-Cosano A, Weirauch MT, Leininger S, Yang A, Torruella G et al (2013) Early evolution of the T-box transcription factor family. Proc Natl Acad Sci U S A 110:16050–16055
Semenza GL (1998) Transcription factors and human disease. (9 Homeodomain proteins). Oxford University Press, Oxford
Shen-Orr SS, Milo R, Mangan S, Alon U (2002) Network motifs in the transcriptional regulation network of Escherichia coli. Nat Genet 31:64–68
Shimeld SM, Degnan B, Luke GN (2010) Evolutionary genomics of the Fox genes: origin of gene families and the ancestry of gene clusters. Genomics 95:256–260
Spagnuolo A, Ristoratore F, Di Gregorio A, Aniello F, Branno M et al (2003) Unusual number and genomic organization of Hox genes in the tunicate Ciona intestinalis. Gene 309:71–79
Sparrow DB, McInerney-Leo A, Gucev ZS, Gardiner B, Marshall M (2013) Autosomal dominant spondylocostal dysostosis is caused by mutation in TBX6. Hum Mol Genet 22(8):1625–1631
Stark AM, Lin F, Kheradpour P, Pedersen JS, Parts L et al (2007) Discovery of functional elements in 12 Drosophila genomes using evolutionary signatures. Nature 450:219–232
The ENCODE Project Consortium (2012) An integrated encyclopedia of DNA elements in the human genome. Nature 489:57–74
Tupler R, Perini G, Green MR (2001) Expressing the human genome. Nature 409:832–833
Vaquerizas JM, Kummerfeld SK, Teichmann SA, Luscombe NM (2009) A census of human transcription factors: function, expression and evolution. Nat Rev Genet 10:252–263
Villar D, Flicek P, Odom DT (2014) Evolution of transcription factor binding in metazoans – mechanisms and functional implications. Nat Rev Genet Nat Rev Genet 15:221–233
Vollmer JY, Clerc RG (1998) Homeobox genes in the developing mouse brain. J Neurochem 71:1–19
Wang M, Wang Q, Zhao H, Zhang X, Pan Y (2009) Evolutionary selection pressure of forkhead domain and functional divergence. Gene 432:19–25
Weigel D, Jäckle H (1990) The fork head domain: a novel DNA binding motif of eukaryotic transcription factors? Cell 63:455–456

Weigel D, Jürgens G, Küttner F, Seifert E, Jäckle H (1989) The homeotic gene fork head encodes a nuclear protein and is expressed in the terminal regions of the Drosophila embryo. Cell 57:645–658

Wolberger C, Vershon AK, Liu B, Johnson AD, Pabo CO (1991) Crystal structure of a MATalpha2 homeodomain-operator complex suggests a general model for homeodomain-DNA interactions. Cell 67:517–536

Wold B, Myers RM (2008) Sequence census methods for functional genomics. Nat Methods 5:19–21

Wolfe SA, Nekludova L, Pabo CO (2000) DNA recognition by Cys2His2 zinc finger proteisn. Annu Rev Biophys Biomol Struct 29:183–212

Zhang H, Ackermann AM, GUSArova GA, Lowe D, Feng X et al (2006) The FoxM1 transcription factor is required to maintain pancreatic beta-cell mass. Mol Endocrinol 20:1853–1866

Zhu C, Byers KJ, McCord RP, Shi Z, Berger MF et al (2009) High-resolution DNA-binding specificity analysis of yeast transcription factors. Genome Res 19:556–566

Zhu X, Ahmad SM, Aboukhalil A, Busser BW, Kim Y et al (2012) Differential regulation of mesodermal gene expression by Drosophila cell type-specific Forkhead transcription factors. Development 139:1457–1466

Chapter 13
Genetics of Diabetes: Are They Thrifty Genotype?

Ituro Inoue and Hirofumi Nakaoka

Abstract Diabetes mellitus: a "thrifty" genotype rendered detrimental by "progress" by James Neel (Am J Hum Genet 14:353–362, 1962).

Type 2 diabetes (T2D) is a complex metabolic disorder in which both genetic and environmental factors play roles in the etiology. Advances in genetic analyses brought tremendous successes in identifying genetic components of common T2D and monogenic form of diabetes such as maturity-onset diabetes of the young (MODY). Large-scale genome-wide association studies identified more than 70 loci of T2D in various populations, which dramatically enhance our understanding of molecular etiology of T2D. In particular, *TCF7L2*, which is the most prominent genetic susceptibility of T2D, is extensively studied including functional impact. MODY is considered to be a monogenic form of diabetes, but complexity of the disorder is revealed; indeed, genetic causalities were identified in only <15% of MODY patients in Japan. More than 50 years ago, Neel made two notions on genetics of T2D: one is calling T2D as geneticist's nightmare and the other is thrifty genotype hypothesis. These notions are challenged, partly solved, with the modern genetic advances and are focused in this chapter.

Keywords Diabetes · GWAS · MODY · Thrifty genotype · TCF7L2 · Natural selection

13.1 Introduction

Genetic basis of metabolic diseases is very complicated because environmental factors including personal lifestyles play substantial and interactive roles in the etiology. Type 2 diabetes (T2D) and other metabolic diseases including obesity are typical examples. James Neel, a pioneer of human genetics, called T2D as "geneticist's nightmare" because of its complexity and also proposed the "thrifty genotype

---

I. Inoue (✉) · H. Nakaoka
Division of Human Genetics, National Institute of Genetics, Mishima, Japan
e-mail: itinoue@nig.ac.jp; hnakaoka@nig.ac.jp

N. Saitou (ed.), *Evolution of the Human Genome I*, Evolutionary Studies,
https://doi.org/10.1007/978-4-431-56603-8_13

hypothesis" claiming genetic etiologies of diabetes mellitus and other metabolic diseases according to population history more than 50 years ago (Neel 1962). The basic concept of the hypothesis is that the genetic background leading to disadvantageous T2D today was advantageous in the past environments. This hypothesis is plausible to understand disease causality regarding natural selection; therefore, it seems to be positively accepted by the research community. In this chapter, genetic components of diabetes mellitus are discussed from the point of view of population history. Meanwhile, the thrifty genotype hypothesis is evaluated or challenged with the most updated genomic information. The metabolic diseases such as T2D are closely related with lifestyle particularly diet of individuals and population history. The human lineage split from the closest lineage, chimpanzee, about six million years ago. Since then, early human ancestor had the hunter and gatherer diet and experienced repeated or presumably continuous feast or famine. Then, about ten thousands years ago, agriculture started and our diet dramatically changed, and more stable diet could be obtained. The metabolic diseases are assumed to be civilized diseases because of the consequence of taking abundant diet, mostly over calories. We discuss here what are the genetic factors of diabetes, how are the factors maintained in human history, and why do the diseases exist.

13.2 Fundamental Problem of Insulin Regulation

Diabetes is a metabolic disease of civilization characterized by high blood glucose (hyperglycemia) resulting from defects in insulin secretion or action. Persistent hyperglycemia results in damage of multiple organs such as the eyes, kidneys, nerves, and blood vessels. Thus diabetes is an insulin problem, and insulin is solely produced by beta cells of the islet of Langerhans in the pancreas. When islet is mistakenly destroyed by autoimmune mechanism accompanied with viral infection, type 1 diabetes (T1D) occurs where glucose control is impaired due to complete loss of insulin. Therefore insulin is absolutely needed for treatment of T1D. If the sufficient insulin is not produced or the insulin action is disabled, T2D occurs showing difficulty in regulation of blood glucose level. T2D has more complicated nature showing either low secretion of insulin or resistant to insulin action. Insulin gene expression is tightly regulated by multiple transcriptional factors. Also secretory pathway of insulin is regulated by many factors. In other word, defects of insulin production and secretion in beta cell by many genetic factors lead to development of common and rare forms of diabetes described below (Maurano et al. 2012; Pasquali et al. 2014).

13.3 Genetics of Diabetes: Is It Still Geneticist's Nightmare?

13.3.1 Genome-Wide Association Studies of T2D

T2D is a common disease affecting 151 million individuals worldwide and will affect 324 millions by 2025 (Zimmet 2003). In Europe, prevalence of diabetes is 6–8% and decreased life expectancy about 11 years when diagnosed at age 40. Heritability of T2D is estimated to range between 30% and 70% based on family studies. Classical linkage and association study identified *TCF7L2* as a susceptibility of T2D in Iceland population (Grant et al. 2006; Helgason et al. 2007). *TCF7L2* is the best replicated susceptibility of T2D to date conferring a relative risk about 1.4. Elevated expression of TCF7L2 in beta cells is implicated in pathophysiology of T2D (Lyssenko et al. 2007). With the advent of genome-wide association study (GWAS), it became a common method, and sample size becomes larger and larger. Many loci, mostly SNPs (>70), were identified as susceptibilities of T2D, although each has relatively small effect showing modest odds ratio. Also, some of the susceptibilities of T2D are population specific, e.g., *TCF7L2*, the most prominent susceptibility in Europeans, is not strongly associated with Asian populations, and *KCNQ1* is opposite for European and Asian populations (Fig. 13.1) (Morris et al. 2012; Groop and Pociot 2014; Hara et al. 2014). With the advent of high-throughput sequencing technology, rare variants have been detected as causalities of T2D (Bonnefond et al. 2012; Steinthorsdottir et al. 2014). In fact, T2D is one of the most extensively analyzed diseases by GWAS, and resequencing studies and genetic factors were most successfully identified among common diseases, which are contradicting to Neel's comment (McCarthy 2010). Probably, we came out of the "nightmare."

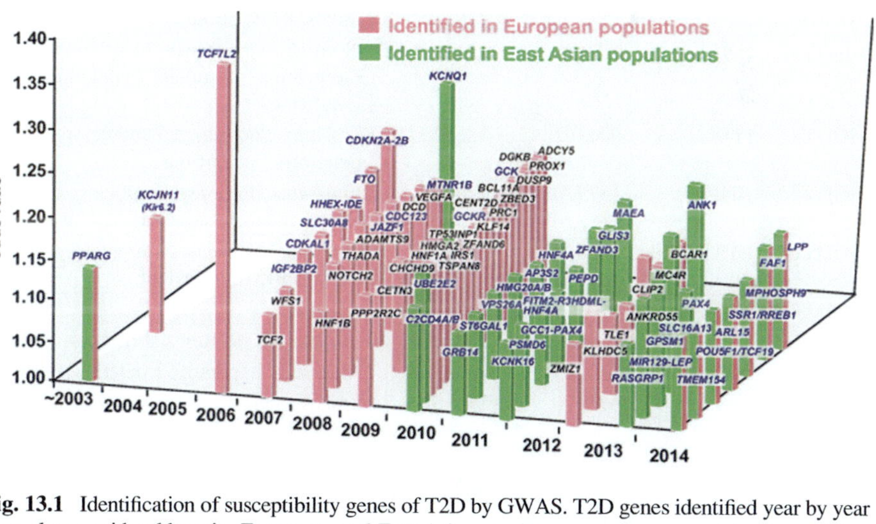

Fig. 13.1 Identification of susceptibility genes of T2D by GWAS. T2D genes identified year by year were shown with odds ratio. European- and East Asian-specific genes were separately shown (Hara et al. 2014)

13.3.2 Complexity of MODY and MODY-Related Genes

Maturity-onset diabetes of the young (MODY) is classically recognized as Mendelian form of diabetes showing an autosomal dominant inheritance, which is distinguished from common type of diabetes (Fajans and Bell 2011). MODY is originally defined as follows: early-onset diabetes diagnosed at <25 years, nonketotic, lean (BMI<25), and family history having at least two generations. It turned out that MODY is not a simple monogenic disease showing locus and allelic heterogeneity with incomplete penetrance and phenocopy, thus now defined as autosomal dominant form of diabetes regardless of onset of age. Also, the term "MODY" has been questioned by clinicians and researchers due to nowaday's ambiguous definition; MODY-type diabetes might be more appropriate in a broad sense. Indeed, the onset age of *HNF1A*-MODY is >25 years in 40% of European American patients. So far at least 13 MODY genes are identified including six major genes, and over 200 mutations have been described in glucokinase gene (MODY2), a glucose sensor, and *HNF1A* (MODY3), a liver and pancreatic transcriptional factor (Table 13.1).

A recent study focused on well-phenotyped population samples to screen MODY genes to survey mutation frequency in the population (Flannick et al. 2013). Seven MODY genes (*HNF1A*, *GCK*, *HNF4A*, *HNF1B*, *PDX1*, *INS*, and *NEUROD1*) were completely sequenced in 4003 population-based cohort samples.

Table 13.1 Identified MODY genes

MODY genes	Gene symbol	OMIM	Location	Reported mutations[a]
MODY1	HNF4A	125850	20q13.12	83 missense/nonsense, 8 splicing, 9 regulatory, 2 indels, 10 deletions, 6 insertions
MODY2	GCK	125851	7p15.3-p15.1	502 missense/nonsense, 57 splicing, 8 indels, 91 deletions, 21 insertions
MODY3	HNF1A	600496	12q24.2	298 missense/nonsense, 31 splicing, 13 regulatory, 13 indels, 73 deletions, 43 insertions
MODY4	PDX1	606392	13q12.1	15 missense/nonsense, 2 regulatory, 2 deletions, 3 insertions
MODY5	HNF1B	137920	17q12	62 missense/nonsense, 6 splicing, 1 regulatory, 1 indel, 13 deletions, 10 insertions
MODY6	NEUROD1	606394	2q32	62 missense/nonsense, 1 regulatory, 13 deletions, 10 insertions
MODY7	KLF11	610508	2p25	5 missense
MODY8	CEL	609812	9q34.3	6 missense, 2 deletions, 1 insertion
MODY9	PAX4	612225	7q32	7 missense, 2 splicing, 1 deletion
MODY10	INS	613370	11p15.5	44 missense/nonsense, 2 splicing, 6 regulatory, 1 Indel, 1 insertion
MODY11	BLK	613,375	8p23-p22	1 missense

[a]The Human Gene Mutation Database

More than 100 nonsynonymous variants across the seven genes were identified, and they are presumably pathogenic variants. One variant carrier harbors typical MODY, and the rest of the carriers remain to have a normal blood glucose level through middle age. Overall, non-symptomatic variant carriers are 0.5–1.5% indicating lifestyle may have an important role in the pathogenesis even for Mendelian form of diabetes.

Gene search for MODY is still underway, because MODY genes have been identified in only <15% of Japanese patients (Yorifuji et al. 2012; Horikawa et al. 2014). In European populations, about >10% of the MODY patients were not genetically diagnosed. Presumably, private mutations, specific to each family, are involved in the etiologies of these patients particularly in Asian patients.

13.3.3 Transcriptional Regulatory Pathways Disrupted by T2D-Associated Variants

Current evidences for causalities of MODY are incomplete due to the abovementioned complexities but provide clues into how we should consider about the genetic components and pathophysiologies of common form of diabetes. MODY is mainly caused by mutations on coding regions of the transcription factors involved in beta-cell function and glucose transport (HNF4α, HNF1α, PDX1, HNF1B, and NEUROD1). That is to say, structural disruptions of these proteins cause Mendelian form of diabetes. On the other hand, SNPs associated with common T2D are localized in noncoding regions (introns and intergenic regions), and therefore, their functional consequences have long been unknown. Recently, large-scale functional genomic studies by ENCODE and Roadmap Epigenomics projects demonstrated that SNPs associated with common T2D are overrepresented in DNase I hypersensitive sites with recognition sequences of the transcription factors whose mutations cause MODY (Maurano et al. 2012). For example, the SNPs associated with T2D on chromosomes 3 and 6 are located in recognition sequences of HNF4α (MODY1). These findings demystify molecular mechanism in which risk variants for common and rare forms of diabetes converge to diabetic phenotypes by perturbing the same regulatory network relevant to insulin production and secretion in beta cell.

13.4 How Do T2D Susceptibilities Survive in Human History?

Here we shed light on the population history regarding the susceptibilities of T2D to provide plausible explanations for the question "why has natural selection not eliminated the T2D susceptibilities from the population?" T2D is basically an adult-onset diabetes, mostly over the age of 30 years old, and has a small effect on reproductivity; therefore, natural selection might not have strong effect. Accordingly, common variant of T2D susceptibility could be explained by genetic drift. Alternatively, there might be an advantage to harbor T2D susceptibilities in the past environments. Neel (1962) proposed a "thrifty genotype" hypothesis in the etiology of diabetes, in which modified regulation of insulin secretion and storage of glucose as glycogen may have provided a survival advantage for some of our hunting-gathering period. In modern industrialized societies where generally abundant food is at hand, specially Westernized food, the thrifty genotype if existing may become detrimental showing elevated level of insulin and leading to insulin resistance eventually causing insulin depletion. Thus the evolutionary adaptation becomes maladaptive during progress of civilization. The thrifty genotype hypothesis fits very well in the etiology of obesity, thereby has been supported by many researchers. However, defect of insulin secretion observed in lean-type T2D frequently observed in Asians is hardly explained by thrifty genotype hypothesis (Okamoto et al. 2010).

Additionally the speculation that newly derived private mutations are involved in the etiologies of MODY is consistent with the recent findings of excess of rare variants in large populations resulting from recent explosive population expansion and weak purifying selection particularly observed in Europeans and Asians. In the case of MODY, modern excess diet is a strong factor together with the genetic factor that causes the disease. This idea is not consistent with Neel's thrifty genotype hypothesis although Neel did not evidently consider about MODY case. The hypothesis for the common diabetes is recently challenged by several reports (Graham 2009; Southam et al. 2009; Chen et al. 2012).

13.5 Reevaluation of "Thrifty" Genotype Hypothesis

According to the "thrifty" genotype hypothesis, the high prevalence of diabetes and its responsible genetic variants have gone positive selection during the hunting-gathering period or early-day agriculture. The recent explosive genetic studies of diabetes identified more than 70 loci with sufficient sample size, which enable to reexamine the "thrifty" genotype hypothesis. According to the original "thrifty" genotype hypothesis corresponding to the Neel's proposal (1962), thrifty allele (i.e., risk) should be ancestral and shared by all populations, furthermore probably observed in great apes. Alternative possibility needs to be considered that thrifty genotype was favored only much later in human evolution, for example, when humans were expanding and faced to new environment such as out of Africa. If this is the case, thrifty allele should be derived and specific to populations or group of populations. According to the second scenario, Southam et al. evaluated systematically 17 loci whether thrifty genotype hypothesis fits or not (Southam et al. 2009). Six of the seventeen loci showed that risk alleles are derived, and only one locus (rs7901695 at *TCF7L2*) showed an elevated F_{st} value. These occurred by chance and evidence of positive selection seems to be weak.

Ayub et al. (2014) extensively evaluated the "thrifty" genotype hypothesis using much more genetic loci (65 index SNPs) considering the above two scenarios as they define thrifty early and thrifty late. Thrifty early hypothesis was not supported because only about half of the risk alleles were ancestral (36/65). Even in comparing Neanderthals (44/65 were called) and Denisovans (65/65), about half of the risk alleles were ancestral. If the thrifty late is the case, positive selection signals around the index SNPs should be observed. Three classes of tests were evaluated: (1) site frequency spectrum (Tajima's D, Fay and Wu's H, Nielsen et al.'s CLR), (2) haplotype characteristics (composite of multiple signals and haplotype diversity), and (3) population differentiation (F_{st}). Despite the large-scale tests, only a small number of loci showed evidence of positive selection. This proportion was similar to the level of positive selection within a set of randomly selected loci. Therefore, overall enrichment of positive selection was not observed.

Although thrifty genotype hypothesis is not supported by the current studies, thrifty late hypothesis needs further considerations. Because genetic association study to detect genetic loci is much more sensitive than detecting natural selection, novel method detecting subtle natural selection otherwise soft sweep is required. As we keep mentioning, T2D is a complex disease and overlapped with other diseases like obesity and hyperlipidemia; interactive genetic factors play roles in the etiology. However, we do not have any tool to detect interactive natural selection. Therefore, thrifty late hypothesis is still an open question.

References

Ayub Q, Moutsianas L, Chen Y et al (2014) Revisiting the thrifty gene hypothesis via 65 loci associated with susceptibility to type 2 diabetes. Am J Hum Genet 94:176–185

Bonnefond A, Clément N, Fawcett K et al (2012) Rare MTNR1B variants impairing melatonin receptor 1B function contribute to type 2 diabetes. Nat Genet 44:297–301

Chen R, Corona E, Sikora M et al (2012) Type 2 diabetes risk alleles demonstrate extreme directional differentiation among human populations, compared to other diseases. PLoS Genet 8:e1002621

Fajans SS, Bell GI (2011) MODY: history, genetics, pathophysiology, and clinical decision making. Diabetes Care 34:1878–1884

Flannick J, Beer NL, Bick AG et al (2013) Assessing the phenotypic effects in the general population of rare variants in genes for a dominant Mendelian form of diabetes. Nat Genet 45:1380–1385

Graham W (2009) The thrifty gene hypothesis: considering the significance of a 47-year-old theory. Quill Scope 2:10–13

Grant SF, Thorleifsson G, Reynisdottir I et al (2006) Variant of transcription factor 7-like 2 (TCF7L2) gene confers risk of type 2 diabetes. Nat Genet 38:320–323

Groop L, Pociot F (2014) Genetics of diabetes – are we missing the genes or the disease? Mol Cell Endocrinol 382:726–739

Hara K, Shojima N, Hosoe J, Kadowaki T (2014) Genetic architecture of type 2 diabetes. Biochem Biophys Res Commun 452:213–220

Helgason A, Pálsson S, Thorleifsson G et al (2007) Refining the impact of TCF7L2 gene variants on type 2 diabetes and adaptive evolution. Nat Genet 39:218–225

Horikawa Y, Enya M, Fushimi N et al (2014) Screening of diabetes of youth for hepatocyte nuclear factor 1 mutations: clinical phenotype of HNF1β-related maturity-onset diabetes of the young and HNF1α-related maturity-onset diabetes of the young in Japanese. Diabet Med 31:721–727

Lyssenko V, Lupi R, Marchetti P et al (2007) Mechanisms by which common variants in the TCF7L2 gene increase risk of type 2 diabetes. J Clin Invest 117:2155–2163

Maurano MT, Humbert R, Rynes E et al (2012) Systematic localization of common disease-associated variation in regulatory DNA. Science 337:1190–1195

McCarthy MI (2010) Genomics, type 2 diabetes, and obesity. N Engl J Med 363:2339–2350

Morris AP, Voight BF, Teslovich TM et al (2012) Large-scale association analysis provides insights into the genetic architecture and pathophysiology of type 2 diabetes. Nat Genet 44:981–990

Neel JV (1962) Diabetes mellitus: a "thrifty" genotype rendered detrimental by "progress". Am J Hum Genet 14:353–362

Okamoto K1, Iwasaki N, Nishimura C et al (2010) Identification of KCNJ15 as a susceptibility gene in Asian patients with type 2 diabetes mellitus. Am J Hum Genet 86:54–64

Pasquali L, Gaulton KJ, Rodríguez-Seguí SA et al (2014) Pancreatic islet enhancer clusters enriched in type 2 diabetes risk-associated variants. Nat Genet 46:136–143

Southam L, Soranzo N, Montgomery SB et al (2009) Is the thrifty genotype hypothesis supported by evidence based on confirmed type 2 diabetes- and obesity-susceptibility variants? Diabetologia 52:1846–1851

Steinthorsdottir V, Thorleifsson G, Sulem P et al (2014) Identification of low-frequency and rare sequence variants associated with elevated or reduced risk of type 2 diabetes. Nat Genet 46:294–298

Yorifuji T, Fujimaru R, Hosokawa Y et al (2012) Comprehensive molecular analysis of Japanese patients with pediatric-onset MODY-type diabetes mellitus. Pediatr Diabetes 13:26–32

Zimmet P (2003) The burden of type 2 diabetes: are we doing enough? Diabete Metab 29(4):6S9–618

Chapter 14
Disease-Related Genes from Population Genetic Aspect and Their Functional Significance

Ituro Inoue and Hirofumi Nakaoka

Abstract Numerous disease- and trait-related genes have been identified owing to advances of modern genomic technologies. Atopic dermatitis (AD) is the most common and complex disease, and filaggrin gene (*FLG*) was identified as a strong genetic component. Numerous variations of *FLG* were identified and European and Asian patients harbor each distinct mutation profiles. Excess non-synonymous variants over synonymous variants of *FLG* in general populations indicate that positive selection may have a substantial role in shaping the mutation profiles. In various disorders, population history is important particularly considering population-specific mutations. Recently, genomes of archaic humans such as Neanderthals and Denisovans had been sequenced and provided new evidence that 1–6% of modern Eurasian genome were introgressed by archaic humans. The introgressions had key roles in several immunological pathways including acquired and innate immunity. For example, parts of human leucocyte antigen (*HLA*) alleles were derived from archaic humans, which provided advantages for survival for ancestry of modern humans after the out-of-Africa expansion. A growing body of evidence supports the importance of variants in noncoding regions on human complex traits and natural selection. Thus, we introduce a topic on understanding of gene regulatory network recently demonstrated by ENCODE project, which provides clues of seeking functional impact of SNPs identified by genome-wide association study. SNPs located on DNase I hypersensitive sites are associated with disease causality through modulating transcriptional regulatory mechanisms. The allele-specific gene regulation is the key to understand the pathophysiology of common diseases.

Keywords Filaggrin · Atopic dermatitis · Copy number variation · Introgression · Archaic genome · Chromatin conformation

---

I. Inoue (✉) · H. Nakaoka
Division of Human Genetics, National Institute of Genetics, Mishima, Japan
e-mail: itinoue@nig.ac.jp; hnakaoka@nig.ac.jp

N. Saitou (ed.), *Evolution of the Human Genome I*, Evolutionary Studies,
https://doi.org/10.1007/978-4-431-56603-8_14

14.1 Introduction

With the widely applied genome-wide association study (GWAS), genetic components of numerous common diseases and phenotypes such as stature, skin, and eye color have been identified with high statistical significance. Some of the traits show signatures of positive selection. We present genetic studies of atopic dermatitis (AD), which is a typical and most common complex disease. In AD, filaggrin gene (*FLG*) plays substantial roles in the pathology. An array of non-synonymous variants was identified and they are population specific. It has been well recognized that population history also plays key roles in the etiologies of common diseases. Particularly, introgression of Neanderthal or Denisovan genome into ancestry of modern humans is highlighted because of disease and trait associations. In the case of human major histocompatibility complex (*MHC*), human leukocyte antigen (*HLA*) genes are highly polymorphic, and parts of *HLA* alleles were derived from archaic humans, which provided advantages for survival for ancestry of modern humans when they arrived in a new environment. Several lines of evidence support importance of variants in noncoding regions on human complex traits and natural selection. Molecular mechanisms of the GWAS-hit SNPs in noncoding regions leading to the diseases or traits are mostly unclarified because of subtle functional significance. The same situation is true for the SNPs that were identified under natural selection. ENCODE project uncovered genome-wide gene regulatory networks in various human cells and tissues, which can be directly applied to GWAS-identified SNPs for seeking the biological mechanisms.

In the first half, we describe the recent progress of genetic analyses of common diseases and adaptive traits related with population history. In the second half, we focus on functional aspect of noncoding variants by introducing functional genomic approaches exploring molecular mechanisms of SNPs that are identified in pigmentation traits. Allele-specific effects of SNPs on chromatin structure and gene regulatory machinery are highlighted.

14.2 Atopic Dermatitis

AD is one of the most common complex disorders in the developed countries affecting 10–25% of children in the USA and UK. AD is also prevalent in Asian and African populations (Asher et al. 2006). Itchy inflammatory skin lesion is the most common symptom of AD that is frequently overlapped with asthma, allergic rhinitis, or urticaria. Several family studies had identified genetic loci involved in the etiology of AD (Lee et al. 2000; Cookson et al. 2001), but AD is recognized as a typical complex disorder where multiple genetic and environmental factors mutually play roles in the disease development (Thomsen 2014).

14.3 Atopic Dermatitis and Filaggrin Gene

Many genes have been identified as susceptibilities of AD mostly through GWAS, particularly genes encoding epidermal structural proteins and key elements of the immune system. Among them, the most prominent discovery is a strong association between *FLG* and AD. It was first reported that two common null mutations of *FLG* are causality of familial ichthyosis vulgaris, a common keratinizing disease (Smith et al. 2006). Because it is well known that many of the ichthyosis vulgaris patients are accompanied with AD, *FLG* was screened for AD and was reported as a susceptibility of AD and secondly allergic diseases (Palmer et al. 2006). The two variants, R501X and 2282del4, were carried by 55% of AD patients and 9% of European-origin populations in the original report. Two association studies showed extremely strong associations between AD and *FLG* mutations (odds ratio = 7.44 at heterozygous state) (Palmer et al. 2006; Sandilands et al. 2007). Most of the mutations are nonsense or frameshift, which result in loss of function of *FLG*.

FLG encoding profilaggrin (filament-aggregating protein) is on chromosome 1q21 where the epidermal differentiation complex locates and plays critical roles in the terminal differentiation of epidermis (Brown and Irvine 2008; Sandilands et al. 2009; Osawa et al. 2010; McGrath 2012). *FLG* consists of three exons and two introns. The initiation codon of *FLG* is located in exon 2, while most of profilaggrin is encoded in exon 3. Exon 3 shows copy number variations encoding nearly identical 10–12 tandem repeats spanning 12.7–14.7 kb. Each of these repeats produces a copy of 324 amino acids of the filaggrin polypeptide. The number of repeats has a relationship with the quantity of filaggrin expressed in the epidermis. A greater number of repeats could be protective against dermatitis, whereas a smaller number of repeats increase the risk.

Profilaggrin is dephosphorylated and proteolyzed into multiple filaggrin monomers, which bind to keratin intermediate filaments (Presland 2009; Brown and McLean 2012). The filaggrin monomer is then further degraded to release its component amino acids (Fig. 14.1). Filaggrin plays a major role in skin barrier, where it maintains skin function against water loss and also minimizes the entry of allergens and microorganisms. Nonsense mutations within the third exon result in a truncated profilaggrin lacking the C-terminal, leading almost complete absence of filaggrin molecules. This is associated with disorganized keratin filaments, impaired lamellar body loading, and abnormal architecture of the lamellar bilayer. *FLG*-null mutations at heterozygous state result in a reduction of the natural moisturizing factors in the stratum corneum, and its haploinsufficiency contributes to several common dermatological disorders.

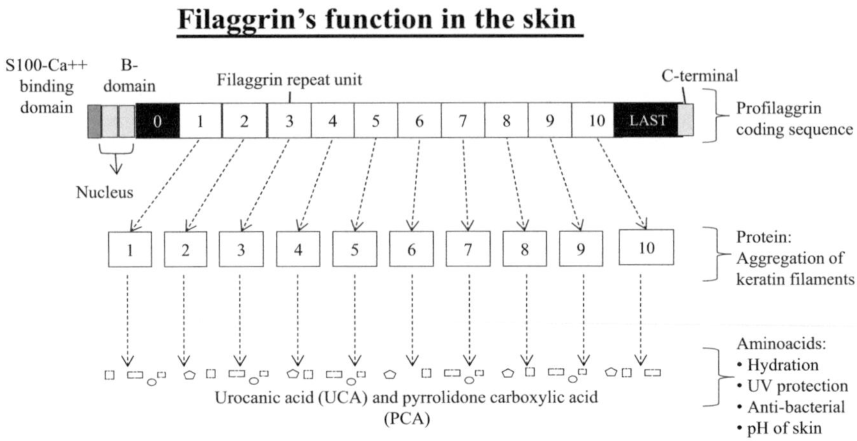

Fig. 14.1 Human filaggrin gene includes three exons and two introns, with the repeat region being found on the third. The filaggrin repeat region consists of complete repeats that are flanked by two partial repeats. *FLG* encodes profilaggrin protein, which is dephosphorylated and degraded to monomeric filaggrin repeats and then further proteolyzed to amino acids. Profilaggrin, filaggrin repeats, and amino acids are localized in the outer layers of the epidermis and have important functions in the skin. FLG unit binds to the keratin fibers and to the cornified envelope forming a barrier to water loss and minimizing the entry of allergens. FLG units are rich in glutamine and histidine into which they are degraded in the outer layer. Glutamine is converted into pyrrolidone-5-carboxylic acid (PCA) and histidine is deiminated to cis-urocanic acid (UCA). PCA and UCA serve important functions as protecting against UV irradiation, maintaining the acidic pH, being involved in the local immune response, and maintaining overall homeostasis. Black = partial repeats and white = complete repeats

14.4 Filaggrin Mutations

FLG-null mutations either nonsense or frameshift have a high prevalence in various populations (Fig. 14.2). Two major null mutations (R501X and 2282del4) are observed in 7–10% of European ancestry population. Furthermore, a total of 420 other rare or family-specific loss-of-function mutations on exon 3 have been thus far discovered in populations of European ancestry. Asian populations had been also shown to have their specific mutation spectra, in Japanese, Han Chinese, and Singaporean Chinese populations; R501X and 2282del4 mutations are rarely observed (Osawa et al. 2010). Interestingly, R501X mutation observed in Japanese was not derived from European population based on the haplotype analysis but occurred de novo (Hamada et al. 2008). These identified causal mutations revealed the complex genetic architecture of this locus, where each ancestral population has its own unique spectrum of mutations.

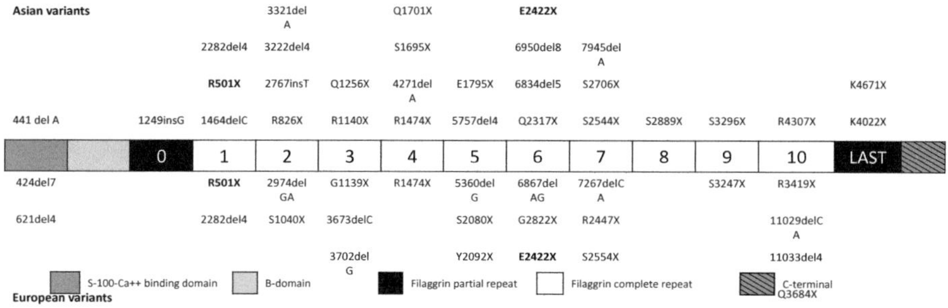

Fig. 14.2 Diagram of profilaggrin and FLG mutations reported in European and Asian populations. R501X and E2422X were detected the most in both European and Asian populations (Ellinghaus et al. 2013 and Osawa et al. 2011)

14.5 Natural Vaccination Hypothesis

One of the intriguing questions about mutation spectrum of *FLG* is why null mutations are so prevalent in various populations. Indeed, non-synonymous variants are more prevalent than synonymous variants. Also, population-specific mutations, mostly rare variants, are observed. Because most of the mutations are heterozygous and considered to be loss of function, heterozygous advantage might exist. These evidence indicate natural selection might shape the mutation spectrum of *FLG*. One might explain with "natural vaccination" hypothesis, in which heterozygous state of *FLG* having "leaky" skin could have an advantage against several pathogens by having vaccination via skin barrier in early days probably during pandemics in our ancient past (Irvine and McLean 2006). Determining the age of these null mutations in human population history could provide interesting clues as to why they are so common.

14.6 Introgression of Archaic Genomes

With the technical advances of sequencing method, archaic human genomes such as Neanderthals and Denisovans have been sequenced, and it has become clear that intercross between Neanderthals or Denisovans and ancestry of modern humans give some features of population history of non-sub-Saharan populations, mostly Eurasians (Green et al. 2010; Prüfer et al. 2014; Fu et al. 2014; Sankararaman et al. 2014). Because intercross with ancestor of modern humans is known, Neanderthal and Denisovan are called as archaic human here. Neanderthal ancestors departed Africa about 400,000 years ago before modern humans were shaped. Thereby, the modern humans in sub-Saharan Africa do not share Neanderthal genome. About 80,000 years ago, ancestors of modern humans in Africa migrated out of Africa and settled in the Middle East. Then part of the groups migrated to north, and part went east spreading over Eurasia.

Archaic genome sequence data have been accumulated and provided thus far unidentified human evolutional history (Lowery et al. 2013; Kelso and Prüfer 2014). Remarkably, 1–6% of the genome of today's Europeans and Asians were originated from the Neanderthals or Denisovans. Also introgression of ancient genomes into modern humans is highlighted in relation with disease causalities and phenotypic characters such as type 2 diabetes in Mexicans and high-altitude adaptation in Tibetans (The SIGMA Type 2 Diabetes Consortium 2014; Huerta-Sanchez et al. 2014).

14.7 Archaic DNA Introgression into Modern Human Immune System

On migration out of Africa, the ancestors of modern humans encountered new environments particularly thus far unfaced pathogens. Because archaic humans resided in Eurasia more than 200,000 years, they had better immune system adapted to the local pathogens. With admixtures with archaic humans, the ancestor of modern humans rapidly acquired immune system fit to the new environments, which could provide major advantages for survival.

Abi-Rached et al. (2011) focused on the human *MHC* (*HLA*) region, which plays a central role of immune defense system, and examined whether archaic *HLA* alleles of Neanderthals or Denisovans were introgressed to modern humans. *HLA-B*73:01* is observed with high frequency in West Asians and rare or absent in other regions. *B*73:01* is basically in linkage disequilibrium (LD) with *HLA-C*15:05* shaping a haplotype. The LD of the two alleles or haplotype is decayed in West Asians; meanwhile, complete LD is maintained in African population even if the haplotype is very rare. Europeans are between them. These evidence indicate that the haplotype originated in West Asia and are also consistent with introgression in West Asia of the archaic *B*73:01-C*15:05* haplotype. However, thus far identified *HLA-B* alleles of Denisovan were not B*73, and further studies are evidently needed. In contrast, *HLA-A* of Denisovan was estimated to be *A*02:01* and *A*11:01*. As shown in Fig. 14.3, *HLA-A*11* is common in West Asia, less common in Europe, and absent in Africa. *HLA-A*02:06* is also specifically observed in West Asia. A simple interpretation is that modern *A*11* allele was derived from Denisovan and maintained with high frequency. Because the haplotype phase cannot be determined in Denisovans, all possible haplotypes of *HLA-A-HLA-C* were taken into account and then compared with the haplotype of modern populations resulting that all types showed high frequency in West Asians and virtually absent in Africans. Because ancestors of modern humans and Denisovans separated >250,000 years ago, i.e., ~10,000 generation, and frequent recombinations occurred over that time, *HLA* haplotype cannot be shared between Denisovans and modern humans if both stayed in Africa. It is likely that the modern humans acquired these haplotypes by introgression from Denisovans. Because *HLA* loci are inherently highly polymorphic and *HLA* loci of archaic humans are not fully determined, it is possible that we observed simply population-specific alleles.

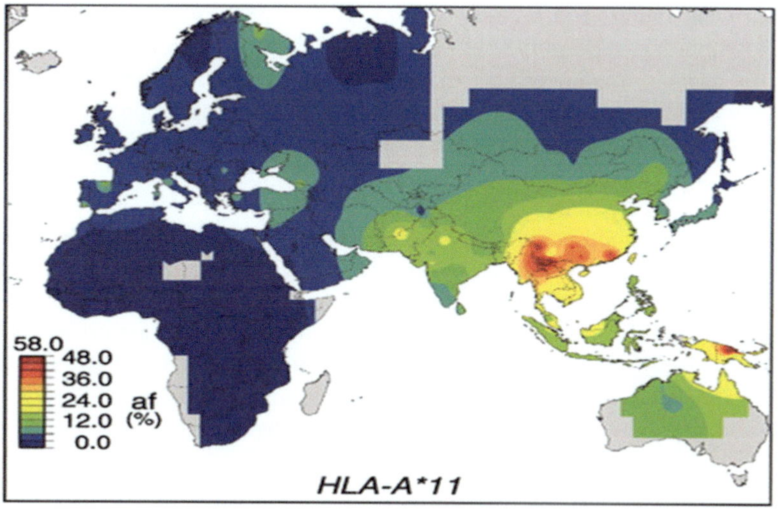

Abi-Rached L et al. 2011. Science

Fig. 14.3 Distribution of *HLA-A*11* in modern humans. *HLA-A*11* shared with Denisovans is mostly observed in East Asians but not in Europeans and Africans

More convincingly, *STAT2*, one of the key molecules of immunity, is another example of introgression of Neanderthal's genome into the Papuans of New Guinea (Mendez et al. 2012). One haplotype comprising 18 sites spanning 8.6 kb, the so-called N linage, completely matches with the Neanderthal sequence. N linages are broadly distributed in Eurasians at relatively low frequencies and are not observed in sub-Saharan African populations. Similar results were reported for 2′-5′ oligoadenylate synthetase (OAS) genes, indispensable components of the immune system having significant antiviral functions (Mendez et al. 2013). A Neanderthal-like haplotype spanning about 180 kb containing *OAS1*, *OAS2*, and *OAS3* was estimated to have diverged from the Neanderthal about 125,000 years and is mostly absent in African populations. Most recently, *TLR6-TLR1-TLR10* gene cluster, which are important for innate immunity, was shaped as a result of archaic admixture (16% in CEPH Europeans and 49% Han Chinese) (Dannemann et al. 2016; Deschamps et al. 2016).

14.8 Chromatin Structure Related with Disease Etiology and Natural Selection Under the ENCODE Project

GWAS is such a powerful method to identify genetic loci of common diseases and traits. Most of the identified SNPs are located on intergenic and intronic regions rather than coding regions, suggesting that these noncoding SNPs are associated with the disease risk through regulation of gene expression (Hindorff et al. 2009). Similarly, SNPs putatively associated with local adaptation in terms of climate variables were reported to be overrepresented in expression quantitative trait loci (eQTL) within cis-regulatory elements (Fraser 2013). Therefore exploration of molecular mechanisms underlying disease- and trait-associated noncoding SNPs is essential for better understanding of genetic architecture shaping present-day human populations. It is generally accepted that identified SNPs by GWAS could be merely markers or surrogates, and functional or causal SNPs in LD with the marker need to be identified and analyzed. Under the ENCODE project, Maurano et al. (2012) systematically surveyed positions of disease- or trait-associated SNPs. They examined 5654 noncoding SNPs, which are identified by GWAS related with diseases and traits. Forty percent of GWAS SNPs were enriched in DHS (DNase I hypersensitive site), and 76.6% of all noncoding SNPs either lie within DHS or are in complete LD with SNPs in nearby DHS. DHSs serve as a functional site for cis-regulatory element including enhancers, promoters, insulators, and silencers. Therefore SNPs in DHS are expected to alter bindings of transcriptional factors then gene regulations.

14.9 Three-Dimensional Chromatin Structure and Phenotypic Status

Skin, eye, and hair colors reflect one of the most striking common phenotypes within and among populations and are associated with local adaptation. It is reported that the SNP (rs12913832) within *HERC2* is strongly associated with human eye, hair, and skin color variation. Particularly strong association is reported in comparison between red hair and black hair phenotypes ($P = 4.3 \times 10^{-39}$) (Han et al. 2008). rs12913832 is located in intron 86 of *HERC2*, which is not known as pigmentation-related gene, but 21 kb upstream of the promoter of *OCA2*. *OCA2* is involved in human pigmentation because it regulates melanosomal pH, and mutations of the gene cause oculocutaneous albinism type II. Therefore, rs12913832 is speculated to function as a distal regulatory element of *OCA2*. In other case, *KITLG* encoding a secreted ligand for the KIT receptor tyrosine kinase is associated with common blond hair color in northern Europeans. The SNP (rs12821256) located in an intergenic region over 350 kb upstream of *KITLG* transcriptional start site is associated with blond hair phenotype in Iceland and Netherland with odds ratio 1.9–2.4 in comparison of blond versus brown hairs in northern European. The blond-associated A>G substitution of rs12821256 is prevalent in Northern Europeans while is virtually absent in African and Asian populations.

Functional involvement of the noncoding SNP, rs12913832, was analyzed with chromatin structure analyses using human relevant cell lines. Visser et al. (2012) used two different cell lines of melanocyte origin homozygous for rs12913832 C-allele (HEMn-LP) and T-allele (HEMn-DP). HEMn-LP cell is a lightly pigmented cell line and HEMn-DP is darkly pigmented. Using formaldehyde-assisted identification of regulatory elements (FAIRE) assay, which identifies regulatory DNA region based on differences in cross-linking ability between enhancer elements and the bulk of chromatin, the region around rs12913832 was shown in open chromatin structure in both cells. In MCF7, a breast cancer cell line using as a negative control, the region showed closed structure suggesting that the SNP site is a melanocyte-specific regulatory element. Furthermore higher enrichment was observed in HEMn-DP cells indicating rs12913832 T-allele has more open structure than C-allele. If the region around rs12913832 functions as a distal enhancer of *OCA2*, the rs12913832 region and the promoter of *OCA2* form a long-range chromatin loop. The long-range chromatin loop can be detected using chromosome conformation capture (3C) method. In 3C, the cross-linked chromatin is digested with an appropriate restriction enzyme followed by intramolecular ligation under diluted conditions. Then ligation product is determined by qPCR and allelic differences in chromatin loop formation are evaluated. As expected, rs12913832 T-allele showed enriched loop formation with *OCA2* promoter resulting in enhanced expression of *OCA2* along with transcription factors binding at the loop.

In general, it is difficult to infer biological mechanism of the GWAS-hit SNPs due to the modest effects. Chromatin structure analyses with aids of the next-generation sequencing technologies would assign functional basis of the genetic susceptibilities of common diseases in a quantitative manner. Evidence are accumulating that trait-associated SNPs are mostly eQTLs regulating gene expression (Nicolae et al. 2010; Nica et al. 2010; Nica and Dermitzakis 2013; Nakaoka et al. 2016). By applying next-generation sequencers, precise measurements of the allele-specific loop formation and bindings of transcriptional factors become possible and provide clues to understand subtle biological impact of the SNPs and mechanisms of the diseases.

References

Abi-Rached L, Jobin MJ, Kulkarni S et al (2011) The shaping of modern human immune systems by multiregional admixture with archaic humans. Science 334:89–94

Asher MI, Montefort S, Björkstén B et al (2006) Worldwide time trends in the prevalence of symptoms of asthma, allergic rhinoconjunctivitis, and eczema in childhood: ISAAC Phases One and Three repeat multicountry cross-sectional surveys. Lancet 368:733–743

Brown SJ, Irvine AD (2008) Atopic eczema and the filaggrin story. Semin Cutan Med Surg 27:128–137

Brown SJ, McLean WH (2012) One remarkable molecule: filaggrin. J Invest Dermatol 132:751–762

Cookson WO, Ubhi B, Lawrence R et al (2001) Genetic linkage of childhood atopic dermatitis to psoriasis susceptibility loci. Nat Genet 27:372–373

Dannemann M, Andres AM, Kelso J (2016) Introgression of Neandertal- and Denisovan-like haplotypes contributes to adaptive variation in human toll-like receptors. Am J Hum Genet 98:22–33

Deschamps M, Lavel G, Fagny M et al (2016) Genomic signatures of selective pressures and introgression from archaic hominins at human innate immunity genes. Am J Hum Genet 98:5–21

Ellinghaus D, Baurecht H, Esparza-Gordillo J et al (2013) High-density genotyping study identifies four new susceptibility loci for atopic dermatitis. Nat Genet 45:808–812

Fraser HB (2013) Gene expression drives local adaptation in humans. Genome Res 23:1089–1096
Fu Q, Li H, Moorjani P et al (2014) Genome sequence of a 45,000-year-old modern human from western Siberia. Nature 514:445–449
Green RE, Krause J, Briggs AW et al (2010) A draft sequence of the Neandertal genome. Science 328:710–722
Hamada T, Sandilands A et al (2008) De novo occurrence of the filaggrin mutation R501X with prevalent mutation c.3321dela in Japanese family with ichthyosis vulgaris complicated by atopic dermatitis. J Invest Dermatol 128:1323–1325
Han J, Kraft P, Nan H et al (2008) A genome-wide association study identifies novel alleles associated with hair color and skin pigmentation. PLoS Genet 4:e1000074
Hindorff LA, Sethupathy P, Junkins HA et al (2009) Potential etiologic and functional implications of genome-wide association loci for human diseases and traits. Proc Natl Acad Sci U S A 106:9362–9367
Huerta-Sánchez E, Jin X, Asan et al (2014) Altitude adaptation in Tibetans caused by introgression of Denisovan-like DNA. Nature 512:194–197
Irvine AD, McLean WH (2006) Breaking the (un)sound barrier: filaggrin is a major gene for atopic dermatitis. J Invest Dermatol 126:1200–1202
Kelso J, Prüfer K (2014) Ancient humans and the origin of modern humans. Curr Opin Genet Dev 29:133–138
Lee YA, Wahn U, Kehrt R et al (2000) A major susceptibility locus for atopic dermatitis maps to chromosome 3q21. Nat Genet 26:470–473
Lowery RK, Uribe G, Jimenez EB et al (2013) Neanderthal and Denisova genetic affinities with contemporary humans: introgression versus common ancestral polymorphisms. Gene 530:83–94
Maurano MT, Humbert R, Rynes E et al (2012) Systematic localization of common disease-associated variation in regulatory DNA. Science 337:1190–1195
McGrath JA (2012) Profilaggrin, dry skin, and atopic dermatitis risk: size matters. J Invest Dermatol 132:10–11
Mendez FL, Watkins JC, Hammer MF (2012) A haplotype at STAT2 Introgressed from Neanderthals and serves as a candidate of positive selection in Papua New Guinea. Am J Hum Genet 91:265–274
Mendez FL, Watkins JC, Hammer MF (2013) Neanderthal origin of genetic variation at the cluster of OAS immunity genes. Mol Biol Evol 30:798–801
Nakaoka H, Gurumurthy A, Hayano T et al (2016) Allelic imbalance in regulation of ANRIL through chromatin interaction at 9p21endometriosis risk locus. PLoS Genet 12:e1005893
Nica AC, Dermitzakis ET (2013) Expression quantitative trait loci: present and future. Philos Trans R Soc Lond Ser B Biol Sci 368:20120362
Nica AC, Montgomery SB, Dimas AS et al (2010) Candidate causal regulatory effects by integration of expression QTLs with complex trait genetic associations. PLoS Genet 6: e1000895
Nicolae DL, Gamazon E, Zhang W et al (2010) Trait-associated SNPs are more likely to be eQTLs: annotation to enhance discovery from GWAS. PLoS Genet 6:e1000888
Osawa R, Konno S, Akiyama M et al (2010) Japanese-specific filaggrin gene mutations in Japanese patients suffering from atopic eczema and asthma. J Invest Dermatol 130:2834–2836
Osawa R, Akiyama M, Shimizu H (2011) Filaggrin gene defects and the risk of developing allergic disorders. Allergol Int 60:1–9
Palmer CN, Irvine AD, Terron-Kwiatkowski A et al (2006) Common loss-of-function variants of the epidermal barrier protein filaggrin are a major predisposing factor for atopic dermatitis. Nat Genet 38:441–446
Presland RB (2009) Function of filaggrin and caspase-14 in formation and maintenance of the epithelial barrier. Dermatol Sin 27:1–14
Prüfer K, Racimo F, Patterson N et al (2014) The complete genome sequence of a Neanderthal from the Altai Mountains. Nature 505:43–49

Sandilands A, Terron-Kwiatkowski A, Hull PR et al (2007) Comprehensive analysis of the gene encoding filaggrin uncovers prevalent and rare mutations in ichthyosis vulgaris and atopic eczema. Nat Genet 39:650–654

Sandilands A, Sutherland C, Irvine AD, McLean WH (2009) Filaggrin in the frontline: role in skin barrier function and disease. J Cell Sci 122:1285–1294

Sankararaman S, Mallick S, Dannemann M et al (2014) The genomic landscape of Neanderthal ancestry in present-day humans. Nature 507:354–357

SIGMA Type 2 Diabetes Consortium, Williams AL, Jacobs SB et al (2014) Sequence variants in SLC16A11 are a common risk factor for type 2 diabetes in Mexico. Nature 506:97–101

Smith FJ, Irvine AD, Terron-Kwiatkowski A et al (2006) Loss-of-function mutations in the gene encoding filaggrin cause ichthyosis vulgaris. Nat Genet 38:337–342

Thomsen SF (2014) Atopic dermatitis: natural history, diagnosis, and treatment. ISRN Allergy 2014:354250

Visser M, Kayser M, Palstra RJ (2012) HERC2 rs12913832 modulates human pigmentation by attenuating chromatin-loop formation between a long-range enhancer and the *OCA2* promoter. Genome Res 22:446–455

Chapter 15
Microbe Genomes Associated with Human Body

Chaochun Wei and Ben Jia

Abstract This chapter introduces the concept of human superorganism first. Microbe genomes associated with the human body have been studied via metagenomics methods. Metagenomics is introduced first. We then move to the Human Microbiome Project. The community and gene composition of the microbiota associated with the human body is outlined. The total number of genes in the human gut symbiota is estimated. Disease-associated microbe genomes in/on the human body are introduced, followed with the discussion about the coevolution of the microbes and the human genome.

Keywords Metagenomics · Human microbiome project · Coevolution · Human superorganism · Human gut microbiota · 16S rRNA

Humans are superorganisms consisting of both human cells and microbial cells. The number of microbial cells (~100 trillion) can be an order of magnitude larger than the number of human cells (~10 trillion), and the number of genes encoded in the human genome itself can be much less than the genes encoded in the microbe genomes associated with the human body (Zhao 2013; Gill et al. 2006). Figure 15.1 showed a diagram of the human superorganism and how microorganisms associated with the human body interact with the human genome. These microorganisms provide human with genetic and metabolic attributes that were not required to evolve in the human genome, including the ability to harvest otherwise indigestible nutrients such as polysaccharides (Backhed et al. 2005). The host genome imposes selective pressure on the microorganisms. Environment elements, such as diets and particular drugs, can impact the structure of the gut microbiota greater than host genetics and thus influence the health of the human superorganism (Zhao 2013).

In order to understand this human superorganism, we need to introduce metagenomics.

C. Wei (✉) · B. Jia
School of Life Sciences and Biotechnology, Shanghai Jiao Tong University, Shanghai, China
e-mail: ccwei@sjtu.edu.cn

N. Saitou (ed.), *Evolution of the Human Genome I*, Evolutionary Studies,
https://doi.org/10.1007/978-4-431-56603-8_15

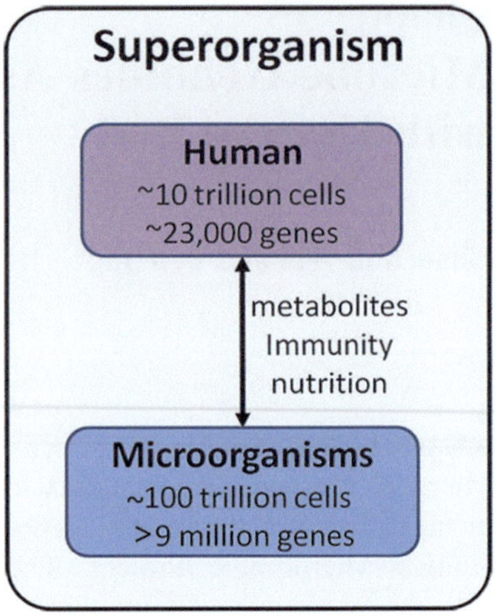

Fig. 15.1 The concept of considering human as a superorganism

15.1 What Is Metagenomics and Why

In a natural environment, microbes exist in a form of community, and microbes in a community interact with each other. In order to understand the microbe community, we need to understand the composition structure of the microbe community, the functional gene composition, etc. However, the traditional isolate-and-culture methods can only analyze a tiny fraction of the total microbes (Suau et al. 1999; Furrie 2006).

A metagenome is all genetic materials recovered directly from an environmental sample. Metagenomics is a research field about metagenomes. It is a new research field full of promising since it can provide insights to a large unexplored field of uncultured microbes (Venter et al. 2004; Tyson et al. 2004).

15.1.1 The Current Progress of Metagenomics

Metagenomics methods have been applied to analyze numerous microbial communities (Kowalchuk et al. 2007), and new organisms or enzymes have been found by using metagenomics methods (Yun et al. 2004; Ferrer et al. 2007). For example, the gut microbiota can be considered as a special metabolite organ of the human body; it is closely related to obesity (Turnbaugh et al. 2006), diabetes (Cani et al. 2007), and hypertension (Holmes et al. 2008). Researches have shown that gut microbiota can impact the food digestion, the nutrition absorption (Turnbaugh et al. 2006; Backhed et al. 2004), and even the consequent metabolism procedures (Backhed et al. 2007). In fact the impact of gut microbiota on the chronic metabolic diseases has attracted more and more attention in the field, and it becomes an important direction for drug target gene finding by integrating genes from gut metagenomes and disease-related gene networks (Zhu et al. 2008). Metagenomics methods have provided a series of new technologies and concepts for the environment problem (Etchebehere et al. 2003; Narihiro and Sekiguchi 2007) and energy problems (Kim et al. 2007; Levin et al. 2007; Zuo et al. 2008). More recently, metagenomics methods have been applied to analyze the impact of gut microbiota on the brain and behavior (Cryan and Dinan 2012), thus revealing the molecular mechanisms of diseases like schizophrenia (Dinan et al. 2014).

Through a series of genomics methods, metagenomics try to answer three questions: (1) who are they (measuring the diversity), (2) what they do (functional annotation), and how to compare them (comparative genomics) (Wooley et al. 2010). The first question is about to measure the diversity or composition structure of the microbial community; the second question is about the function or gene composition of the microbiome; and the last question is about how to compare different metagenomes. They have been reviewed briefly below.

15.1.2 Microbial Community Composition Structure Analysis and Metagenome Functional Annotation

Determining the community composition structure is the first important step in metagenomics analysis. There are two types of methods to represent the community structure. One is organism composition, i.e., the compositional organisms and their corresponding proportions; and the other is the composition of functional genes, i.e., the functional groups of genes and their abundances.

The community composition structure can be determined by sequencing the conserved genes such as 16S rRNA genes. 16S rRNA is common in all prokaryotes while the sequence specificity is diverse enough to discriminate organism in genus or species level (Pace 1997). This method is becoming a gold standard for community diversity, especially for large sample size research. By using high-throughput sequencing and barcode technology, a single sequencing run can process many samples or even hundreds of samples (Ling et al. 2010; Zhang et al. 2010).

However, the copy numbers of 16S rRNA genes vary in different bacterial genomes. Figure 15.2 shows the numbers of 16S rRNA genes varied from 1 to 15, with an average of 4.4 copies per genome. The number of 16S copies was also related to its genome size. In general, larger genomes had more copies of 16S rRNA genes. Therefore, 16S rRNA gene-based methods have a strong bias toward species with high copy numbers of 16S rRNAs (Fig. 15.3).

Functional analysis is another aspect of microbe community structure. It is to study the composition structure of the functional genes. The shotgun sequencing of metagenomes can be used to find all functional genes including those new functional genes. Gill et al. (2006) studied the 72 Mbp DNA sequences and 16S rRNA sequences from two fecal samples of two healthy individuals. They found many nutrition metabolism-related genes, especially for polysaccharide metabolism, and the frequencies of these genes were even higher than those of related human genes

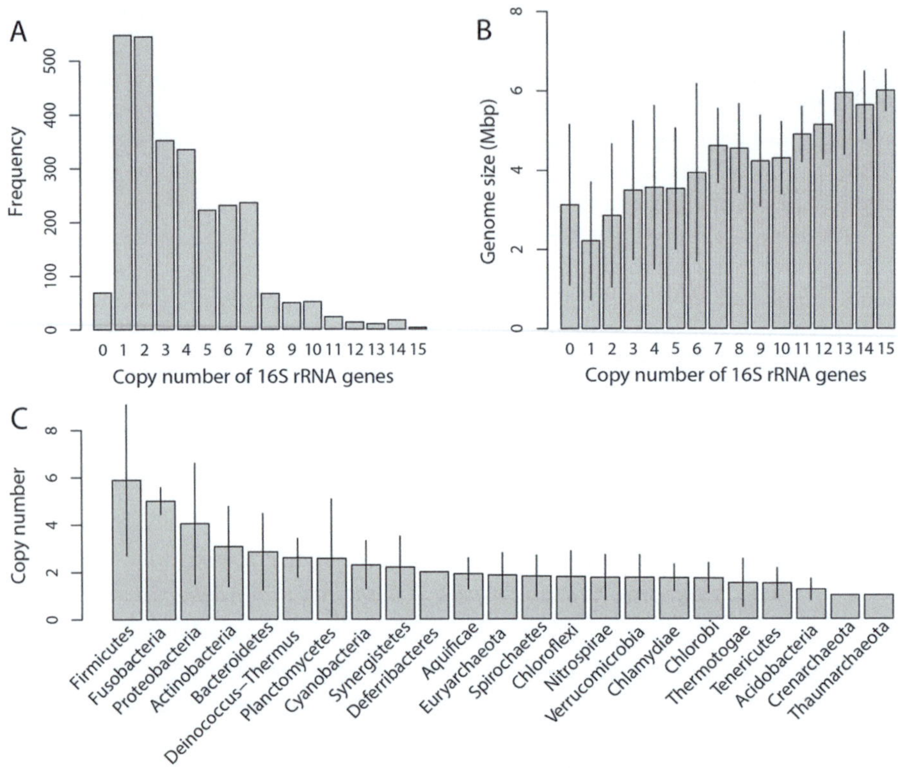

Fig. 15.2 The copy numbers of 16S rRNA genes in bacterial genomes. The copy numbers of 16S rRNA genes from 2787 fully sequenced genomes with 16S rRNA genes annotated (from ftp://ftp.ncbi.nlm.nih.gov/genomes/Bacteria/all.gbk.tar.gz, November, 2014). (**A**) The copy number of 16S rRNA genes in a genome varies in an order of magnitude between different genomes. (**B**) The copy number of 16S rRNA genes in a genome increases as the genome size increases. (**C**) Different phyla have different numbers of 16S rRNA gene copies

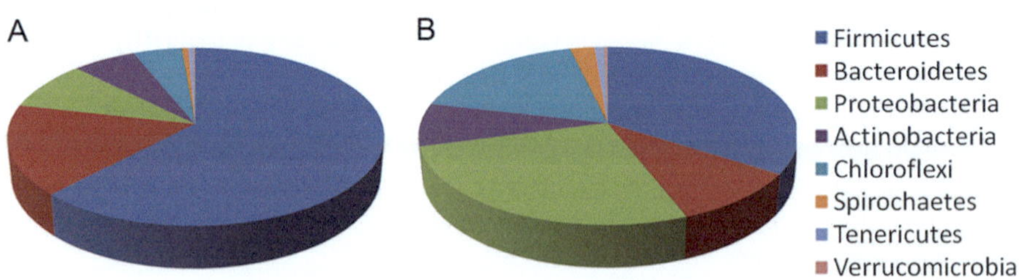

Fig. 15.3 Impact of the number of 16S rRNAs in different organisms on metagenomics analysis. This figure shows the same composition structure estimated without (**A**) and with (**B**) considering the copy numbers of 16S rRNA genes (The 16S rRNA data were from Zhang et al. 2010)

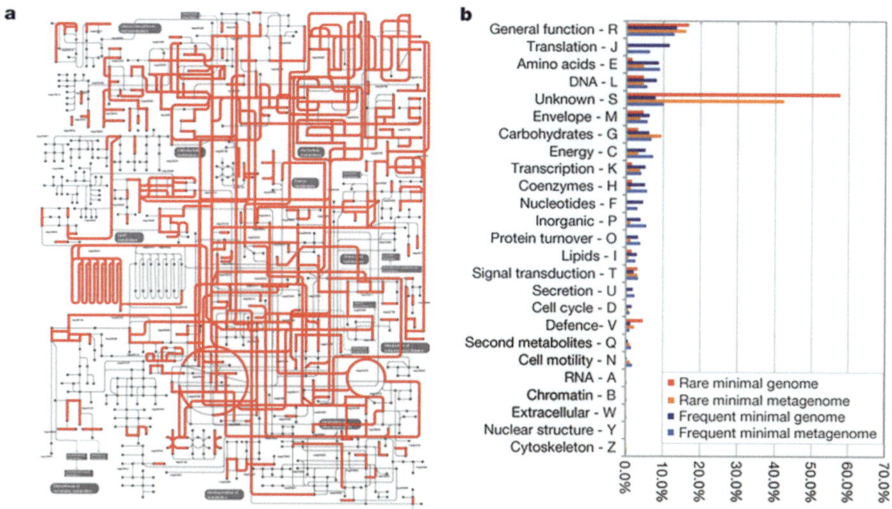

Fig. 15.4 Human gut microbial gene catalogue from metagenomics sequencing. (**a**) Projection of minimal gut genome on KEGG pathways; (**b**) functional composition of the minimal gut genomes shared by all samples and metagenomes (Figure adapted by permission from Macmillan Publishers Ltd.: Nature (Qin et al. 2010), copy right (2010))

(Gill et al. 2006). Gordon lab demonstrated that gut microbiota contributed significantly to obesity because the types of genes related to energy absorption in fat mouse individuals were more rich than those in lean individuals (Turnbaugh et al. 2006). Qin et al. (2010) sequenced gut metagenomes from 124 European individuals and generated about 3.3 million unique microbial genes. By using tools like iPATH (Letunic et al. 2008), COGs (the Clusters of Orthologous Groups of protein) (Tatusov et al. 2001), and KEGG (Kyoto Encyclopedia of Genes and Genomes) (Kanehisa and Goto 2000), the functional categorization of these microbial genes is shown in Fig. 15.4.

The functional analysis of a metagenome, such as gene prediction and gene functional analysis, needs public databases. A few popular public databases are listed here, including COGs (the Clusters of Orthologous Groups of protein) (Tatusov et al. 2001), Pfam (Protein family) (Finn et al. 2008), Kyoto Encyclopedia of Genes and Genomes (Kanehisa and Goto 2000), etc.

15.1.3 Current Progress in Metagenome Sequencing, Assembly, Gene Prediction, Functional Analysis, and Comparative Metagenomics

High-throughput sequencing technologies play a key role in metagenomics, especially for shotgun sequencing-based metagenomics analysis (Venter et al. 2004;

Tyson et al. 2004). Overall, these new sequencing platforms, which are called next-generation sequencing (NGS) technologies in general, have revolutionized the sequencing landscape in the past few years (Margulies et al. 2005; Mardis 2008). Due to their high-throughput and cost performance efficiency, NGS platforms have made it possible to apply metagenomics methods to a wide range of research areas.

Similar to traditional genome sequence analysis, metagenome sequencing data analysis includes quality control (QC), sequence assembly, sequence binning, gene prediction, and gene functional analysis using public databases.

In the end, comparative genomics methods and statistical analysis methods were applied to find environmental-specific genes or the dynamic change of the gene composition structure of metagenomes in the same environment.

The above analysis (including sequence assembly, gene prediction, functional analysis, and comparative metagenomics analysis) requires to integrate methods and public databases. The current databases for metagenomics research included metagenomic-specific databases such as MG_RAST (Meyer et al. 2008) (http://metagenomics.anl.gov/, metagenome data), CAMERA (http://camera.calit2.net/, metagenome data), MMCD (http://mmcd.nmrfam.wisc.edu/, metabolism data), and IMG (http://img.jgi.doe.gov/, genomic data) and general database such as NCBI (http://www.ncbi.nlm.nih.gov/, metagenomic and single cell genomic data).

15.1.4 Single Cell Sequencing Technology for Metagenome Analysis

Traditional metagenome sequencing data measures the composition of a community, and it is difficult to reveal the in vivo mechanisms and activities of some procedures in one cell. However, the phenotypes of a population of cells with the same genotypes can be different. Single cell difference can be used to reveal organism interactions in uncultivated microbes (Yoon et al. 2011), mechanism of cell proliferation, tissue development, and cancer onset (Navin et al. 2011; Tay et al. 2010; Sato et al. 2009; Bendall et al. 2011). Single cell sequencing has greatly improved the metagenomics methods.

15.2 The Human Microbiome Project

The microorganisms that live inside and on humans are known as microbiota. The Human Microbiome Project (HMP, http://commonfund.nih.gov/hmp/index) was initialized as a strategy to understand the microbe components of the human genetics, and it aimed to develop tools and datasets for the research community to understand the range of human genetic and physiological diversity in terms of microbiota and their impacts on the human health and disease (Turnbaugh et al.

2007). In the first stage of HMP, the community composition and diversity as well as microbe genomes from five sites on the human body were investigated. The five body sites included nasal passages, oral cavities, skin, gastrointestinal tract, and urogenital tract. The current stage of the HMP focused on cohort studies of microbiome-associated diseases.

Metagenomics analysis methods can be applied to the human microbiome data. Although large variation in human microbiomes between people exists, it is a novel direction to improve our understanding about ourselves as a superorganism. Metagenomics methods have been applied to research areas including the diagnosis, therapy, and prevention of chronic diseases.

With the progress of molecular biotechnology and data accumulation, our understanding about gene and disease is under a revolution: diseases are not only associated with genes in human genome but also related to genomes from the environment around and inside human body. The human microbiome was speculated to have important functions in health and disease. A notable example of what environmental genome changes can result in is human gut bacterial community. The change of human gut bacterial community is associated with obesity (Turnbaugh et al. 2006), diabetes (Cani et al. 2007), hypertension (Holmes et al. 2008), and so on. Therefore gut bacteria has become one of the hot research areas especially for those researches about chronic and metabolism-related diseases (Turnbaugh et al. 2009). Also, it is becoming an important research direction for finding drug target genes using human gene networks (Zhu et al. 2008). Drug target genes can also be identified by studying similar gene networks containing mixture of both human genes and genes from human gut bacteria (Chen et al. 2008).

15.2.1 The Gut Microbiota

The gut microbiota is among the most investigated microbiota inside/on the human body due to its extreme importance. From fecal samples of 124 European individuals, Qin et al. (2010) sequenced and categorized about 3.3 million unique microbial genes from 576 gigabases of sequence. Overall, the human gut microbiota contains 1,000–1,5000 prevalent bacterial species, and each individual contains at least 160 such species (Qin et al. 2010).

Obesity and diabetes are the most common chronic metabolic diseases and are becoming worldwide problem in health. Recent researches first showed that gut microbiota was associated with obesity (Turnbaugh et al. 2006), diabetes (Cani et al. 2007), and hypertension (Holmes et al. 2008). Without changing the intake of food, the gut microbiota can impact the food digestion, nutrition absorption (Turnbaugh et al. 2006; Backhed et al. 2004), and consequent metabolic processes (Backhed et al. 2007). Then, more recent researches showed that the gut microbiota might play an essential or even causal role in some chronic metabolic diseases (Zhao 2013). Therefore, the human gut microbiota can be considered as a special organ of the human body, and combining gut microbiome and the human gene network can be a potential approach for disease target gene finding (Zhu et al. 2008; Xu et al. 2014; Xiao and Zhao 2014). This can also be a new research area for personalized medicine.

15.2.2 The Oral Microbiota

Oral microbiota plays a vital role in maintaining the homeostasis of oral cavity. Dental diseases like caries are common in human. By high-throughput barcoded pyrosequencing of the oral microbiota, Ling et al. (2010) found that bacterial diversity was far more complex than we expected previously. There were more than 200 genera belonging to ten phyla, and the genera of *Streptococcus*, *Veillonella*, *Actinomyces*, *Granulicatella*, *Leptotrichia*, and *Thiomonas* were significantly associated with dental caries (Ling et al. 2010). The results showed that there was no one specific pathogen but rather pathogenic populations in plaque that significantly correlated with dental caries. The enormous diversity of oral microbiota allowed for a better understanding of oral microecosystem, and these pathogenic populations in plaque provide new insights into the etiology of dental caries and suggest new targets for interventions of the disease.

Yang et al. (2010) also reported that saliva microbiomes can distinguish caries active from healthy human populations and caries is a polymicrobial disease caused by a complex community formed by tens of bacterial species. By comparing to the gut microbiomes, they concluded that the oral cavity is functionally a different environment from the gut (Yang et al. 2012).

15.2.3 Microbiota in/on Different Body Sites

The microbiota occupying different body sites vary remarkably within or among individuals in terms of composition structure, function, and diversity. This holds even for healthy individuals (Human_Microbiome_Project_Consortium 2012).

As we mentioned above, the human gut microbiota plays an important role in human health, especially for those chronic and metabolism-related diseases (Turnbaugh et al. 2009). In addition, it is becoming an important research direction for finding drug target genes using human gene networks (Zhu et al. 2008). Drug target genes can also be identified by studying similar gene networks containing mixture of both human genes and genes from human gut bacteria (Chen et al. 2008).

Further studies in these directions require that we understand the gene composition of the human gut microbe community. Estimating the number of genes in it is one of the most important steps to understand the scale of the problem that we are dealing with.

15.3 Genes in the Microbial Genomes Associated with Human Body

Estimating the number of genes in human genome was one of the most fundamental problems in computational biology. The number of genes estimated for human genome dropped from more than 100,000 (Liang et al. 2000) to 60,000 (Fields et al. 1994), 40,000 (Das et al. 2001), and 30,000 after the draft human genome came out in 2001 (Lander et al. 2001), then to the current estimation of about 23,000 (Wei and Brent 2006). This number is not much larger than 17,000, the number of genes in C. *elegans* (Wei et al. 2005), a model organism about 1 mm in length and about 1,000 cells in total. The methods used to estimate gene numbers include transcript-based methods, CpG island counting methods, and, ultimately, gene prediction methods. Transcript-based methods contain cDNA counting, EST clustering and Refseq gene counting, etc. Although the accurate number of genes in a human genome is still not determined, the scale of the gene number has been set to be about 20,000. However, the number of genes in the microbial genomes associated with human body is unknown, and estimation of the number is also an open problem.

A human body contains not only the human genome. A new concept is to consider human as a superorganism containing those microbes in or on human body as well (Lederberg 2000). There are more than 100 trillion bacterial cells in human gut, which are about ten times more than cells in human itself (Turnbaugh et al. 2007). Those bacteria can help digest food and harvest nutrition and energy that otherwise cannot be collected by the human body directly (Turnbaugh et al. 2006; Backhed et al. 2004, 2007), i.e., human has obtained many genes needed for itself though these genes did not evolve in the human genome.

Based on the diversity of gut microbes and the average number of genes contained in a microbe genome, the number of genes in human gut microbiota was guessed to be 100 times greater than that of our human genome (Backhed et al. 2005). Since the scale of total number of genes in a human genome is about 20,000, this makes the guess of human gut microbiota genes to be at least 2,000,000. However, the exact number is unknown.

Yang et al. (2009) presented a model (Fig. 15.5) to estimate the number of genes in human gut bacterial community, the largest microbe community in or on the human body, and estimated the number of genes in the microbial genomes in the human gut was more than nine million (Yang et al. 2009). Qin et al. (2010) sequenced gut metagenomes from 124 European individuals and generated about 3.3 million unique microbial genes, which is about 150 times larger than the number of human genes, and claimed that they may have missed 15 % of the genes, which make the estimated number of about four million genes based on a real large-scale sequencing (Qin et al. 2010). More recent large-scale human gut metagenome sequencing result showed that this number can be more than eight million (Franzosa et al. 2014).

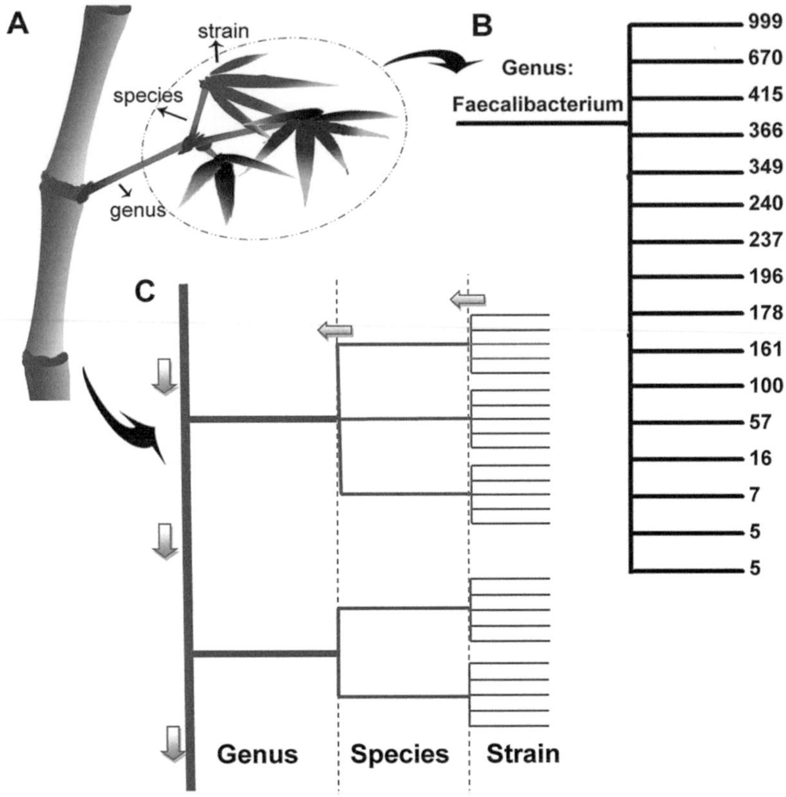

Fig. 15.5 The gene counting model for the human gut bacterial community. A bamboo-like tree structure represented the human gut bacterial community. The estimated number of unique genes in the human gut bacterial community is more than nine million (Figure is from Yang et al. 2009)

15.4 Disease-Associated Microbes in/on Human Body

Metagenomics and metatranscriptomics methods can be applied to find unknown pathogens. They have been applied to find disease-associated microbes in/on animal bodies and then human bodies. In 2007, Cox-Foster et al. reported a metagenomics survey of microbes in honeybee colony collapse disorder and proposed that Israeli acute paralysis virus of mites was strongly correlated with the disease (Cox-Foster et al. 2007). Feng et al. (2008) reported a previously undescribed polyomavirus in human skin cancer cell using metatranscriptomics method. Palacios et al. (2008) reported the identification of a new arenavirus in a cluster of fatal transplant-associated diseases with metagenomics methods. New microbial pathogen identification with metagenomics methods has been reviewed by MacConaill and Meyerson (2008).

Later, researchers from the USA and Japan started metagenomics studies for clinical samples of the gut (Nakamura et al. 2008) and respiratory track (Nakamura et al. 2009; Willner et al. 2009; Greninger et al. 2010). These researches did found not only disease-associated novel pathogens (such as small rRNA virus), but they also revealed the pathogen spectrum of the human gut and respiratory track. Fan et al. also used metagenomics method to detect H1N1A flu virus (Yongfeng et al. 2011).

More recently, metagenomics methods have been applied to analyze the impact of gut microbiota on the brain and behavior (Cryan and Dinan 2012), thus to reveal the molecular mechanisms of diseases like schizophrenia (Dinan et al. 2014).

15.5 The Coevolution of the Human Genome and the Microbial Genomes

The human body harbors not only the human genome but also microbiota, such as gut microbiota. The pathogen-host interactions and coevolution have been one of the research focus. Several highly human-adapted pathogen bacterial genomes were sequenced to prove novel insights to the pathogen-host interaction and coevolution in the early age of genomics (Field et al. 1999). Xu et al. (2003) show the symbiosis of human and a Gram-negative anaerobe *Bacteroides thetaiotaomicron*, which is a dominant member of human normal distal intestinal microbiota, by sequencing the full genome and proteomic methods. They showed that the bacterial genome contains genes to encode proteins to digest dietary polysaccharides, which are indigestible to human (Xu et al. 2003). Recent study showed that the human microbiota has strong flexibility to optimize host physiology state from daily life to lifespan scales and they coevolved with the human genome (Quercia et al. 2014). Turroni et al. reported an example of possible human-microbe coevolution by studying the genetic strategies for mucin metabolism in *Bifidobacterium bifidum* PRL2010 (Turroni et al. 2011). In addition, it was proposed that some diseases, especially infectious diseases, can be modeled as a disruption to the human-pathogen coevolution (Kodaman et al. 2014).

References

Backhed F, Ding H, Wang T, Hooper LV, Koh GY, Nagy A, Semenkovich CF, Gordon JI (2004) The gut microbiota as an environmental factor that regulates fat storage. Proc Natl Acad Sci U S A 101(44):15718–15723

Backhed F, Ley RE, Sonnenburg JL, Peterson DA, Gordon JI (2005) Host-bacterial mutualism in the human intestine. Science 307(5717):1915–1920

Backhed F, Manchester JK, Semenkovich CF, Gordon JI (2007) Mechanisms underlying the resistance to diet-induced obesity in germ-free mice. Proc Natl Acad Sci U S A 104 (3):979–984

Bendall SC, Simonds EF, Qiu P, Amir el AD, Krutzik PO, Finck R, Bruggner RV, Melamed R, Trejo A, Ornatsky OI, Balderas RS, Plevritis SK, Sachs K, Pe'er D, Tanner SD, Nolan GP (2011) Single-cell mass cytometry of differential immune and drug responses across a human hematopoietic continuum. Science 332(6030):687–696

Cani PD, Amar J, Iglesias MA, Poggi M, Knauf C, Bastelica D, Neyrinck AM, Fava F, Tuohy KM, Chabo C, Waget A, Delmee E, Cousin B, Sulpice T, Chamontin B, Ferrieres J, Tanti JF, Gibson GR, Casteilla L, Delzenne NM, Alessi MC, Burcelin R (2007) Metabolic endotoxemia initiates obesity and insulin resistance. Diabetes 56(7):1761–1772

Chen Y, Zhu J, Lum PY, Yang X, Pinto S, MacNeil DJ, Zhang C, Lamb J, Edwards S, Sieberts SK, Leonardson A, Castellini LW, Wang S, Champy MF, Zhang B, Emilsson V, Doss S, Ghazalpour A, Horvath S, Drake TA, Lusis AJ, Schadt EE (2008) Variations in DNA elucidate molecular networks that cause disease. Nature 452(7186):429–435

Cox-Foster DL, Conlan S, Holmes EC, Palacios G, Evans JD, Moran NA, Quan PL, Briese T, Hornig M, Geiser DM, Martinson V, vanEngelsdorp D, Kalkstein AL, Drysdale A, Hui J, Zhai J, Cui L, Hutchison SK, Simons JF, Egholm M, Pettis JS, Lipkin WI (2007) A metagenomic survey of microbes in honey bee colony collapse disorder. Science 318 (5848):283–287

Cryan JF, Dinan TG (2012) Mind-altering microorganisms: the impact of the gut microbiota on brain and behaviour. Nat Rev Neurosci 13(10):701–712

Das M, Burge CB, Park E, Colinas J, Pelletier J (2001) Assessment of the total number of human transcription units. Genomics 77(1–2):71–78

Dinan TG, Borre YE, Cryan JF (2014) Genomics of schizophrenia: time to consider the gut microbiome? Mol Psychiatry 19(12):1252–1257

Etchebehere C, Cabezas A, Dabert P, Muxi L (2003) Evolution of the bacterial community during granules formation in denitrifying reactors followed by molecular, culture-independent techniques. Water Sci Technol 48(6):75–79

Feng H, Shuda M, Chang Y, Moore PS (2008) Clonal integration of a polyomavirus in human Merkel cell carcinoma. Science 319(5866):1096–1100

Ferrer M, Beloqui A, Golyshina OV, Plou FJ, Neef A, Chernikova TN, Fernandez-Arrojo L, Ghazi I, Ballesteros A, Elborough K, Timmis KN, Golyshin PN (2007) Biochemical and structural features of a novel cyclodextrinase from cow rumen metagenome. Biotechnol J 2 (2):207–213

Field D, Hood D, Moxon R (1999) Contribution of genomics to bacterial pathogenesis. Curr Opin Genet Dev 9(6):700–703

Fields C, Adams MD, White O, Venter JC (1994) How many genes in the human genome? Nat Genet 7(3):345–346

Finn RD, Tate J, Mistry J, Coggill PC, Sammut SJ, Hotz HR, Ceric G, Forslund K, Eddy SR, Sonnhammer EL, Bateman A (2008) The Pfam protein families database. Nucleic Acids Res 36(Database issue):D281–D288

Franzosa EA, Morgan XC, Segata N, Waldron L, Reyes J, Earl AM, Giannoukos G, Boylan MR, Ciulla D, Gevers D, Izard J, Garrett WS, Chan AT, Huttenhower C (2014) Relating the metatranscriptome and metagenome of the human gut. Proc Natl Acad Sci U S A 111(22): E2329–E2338

Furrie E (2006) A molecular revolution in the study of intestinal microflora. Gut 55(2):141–143

Gill SR, Pop M, Deboy RT, Eckburg PB, Turnbaugh PJ, Samuel BS, Gordon JI, Relman DA, Fraser-Liggett CM, Nelson KE (2006) Metagenomic analysis of the human distal gut microbiome. Science 312(5778):1355–1359

Greninger AL, Chen EC, Sittler T, Scheinerman A, Roubinian N, Yu G, Kim E, Pillai DR, Guyard C, Mazzulli T, Isa P, Arias CF, Hackett J, Schochetman G, Miller S, Tang P, Chiu CY (2010) A metagenomic analysis of pandemic influenza A (2009 H1N1) infection in patients from North America. PLoS One 5(10):e13381

Holmes E, Loo RL, Stamler J, Bictash M, Yap IK, Chan Q, Ebbels T, De Iorio M, Brown IJ, Veselkov KA, Daviglus ML, Kesteloot H, Ueshima H, Zhao L, Nicholson JK, Elliott P (2008)

Human metabolic phenotype diversity and its association with diet and blood pressure. Nature 453(7193):396–400
Human_Microbiome_Project_Consortium (2012) Structure, function and diversity of the healthy human microbiome. Nature 486(7402):207–214
Kanehisa M, Goto S (2000) KEGG: Kyoto encyclopedia of genes and genomes. Nucleic Acids Res 28(1):27–30
Kim BH, Chang IS, Gadd GM (2007) Challenges in microbial fuel cell development and operation. Appl Microbiol Biotechnol 76(3):485–494
Kodaman N, Sobota RS, Mera R, Schneider BG, Williams SM (2014) Disrupted human-pathogen co-evolution: a model for disease. Front Genet 5:290
Kowalchuk GA, Speksnijder AG, Zhang K, Goodman RM, van Veen JA (2007) Finding the needles in the metagenome haystack. Microb Ecol 53(3):475–485
Lander ES, Linton LM, Birren B, Nusbaum C, Zody MC, Baldwin J, Devon K, Dewar K, Doyle M, FitzHugh W, Funke R, Gage D, Harris K, Heaford A, Howland J, Kann L, Lehoczky J, LeVine R, McEwan P, McKernan K, Meldrim J, Mesirov JP, Miranda C, Morris W, Naylor J, Raymond C, Rosetti M, Santos R, Sheridan A, Sougnez C, Stange-Thomann N, Stojanovic N, Subramanian A, Wyman D, Rogers J, Sulston J, Ainscough R, Beck S, Bentley D, Burton J, Clee C, Carter N, Coulson A, Deadman R, Deloukas P, Dunham A, Dunham I, Durbin R, French L, Grafham D, Gregory S, Hubbard T, Humphray S, Hunt A, Jones M, Lloyd C, McMurray A, Matthews L, Mercer S, Milne S, Mullikin JC, Mungall A, Plumb R, Ross M, Shownkeen R, Sims S, Waterston RH, Wilson RK, Hillier LW, McPherson JD, Marra MA, Mardis ER, Fulton LA, Chinwalla AT, Pepin KH, Gish WR, Chissoe SL, Wendl MC, Delehaunty KD, Miner TL, Delehaunty A, Kramer JB, Cook LL, Fulton RS, Johnson DL, Minx PJ, Clifton SW, Hawkins T, Branscomb E, Predki P, Richardson P, Wenning S, Slezak T, Doggett N, Cheng JF, Olsen A, Lucas S, Elkin C, Uberbacher E, Frazier M, Gibbs RA, Muzny DM, Scherer SE, Bouck JB, Sodergren EJ, Worley KC, Rives CM, Gorrell JH, Metzker ML, Naylor SL, Kucherlapati RS, Nelson DL, Weinstock GM, Sakaki Y, Fujiyama A, Hattori M, Yada T, Toyoda A, Itoh T, Kawagoe C, Watanabe H, Totoki Y, Taylor T, Weissenbach J, Heilig R, Saurin W, Artiguenave F, Brottier P, Bruls T, Pelletier E, Robert C, Wincker P, Smith DR, Doucette-Stamm L, Rubenfield M, Weinstock K, Lee HM, Dubois J, Rosenthal A, Platzer M, Nyakatura G, Taudien S, Rump A, Yang H, Yu J, Wang J, Huang G, Gu J, Hood L, Rowen L, Madan A, Qin S, Davis RW, Federspiel NA, Abola AP, Proctor MJ, Myers RM, Schmutz J, Dickson M, Grimwood J, Cox DR, Olson MV, Kaul R, Shimizu N, Kawasaki K, Minoshima S, Evans GA, Athanasiou M, Schultz R, Roe BA, Chen F, Pan H, Ramser J, Lehrach H, Reinhardt R, McCombie WR, de la Bastide M, Dedhia N, Blocker H, Hornischer K, Nordsiek G, Agarwala R, Aravind L, Bailey JA, Bateman A, Batzoglou S, Birney E, Bork P, Brown DG, Burge CB, Cerutti L, Chen HC, Church D, Clamp M, Copley RR, Doerks T, Eddy SR, Eichler EE, Furey TS, Galagan J, Gilbert JG, Harmon C, Hayashizaki Y, Haussler D, Hermjakob H, Hokamp K, Jang W, Johnson LS, Jones TA, Kasif S, Kaspryzk A, Kennedy S, Kent WJ, Kitts P, Koonin EV, Korf I, Kulp D, Lancet D, Lowe TM, McLysaght A, Mikkelsen T, Moran JV, Mulder N, Pollara VJ, Ponting CP, Schuler G, Schultz J, Slater G, Smit AF, Stupka E, Szustakowski J, Thierry-Mieg D, Thierry-Mieg J, Wagner L, Wallis J, Wheeler R, Williams A, Wolf YI, Wolfe KH, Yang SP, Yeh RF, Collins F, Guyer MS, Peterson J, Felsenfeld A, Wetterstrand KA, Patrinos A, Morgan MJ, Szustakowki J, de Jong P, Catanese JJ, Osoegawa K, Shizuya H, Choi S, Chen YJ (2001) Initial sequencing and analysis of the human genome. Nature 409(6822):860–921
Lederberg J (2000) Infectious history. Science 288(5464):287–293
Letunic I, Yamada T, Kanehisa M, Bork P (2008) iPath: interactive exploration of biochemical pathways and networks. Trends Biochem Sci 33(3):101–103
Levin DB, Zhu H, Beland M, Cicek N, Holbein BE (2007) Potential for hydrogen and methane production from biomass residues in Canada. Bioresour Technol 98(3):654–660
Liang F, Holt I, Pertea G, Karamycheva S, Salzberg SL, Quackenbush J (2000) Gene index analysis of the human genome estimates approximately 120,000 genes. Nat Genet 25 (2):239–240

Ling Z, Kong J, Jia P, Wei C, Wang Y, Pan Z, Huang W, Li L, Chen H, Xiang C (2010) Analysis of oral microbiota in children with dental caries by PCR-DGGE and barcoded pyrosequencing. Microb Ecol 60(3):677–690

MacConaill L, Meyerson M (2008) Adding pathogens by genomic subtraction. Nat Genet 40 (4):380–382

Mardis ER (2008) The impact of next-generation sequencing technology on genetics. Trends Genet 24(3):133–141

Margulies M, Egholm M, Altman WE, Attiya S, Bader JS, Bemben LA, Berka J, Braverman MS, Chen YJ, Chen Z, Dewell SB, Du L, Fierro JM, Gomes XV, Godwin BC, He W, Helgesen S, Ho CH, Irzyk GP, Jando SC, Alenquer ML, Jarvie TP, Jirage KB, Kim JB, Knight JR, Lanza JR, Leamon JH, Lefkowitz SM, Lei M, Li J, Lohman KL, Lu H, Makhijani VB, McDade KE, McKenna MP, Myers EW, Nickerson E, Nobile JR, Plant R, Puc BP, Ronan MT, Roth GT, Sarkis GJ, Simons JF, Simpson JW, Srinivasan M, Tartaro KR, Tomasz A, Vogt KA, Volkmer GA, Wang SH, Wang Y, Weiner MP, Yu P, Begley RF, Rothberg JM (2005) Genome sequencing in microfabricated high-density picolitre reactors. Nature 437(7057):376–380

Meyer F, Paarmann D, D'Souza M, Olson R, Glass EM, Kubal M, Paczian T, Rodriguez A, Stevens R, Wilke A, Wilkening J, Edwards RA (2008) The metagenomics RAST server – a public resource for the automatic phylogenetic and functional analysis of metagenomes. BMC Bioinforma 9:386

Nakamura S, Maeda N, Miron IM, Yoh M, Izutsu K, Kataoka C, Honda T, Yasunaga T, Nakaya T, Kawai J, Hayashizaki Y, Horii T, Iida T (2008) Metagenomic diagnosis of bacterial infections. Emerg Infect Dis 14(11):1784–1786

Nakamura S, Yang CS, Sakon N, Ueda M, Tougan T, Yamashita A, Goto N, Takahashi K, Yasunaga T, Ikuta K, Mizutani T, Okamoto Y, Tagami M, Morita R, Maeda N, Kawai J, Hayashizaki Y, Nagai Y, Horii T, Iida T, Nakaya T (2009) Direct metagenomic detection of viral pathogens in nasal and fecal specimens using an unbiased high-throughput sequencing approach. PLoS One 4(1):e4219

Narihiro T, Sekiguchi Y (2007) Microbial communities in anaerobic digestion processes for waste and wastewater treatment: a microbiological update. Curr Opin Biotechnol 18(3):273–278

Navin N, Kendall J, Troge J, Andrews P, Rodgers L, McIndoo J, Cook K, Stepansky A, Levy D, Esposito D, Muthuswamy L, Krasnitz A, McCombie WR, Hicks J, Wigler M (2011) Tumour evolution inferred by single-cell sequencing. Nature 472(7341):90–94

Pace NR (1997) A molecular view of microbial diversity and the biosphere. Science 276 (5313):734–740

Palacios G, Druce J, Du L, Tran T, Birch C, Briese T, Conlan S, Quan PL, Hui J, Marshall J, Simons JF, Egholm M, Paddock CD, Shieh WJ, Goldsmith CS, Zaki SR, Catton M, Lipkin WI (2008) A new arenavirus in a cluster of fatal transplant-associated diseases. N Engl J Med 358 (10):991–998

Qin J, Li R, Raes J, Arumugam M, Burgdorf KS, Manichanh C, Nielsen T, Pons N, Levenez F, Yamada T, Mende DR, Li J, Xu J, Li S, Li D, Cao J, Wang B, Liang H, Zheng H, Xie Y, Tap J, Lepage P, Bertalan M, Batto JM, Hansen T, Le Paslier D, Linneberg A, Nielsen HB, Pelletier E, Renault P, Sicheritz-Ponten T, Turner K, Zhu H, Yu C, Jian M, Zhou Y, Li Y, Zhang X, Qin N, Yang H, Wang J, Brunak S, Dore J, Guarner F, Kristiansen K, Pedersen O, Parkhill J, Weissenbach J, Bork P, Ehrlich SD (2010) A human gut microbial gene catalogue established by metagenomic sequencing. Nature 464(7285):59–65

Quercia S, Candela M, Giuliani C, Turroni S, Luiselli D, Rampelli S, Brigidi P, Franceschi C, Bacalini MG, Garagnani P, Pirazzini C (2014) From lifetime to evolution: timescales of human gut microbiota adaptation. Front Microbiol 5:587

Suau A, Bonnet R, Sutren M, Godon JJ, Gibson GR, Collins MD, Dore J (1999) Direct analysis of genes encoding 16S rRNA from complex communities reveals many novel molecular species within the human gut. Appl Environ Microbiol 65(11):4799–4807

Tatusov RL, Natale DA, Garkavtsev IV, Tatusova TA, Shankavaram UT, Rao BS, Kiryutin B, Galperin MY, Fedorova ND, Koonin EV (2001) The COG database: new developments in

phylogenetic classification of proteins from complete genomes. Nucleic Acids Res 29 (1):22–28
Tay S, Hughey JJ, Lee TK, Lipniacki T, Quake SR, Covert MW (2010) Single-cell NF-kappaB dynamics reveal digital activation and analogue information processing. Nature 466 (7303):267–271
Turnbaugh PJ, Ley RE, Mahowald MA, Magrini V, Mardis ER, Gordon JI (2006) An obesity-associated gut microbiome with increased capacity for energy harvest. Nature 444 (7122):1027–1031
Turnbaugh PJ, Ley RE, Hamady M, Fraser-Liggett CM, Knight R, Gordon JI (2007) The human microbiome project. Nature 449(7164):804–810. 4
Turnbaugh PJ, Hamady M, Yatsunenko T, Cantarel BL, Duncan A, Ley RE, Sogin ML, Jones WJ, Roe BA, Affourtit JP, Egholm M, Henrissat B, Heath AC, Knight R, Gordon JI (2009) A core gut microbiome in obese and lean twins. Nature 457(7228):480–484
Turroni F, Milani C, van Sinderen D, Ventura M (2011) Genetic strategies for mucin metabolism in Bifidobacterium bifidum PRL2010: an example of possible human-microbe co-evolution. Gut Microbes 2(3):183–189
Tyson GW, Chapman J, Hugenholtz P, Allen EE, Ram RJ, Richardson PM, Solovyev VV, Rubin EM, Rokhsar DS, Banfield JF (2004) Community structure and metabolism through reconstruction of microbial genomes from the environment. Nature 428(6978):37–43
Venter JC, Remington K, Heidelberg JF, Halpern AL, Rusch D, Eisen JA, Wu D, Paulsen I, Nelson KE, Nelson W, Fouts DE, Levy S, Knap AH, Lomas MW, Nealson K, White O, Peterson J, Hoffman J, Parsons R, Baden-Tillson H, Pfannkoch C, Rogers YH, Smith HO (2004) Environmental genome shotgun sequencing of the Sargasso Sea. Science 304(5667):66–74
Wei C, Brent MR (2006) Using ESTs to improve the accuracy of de novo gene prediction. BMC Bioinforma 7:327
Wei C, Lamesch P, Arumugam M, Rosenberg J, Hu P, Vidal M, Brent MR (2005) Closing in on the *C. elegans* ORFeome by Cloning TWINSCAN predictions. Genome Res 15:577–582
Willner D, Furlan M, Haynes M, Schmieder R, Angly FE, Silva J, Tammadoni S, Nosrat B, Conrad D, Rohwer F (2009) Metagenomic analysis of respiratory tract DNA viral communities in cystic fibrosis and non-cystic fibrosis individuals. PLoS One 4(10):e7370
Wooley JC, Godzik A, Friedberg I (2010) A primer on metagenomics. PLoS Comput Biol 6(2): e1000667
Xiao S, Zhao L (2014) Gut microbiota-based translational biomarkers to prevent metabolic syndrome via nutritional modulation. FEMS Microbiol Ecol 87(2):303–314
Xu J, Bjursell MK, Himrod J, Deng S, Carmichael LK, Chiang HC, Hooper LV, Gordon JI (2003) A genomic view of the human-Bacteroides thetaiotaomicron symbiosis. Science 299 (5615):2074–2076
Xu J, Lian F, Zhao L, Zhao Y, Chen X, Zhang X, Guo Y, Zhang C, Zhou Q, Xue Z, Pang X, Tong X (2014) Structural modulation of gut microbiota during alleviation of type 2 diabetes with a Chinese herbal formula. ISME J 9:552
Yang X, Xie L, Li Y, Wei C (2009) More than 9,000,000 unique genes in human gut bacterial community: estimating gene numbers inside a human body. PLoS One 4(6):e6074. https://doi.org/10.1371/journal.pone.0006074
Yang F, Zeng X, Ning K, Liu KL, Lo CC, Wang W, Chen J, Wang D, Huang R, Chang X, Chain PS, Xie G, Ling J, Xu J (2012) Saliva microbiomes distinguish caries-active from healthy human populations. ISME J 6(1):1–10
Yongfeng H, Fan Y, Jie D, Jian Y, Ting Z, Lilian S, Jin Q (2011) Direct pathogen detection from swab samples using a new high-throughput sequencing technology. Clin Microbiol Infect 17 (2):241–244
Yoon HS, Price DC, Stepanauskas R, Rajah VD, Sieracki ME, Wilson WH, Yang EC, Duffy S, Bhattacharya D (2011) Single-cell genomics reveals organismal interactions in uncultivated marine protists. Science 332(6030):714–717

Yun J, Kang S, Park S, Yoon H, Kim MJ, Heu S, Ryu S (2004) Characterization of a novel amylolytic enzyme encoded by a gene from a soil-derived metagenomic library. Appl Environ Microbiol 70(12):7229–7235

Zhang C, Zhang M, Wang S, Han R, Cao Y, Hua W, Mao Y, Zhang X, Pang X, Wei C, Zhao G, Chen Y, Zhao L (2010) Interactions between gut microbiota, host genetics and diet relevant to development of metabolic syndromes in mice. ISME J 4(2):232–241

Zhao L (2013) The gut microbiota and obesity: from correlation to causality. Nat Rev Microbiol 11 (9):639–647

Zhu J, Zhang B, Schadt EE (2008) A systems biology approach to drug discovery. Adv Genet 60:603–635

Zuo Y, Xing D, Regan JM, Logan BE (2008) Isolation of the exoelectrogenic bacterium Ochrobactrum anthropi YZ-1 by using a U-tube microbial fuel cell. Appl Environ Microbiol 74(10):3130–3137

Index

N. Saitou (ed.), *Evolution of the Human Genome I*, Evolutionary Studies,
https://doi.org/10.1007/978-4-431-56603-8

O

P

Q

R

S

T

MIX
Papier aus verantwortungsvollen Quellen
Paper from responsible sources
FSC® C105338

If you have any concerns about our products,
you can contact us on
ProductSafety@springernature.com

In case Publisher is established outside the EU,
the EU authorized representative is:
Springer Nature Customer Service Center GmbH
Europaplatz 3, 69115 Heidelberg, Germany

Printed by Libri Plureos GmbH
in Hamburg, Germany